# The Black spotted, Yellow Borer, Conogethes punctiferalis Guenée and Allied Species

Akshay Kumar Chakravarthy
Editor

# The Black spotted, Yellow Borer, Conogethes punctiferalis Guenée and Allied Species

Springer

*Editor*
Akshay Kumar Chakravarthy
Division of Entomology and Nematology
ICAR-Indian Institue of Horticultural Research
Bengaluru, Karnataka, India

ISBN 978-981-13-0389-0 ISBN 978-981-13-0390-6 (eBook)
https://doi.org/10.1007/978-981-13-0390-6

Library of Congress Control Number: 2018950566



Printed on acid-free paper

This Springer imprint is published by the registered company Springer Nature Singapore Pte Ltd.
The registered company address is: 152 Beach Road, #21-01/04 Gateway East, Singapore 189721, Singapore

# Foreword

*The Black spotted, Yellow Borer, Conogethes punctiferalis Guenée and Allied Species* provides exciting insights into the biology of this economically important group of moths with beautiful yellow wings and black dots or patches. For example, males of *Conogethes* use ultrasonic courtship calls to attract females, as well as to ward off competing males. These and other interesting and useful facts about *Conogethes* moths are found in this book.

The genus *Conogethes* is a group of Southeast Asian moths of great economic importance because the larvae of one species, *C. punctiferalis*, also commonly called the yellow peach moth, are borers of fruit trees. This book is well illustrated with 22 chapters and offers insights into the life history, ecology, and behavior of a group of moths for the general enthusiast of moth biology. For practitioners, it provides in-depth information about identification, occurrence, pest risk analysis, reproductive physiology, prediction models, mass rearing, host-plant relationships, biological control, pheromones, and integrated pest management. The information in this book will be useful to those that design experiments to address challenging questions about the biology of *Conogethes* and other borers.

In 1896, Sir George Hampson collated various generic and species names applied to this group of moths under the genus *Dichocrocis* in the fourth volume about Pyraloidea in *The Fauna of British India Including Ceylon and Burma*. He created this group based on external structures, color patterns, and wing venation, but advances in morphological, molecular, and phylogenetic analytical techniques in the last 120+ years have brought considerable changes to the higher classification and composition of this group of moths. In 1896, Hampson placed *Dichocrocis* in the family Pyralidae and subfamily Pyraustinae, but they are now in the family Crambidae and subfamily Spilomelinae, in the genus *Conogethes* Meyrick. Achille Guenée first described *C. punctiferalis* in 1854, and unfortunately the species identification problems have endured. Pyraloid workers have known for many decades that *C. punctiferalis* is a complex of species, with undescribed species new to science. Recently, host-related moths previously identified as *C. punctiferalis* have been registered as new and different species.

This book has benefited from the expertise of a number of entomologists and researchers who generously contributed relevant information and research material. Most of the contributors are Indians because *Conogethes* and a majority of its host plants are native to India. *Conogethes* is a pest of quarantine importance worldwide.

*Conogethes* populations are expected to expand geographically, intensifying potential damage because of global trade, frequent exchange of horticultural produce, human transit, porous boundaries between countries, and inadequate quarantine and phytosanitary protocols for this particular group of moths. This book will contribute to the realistic management of the pest, provide useful data for use by policy makers, and, ultimately, spark interest in these biologically interesting pyraloid moths and their larvae, and propel further studies by researchers, students, and teachers.

SEL, USDA, Smithsonian Institution
National Museum Natural History,
Washington, DC, USA

M. Alma Solis
Alma.Solis@ars.usda.gov

# Preface

In early 1980s, attention was drawn to the shoot and fruit-borer, *Conogethes* spp. infesting cardamom (*Elettaria cardamomum*) – the Queen of spices in the picturesque Western Ghats region, South India. Western Ghats represents evergreen tropical forests tract with deep green valleys and blue-clad rolling mountains interspersed with twining narrow lanes. *Conogethes* moths infesting cardamom appeared different from the similar-looking moths infesting castor beans, *Ricinus communis*. Interestingly, Japanese workers too observed two types of moths, as also workers from few other countries. It is only at the beginning of the twenty-first century that the species branded *C. punctiferalis* were delineated into new species using molecular and integrative taxonomic tools. One of the most recent examples has been *Conogethes sahyadriensis* from India as founded by Shashank and others in 2018.

According to Richard Mally (Chap. 1), *C. punctiferalis* represents a species complex and little is known about the ecology and phylogeny of these moths. Molecular tools–assisted diagnoses are paving way for comprehensive identification of host-related *Conogethes* moth populations (Vasudev Kammar et al., Chap. 2). The Southeast Asia is rich in species richness of *Conogethes* and it needs extensive and intensive research (Kumar et al., Chap. 4). Du, Li, and Wang from China discuss the status of *Conogethes* populations on major crops in China with emphasis on *Wolbachia* infection, genetic diversity, and gene flow. Loc and Chakravarthy have reviewed and updated the literature on *Conogethes* in South Vietnam where it is a major pest on durian, longan, rambutan, soursop, and sugar apple. Interestingly, *Conogethes* is not a major problem in North Vietnam. But it is so on teak and durian in Srilanka. Sivapragasam and others have discussed the status of *Conogethes* in Malaysia and other Southeast Asian countries, where it is a major pest on most of the tropical fruits of commercial importance.

Sridhar and Darren Kriticos projected the potential areas where *Conogethes* can occur and damage crops worldwide. The *Conogethes* moths are of a major concern in South- and Northeast India, Coastal Australia, and in Southeast Asian countries. The book contains four chapters on the status of *Conogethes* in four major parts of India. Sandeep Singh and others have contributed a chapter on the economic and quarantine importance of the pest in the Punjab, India. Mr. Alagar has specifically focused on cocoa where *Conogethes* is an emerging pest in Karnataka, South India, with records up to 80% fruit damage in fields. Devasahayam and coworkers have elaborated on the damage potential of *Conogethes* on spice crops, *viz.*, cardamom, ginger, and turmeric that are economically important. The book also covers aspects like pest risk analysis, mass rearing methods, reproductive physiology, biological control, host-plant relationships, pheromones, integrated pest management, and a new species of *Conogethes* from South India identified as *Conogethes sahyadriensis* sp. nov. The contributors contend that *Conogethes* sp. group forms an interesting material for research in biosystematics, biogeography, evolution, adaptive strategies, and management. It is in cultivated ecosystems that *Conogethes* sp. complex is going to be an increasingly important pest on a wide and new range of crops. Compilation of information on a pest species as *Conogethes* was a daunting task given that only few entomologists have spent time working on the species. All the contributors have shared valuable material that will prove useful in the future not only for *Conogethes* but also for other borers.

Bengaluru, Karnataka, India A. K. Chakravarthy

# Acknowledgment

I began my career working on cardamom pests at the hill research station of the Western Ghats, Mudigere, Chikmagalur, Karnataka, South India. One of the pests I had to research on was the cardamom borer, *Conogethes* species. Since 1983, these bright yellow moths attracted my attention. Native to India, the cardamom borer moths looked alike to the castor shoot and fruit borer, *Conogethes punctiferalis*. This raised inquisitiveness in my mind. I was lucky in having the support of stalwarts like Dr. C. A Virakthamath and late Dr. G. P. Channabasavanna, University of Agricultural sciences, Bengaluru, and Dr. Hiroshi Honda, University of Tokyo, Japan. They made helpful suggestions and encouraged me to work on species of *Conogethes*. I sincerely thank them with gratitude. Subsequently, it came to my knowledge that *Conogethes punctiferalis* is a species complex. I came across works of other entomologists on this group of moths, like Mitsuhashi, Inoue, and Yamanaka; Wang JB and Jing Li from China; H. T. R. Wijesekara from Sri Lanka; and others.

This book is the result of research work carried out by workers from different parts of the world: Richard Mally from Norway, Yang Li and coworkers from China, H. T. Loc from Vietnam, Amani Mannakara from Srilanka, A. Sivapragasam and coworkers from Malaysia, Hiroshi Honda from Japan, and a host of workers from India. Dr. Darren Kriticos from Australia facilitated a chapter on global distribution pattern of *Conogethes punctiferalis* with Dr. V. Sridhar. I am indebted to the many entomologists from India who have contributed information on several aspects of *Conogethes* – Chandish Ballal, N.E. Thyagaraj, S. R. Kulkarni, Sandeep Singh, V. Sridhar, P. V. R Reddy, L. Hanumantharaya, Y. J. Tambe, P. Thiyagarajan, S. S. Bora, K. Dhanapal, B. A. Gudade, A. B. Rema Shree, Gurlaz Kaur, M. Alagar, V. Selvanarayanan, H. Khader Khan, S. Devasahayam, Senthil Kumar, T. K. Jacob, A. R. N. S Subbanna, and G. Preetha. I express sincere thanks to them all. Postgraduate students (most of them scientists now!): P. R. Shashank, Vasudev Kammar, Gundappa, K. S. Nitin, S. Onkara Naik, G. P. Mutturaj, S. Subhash, T. Ambanna, M. A. Rashmi, B. Doddabasappa, P. Swathi, Richa Varshney, G. Stanley, G. C. Ankush, A. T. Rani, Kiran S. Kasareddy, P. N. Guru, Naveen Kumar, Yatish, and others evinced a deep interest in understanding biology of *Conogethes*. I owe them a great deal. I could not think of an entomologist other than Dr. (Ms.) Alma Solis, Smithsonian Institution, USA, to have accepted to write a foreword for this book. I remain indebted to her.

Mr. K. P. Kumar especially evinced keen interest in studying *Conogethes* and is an ardent *Conogethes* researcher. He did Ph.D. on *Conogethes* and wish to lifelong continue researching on *Conogethes*! He remained very helpful not only throughout the production of this book but at all the times. He was instrumental in bringing out for the first time a monograph on *Conogethes*.

I see that an increasing number of biologists are taking interest in studying *Conogethes*, in the hope that the research approach adopted in studying *Conogethes* may be deployed to manage pests on other crops particularly the lepidopterous borer moths.

Last but not the least, I and other scientists owe a debt of gratitude to our publisher Springer for the deep interest and excellent support.

Bengaluru, Karnataka, India A. K. Chakravarthy

# Contents


1 **Moths of the Genus *Conogethes*: Taxonomy, Systematics, and Similar Species** . . . . . 1
Richard Mally

2 **Molecular Status of *Conogethes* spp.: An Overview** . . . . . 13
Vasudev Kammar, P. R. Shashank, A. T. Rani, and V. Selvanarayanan

3 ***Conogethes sahyadriensis*: A New Borer on Zingiberaceous Crop Plants from India** . . . . . 23
A. K. Chakravarthy, P. R. Shashank, K. P. Kumar, and Vasudev Kammar

4 **Status of Shoot and Fruit Borer, *Conogethes* spp. (Crambidae: Lepidoptera) in Asia: Central, South, and the Southeast** . . . . . 35
K. P. Kumar, Naveen Kumar, and A. K. Chakravarthy

5 **Research Progress of *Conogethes punctiferalis* (Lepidoptera: Crambidae) in China** . . . . . 45
Yan-Li Du, Jing Li, and Zhen-Ying Wang

6 **Status of *Conogethes punctiferalis* (Guenée) in South of Vietnam** . . . . . 67
H. T. Loc, K. P. Kumar, and A. K. Chakravarthy

7 **Status of Shoot and Fruit Borer, *Conogethes punctiferalis*, in Sri Lanka** . . . . . 81
Amani Mannakkara, A. D. N. T. Kumara, N. I. Suwandharathne, I. Hettiarachchi, and H. T. R. Wijesekara

8 **Status and Management of *Conogethes* spp. in Malaysia** . . . . . 89
A. Sivapragasam, A. Badrulhadza, and M. N. Mohamad Roff

**9 Predicting Potential Global Distribution of The Black Spotted Yellow Borer, *Conogethes punctiferalis* Guenée (Crambidae: Lepidoptera) by CLIMEX Modelling** . . . . . . . . . . . . . . 101
V. Sridhar, Darren Kriticos, K. S. Nitin, A. K. Chakravarthy, and P. Swathi

**10 Bioecological Studies on *Conogethes sahyadriensis* in South India** . . 115
Kiran S. Kasareddy, L. Hanumantharaya, and K. P. Kumar

**11 Status of Shoot and Fruit Borer, *Conogethes punctiferalis* Guenee (Lepidoptera: Crambidae), in Central India** . . . . . . . . . . . . . 131
K. R. Yatish, V. J. Tambe, J. C. Ankush, and P. N. Guru

**12 Shoot and Fruit Borer, *Conogethes* spp. (Crambidae: Lepidoptera) in Northeast India**. . . . . . . . . . . . . . . . . . . 149
P. Thiyagarajan, S. S. Bora, K. Dhanapal, B. A. Gudade, and A. B. Rema Shree

**13 Bioecology and Management of Fruit and Shoot Borer, *Conogethes punctiferalis* Guenée (Crambidae: Lepidoptera), on Fruit Crops in Central India** . . . . . . . . . . . . . . . . . . . . . . . . . . . . . 157
S. Gundappa, H. Khader Khan, and A. K. Chakravarthy

**14 The Shoot and Fruit Borer, *Conogethes punctiferalis* (Guenee): An Important Pest of Tropical and Subtropical Fruit Crops** . . . . . . . 165
Sandeep Singh, Gurlaz Kaur, S. Onkara Naik, and P. V. Rami Reddy

**15 Bio-ecology, Damage Potential and Management of *Conogethes punctiferalis* Guenee in Plantation Crops** . . . . . . . . . . . . . . . . . . . . . . . 193
M. Alagar

**16 Status of Shoot and Fruit Borer, *Conogethes sahyadriensis*, on Spice Crops**. . . . . . . . . . . . . . . . . . . . . . . . . . . . . . . . . . . . . . . . 205
S. Devasahayam, C. M. Senthil Kumar, and T. K. Jacob

**17 Pest Risk Analysis for the Shoot and Fruit Borer, *Conogethes* spp. (Crambidae: Lepidoptera)**. . . . . . . . . . . . . . . . . . . . . . . . . . . . . . 219
G. P. Mutturaj, S. Subhash, Sandeep Singh, and A. K. Chakravarthy

**18 Rearing of *Conogethes punctiferalis* Guenée (Lepidoptera: Crambidae) and Feasibility of Its Biological Control** . . . . . . . . . . . . . 235
Chandish R. Ballal, K. P. Kumar, P. Ambanna, A. K. Chakravarthy, Richa Varshney, and H. Khader Khan

**19 Reproduction in the Shoot and Fruit Borer, *Conogethes* spp. (Crambidae: Lepidoptera): Strategizing Survival?**. . . . . . . . . . . . . . . 257
M. A. Rashmi, A. K. Chakravarthy, and S. R. Kulkarni

**20 Host Plant Relationships of the Shoot and Fruit Borer, *Conogethes* spp. (Crambidae: Lepidoptera): Mechanisms and Determinants** . . . . . . . . 279
B. Doddabasappa, K. R. M. Bhanu, and S. Subhash

**21 Extraction and Identification of Pheromones of the Borer, *Conogethes punctiferalis* (Crambidae: Lepidoptera)** . . . . . . . . 307
J. Stanley, A. R. N. S. Subbanna, and G. Preetha

**22 Novel Tools for the Management of *Conogethes punctiferalis* Guenée (Crambidae: Lepidoptera)** . . . . . . . . 333
N. E. Thyagaraj, K. S. Jagadish, and Naveen Kumar

# Editor and Contributors

## About the Editor

**Dr. A. K Chakravarthy** Head and Principal Scientist, Division of Entomology and Nematology, is an author of many books and over 300 scientific papers and 30 chapters in entomology and natural history. His interest includes insects, birds, bats, rodents, and mammals. With over three decades of experience of teaching, research, and extension, Dr. Chakravarthy has been the investigator for over 35 research projects and has guided more than 25 students for postgraduation. A PhD graduate from Punjab Agricultural University and Fellow from IARI, New Delhi, he is a member of several national and international scientific academia, an advisor, a panelist, a referee, reviewer, and editor, and he is associated with publication of several national and international journals worldwide. A field orientated, widely traveled biologist, he is actively working on novel approaches in integrated pest management, host plant interaction, vertebrate pest management, and biodiversity and environmental conservation issues. Recently, he has published books titled *New Horizons in Insect Science: Towards Sustainable Pest Management*, *Economic and Ecological Significance of Arthropods in Diversified Ecosystems – Sustaining Regulatory Mechanisms*, and *Arthropod Diversity and Conservation in the Tropics and Sub-tropics* by Springer.

## Contributors

**M. Alagar** ICAR-Krish Vigyan Kendra, Nagapattinam, Tamil Nadu, India

**P. Ambanna** Department of Agricultural Entomology, University of Agricultural Sciences, Gandhi Krishi Vignan Kendra (GKVK), Bengaluru, Karnataka, India

**J. C. Ankush** Department of Entomology, Mahatma Phule Krishi Vidyapeeth (MPKV), Ahmednagar, Maharashtra, India

**A. Badrulhadza** Crops and Soil Science Research Center, Malaysian Agricultural Research and Development Institute (MARDI), Serdang, Malaysia

**Chandish R. Ballal** ICAR-National Bureau of Agricultural Insect Resources (NBAIR), Bengaluru, Karnataka, India

**K. R. M. Bhanu** Bio-Control Research Laboratories, A Division of Pest Control (India) Pvt. Ltd., Bangalore, Karnataka, India

**S. S. Bora** Regional Research Station, ICAR-Indian Cardamom research Institute, Spices Board, Tadong, Sikkim, India

**A. K. Chakravarthy** Division of Entomology and Nematology, ICAR-Indian Institute of Horticultural Research, Bengaluru, Karnataka, India

**S. Devasahayam** ICAR-Indian Institute of Spices Research, Kozhikode, Kerala, India

**K. Dhanapal** Regional Research Station, ICAR-Indian Cardamom research Institute, Spices Board, Tadong, Sikkim, India

**B. Doddabasappa** Department of Entomology, College of Horticulture, Kolar, Karnataka, India

**Yan-Li Du** Plant Science and Technology College, Beijing University of Agriculture, Beijing, China

**B. A. Gudade** Indian Cardamom Research Institute, Regional Research Station, Spices Board, (Ministry of Commerce & Industry, Govt. of India), Tadong Gangtok, Sikkim, India

**S. Gundappa** ICAR-Central Institute for Subtropical Horticulture, Lucknow, Uttar Pradesh, India

**P. N. Guru** Department of Entomology, Mahatma Phule Krishi Vidyapeeth (MPKV), Ahmednagar, Maharashtra, India

**L. Hanumantharaya** College of Horticulture, University of Agricultural and Horticultural Sciences -Shimoga, Mudigere, Karnataka, India

**I. Hettiarachchi** Department of Agriculture, Fruit Crop Research and Development Institute, Horana, Sri Lanka

**T. K. Jacob** ICAR-Indian Institute of Spices Research, Kozhikode, Kerala, India

**K. S. Jagadish** Department of Agricultural Entomology, University of Agricultural Sciences, Gandhi Krishi Vignan Kendra (GKVK), Bengaluru, Karnataka, India

**Vasudev Kammar** Department of Food and Public Distribution, Ministry of Consumer Affairs and Food and Public Distribution, Government of India, New Delhi, India

**Gurlaz Kaur** Department of Fruit Science, Punjab Agricultural University, Ludhiana, Punjab, India

**H. Khader Khan** Department of Agricultural Entomology, University of Agricultural Sciences, Gandhi Krishi Vignan Kendra (GKVK), Bengaluru, Karnataka, India

**Darren Kriticos** Health and Biosecurity CSIRO, Canberra, ACT, Australia

**S. R. Kulkarni** Department of Entomology, Mahatma Phule Krishi Vidyapeeth (MPKV), Rahuri, Ahmednagar, Maharashtra, India

**A. D. N. T. Kumara** Crop Protection Division, Coconut Research Institute, Lunuwila, Sri Lanka

**K. P. Kumar** Department of Entomology, University of Agricultural Sciences, Gandhi Krishi Vignan Kendra (GKVK), Bengaluru, Karnataka, India

**Naveen Kumar** Denthottu House, Ballamanja, Manchina, Belthangady, Mangaluru, Karnataka, India

**Jing Li** School of Biological and Environmental Engineering, Xi'an University, Xi'an, Shaanxi Province, China

**H. T. Loc** Plant Protection Division, Southern Horticultural Research Institute (SOFRI), My Tho, Tien Giang, Vietnam

**Richard Mally** University Museum of Bergen, Natural History Collections Realfagbygget, Bergen, Norway

**Amani Mannakkara** Department of Agricultural Biology, Faculty of Agriculture, University of Ruhuna, Matara, Sri Lanka

**M. N. Mohamad Roff** Director General Office, Malaysian Agricultural Research and Development Institute (MARDI), Serdang, Malaysia

**G. P. Mutturaj** College of Horticulture, Yalachahalli Horticulture Farm, Yelwala Hobli, Mysore, Karnataka, India

**K. S. Nitin** Division of Entomology and Nematology, ICAR-Indian Institute of Horticultural Research (IIHR), Bengaluru, Karnataka, India

**S. Onkara Naik** Division of Entomology and Nematology, ICAR-Indian Institute of Horticultural Research, Bengaluru, Karnataka, India

**G. Preetha** Department of Entomology, Tamil Nadu Agricultural University, Coimbatore, Tamil Nadu, India

**P. V. Rami Reddy** Division of Entomology and Nematology, ICAR-Indian Institute of Horticultural Research, Bengaluru, Karnataka, India

**A. T. Rani** Division of Crop protection, ICAR-Indian Institute of Vegetable Research (IIVR), Varanasi, Uttar Pradesh, India

**M. A. Rashmi** Division of Entomology and Nematology, ICAR-Indian Institute of Horticultural Research, Bengaluru, Karnataka, India

**Kiran S. Kasareddy** College of Horticulture, University of Agricultural and Horticultural Sciences -Shimoga, Mudigere, Karnataka, India

**A. B. Rema Shree** ICAR-Indian Cardamom Research Institute, Spices Board, Myladumpara, Idukki, Kerala, India

**V. Selvanarayanan** Department of Entomology, Faculty of Agriculture, Annamalai University, Chidambaram, Tamil Nadu, India

**C. M. Senthil Kumar** ICAR-Indian Institute of Spices Research, Kozhikode, Kerala, India

**P. R. Shashank** Insect Taxonomy Laboratory, Division of Entomology, ICAR-Indian Agricultural Research Institute, New Delhi, India

**Sandeep Singh** Department of Fruit Science, Punjab Agricultural University, Ludhiana, Punjab, India

**A. Sivapragasam** CABI South East Asia, Building A19, Malaysian Agricultural Research and Development Institute (MARDI), Serdang, Selangor, Malaysia

**V. Sridhar** Division of Entomology and Nematology, ICAR-Indian Institute of Horticultural Research (IIHR), Bengaluru, Karnataka, India

**J. Stanley** Vivekananda Institute of Hill Agriculture, Indian Council of Agricultural Research, Almora, Uttarakhand, India

**A. R. N. S. Subbanna** Vivekananda Institute of Hill Agriculture, Indian Council of Agricultural Research, Almora, Uttarakhand, India

**S. Subhash** Department of Agricultural Entomology, University of Agricultural Sciences, Gandhi Krishi Vignan Kendra (GKVK), Bengaluru, Karnataka, India

**N. I. Suwandharathne** Crop Protection Division, Coconut Research Institute, Lunuwila, Sri Lanka

**P. Swathi** Division of Entomology and Nematology, ICAR-Indian Institute of Horticultural Research (IIHR), Bengaluru, Karnataka, India

**V. J. Tambe** Department of Agricultural Entomology, Dr. Panjabrao Deshmukh Krishi Vidyapeeth (PDKV), Akola, Maharashtra, India

**P. Thiyagarajan** ICAR-Indian Cardamom Research Institute, Spices Board, Myladumpara, Idukki, Kerala, India

**N. E. Thyagaraj** Department of Entomology, College of Agriculture, Hassan, Karnataka, India

**Richa Varshney** ICAR-National Bureau of Agricultural Insect Resources (NBAIR), Bengaluru, Karnataka, India

**Zhen-Ying Wang** State Key Laboratory for Biology of Plant Diseases and Insect Pests, Institute of Plant Protection, Chinese Academy of Agricultural Sciences, Beijing, China

**H. T. R. Wijesekara** Crop Protection Division, Coconut Research Institute, Lunuwila, Sri Lanka

**K. R. Yatish** Division of Entomology and Nematology, ICAR-Indian Institute of Horticultural Research, Bengaluru, Karnataka, India

# Moths of the Genus *Conogethes*: Taxonomy, Systematics, and Similar Species

1

Richard Mally

**Abstract**

The main characteristics of the genus *Conogethes* in Spilomelinae of Crambidae are described and updated. To date, the genus comprises 15 species. Little is known about habitat preference, host plants and distributions. Adults of *Conogethes* exhibit similarity with moths of the genera *Marwitzia* and *Polygrammodes*. The phylogenetic relationships both within *Conogethes* and among Spilomelinae are so far unknown. Recent research has focused on the acoustic and chemical communication of *C. punctiferalis*. Several morphological characters of the male and female genitalia have proved useful for species distinction within *Conogethes*.

**Keywords**

*Conogethes* · Diagnosis · Phylogenetics · Systematics

## 1.1 Introduction

*Conogethes* Meyrick 1884, is a genus of Spilomelinae in the family Crambidae and currently comprises 15 species (Table 1; Nuss et al. 2003–2017) distributed in the Austral-Asian region from India, China and Japan through the Indonesian archipelago to New Guinea, the Solomon Islands and Australia. *Conogethes* has also been observed outside of its natural range, with findings in Hawaii (Munroe 1989) and Great Britain (Truscott 2007). Little is known about the habitat preferences, and maybe the availability of host plants is the main determining factor.

R. Mally (✉)
University Museum of Bergen, Natural History Collections, Realfagbygget, Bergen, Norway
e-mail: richard.mally@uib.no



A. K. Chakravarthy (ed.), *The Black spotted, Yellow Borer, Conogethes punctiferalis Guenée and Allied Species*,
https://doi.org/10.1007/978-981-13-0390-6_1

*Conogethes* is mostly known for the economic impact of its larvae on agricultural crops. Substantial research has been undertaken in order to understand the biology of these harmful insects and to develop strategies to confine their impact and especially that of the yellow peach moth *Conogethes punctiferalis* (Guenée 1854), which is reportedly the most harmful species in this genus but actually represents a species complex. Honda and Mitsuhashi (1989) and Doddabasappa et al. (2014) studied the immature stages of *Conogethes* and provided detailed information on them. The copulatory behaviour of the adult moths has been studied by Kaneko (1978) and Konno et al. (1980). This is tightly connected to the ongoing research on the chemical communication between males and females via sex pheromones. These volatile substances are released from the coremata, structures near the male genitalia that can be everted for that purpose. The microstructure and functional morphology of these coremata have been studied by Kimura et al. (2002). So far, sex pheromone composition is known for only two species, both of economic importance: *C. punctiferalis* (e.g. Konno et al. 1982; see also list of publications in El-Sayed 2017) and *C. pluto* (Butler, 1887) (El-Sayed et al. 2012).

Recent research also focused on the acoustic communication of *C. punctiferalis* (Nakano et al. 2012a, b). Nakano et al. (2012a) observed and reported on the ultrasonic courtship song which seems necessary to initiate copulation, and they identified a mesothoracic tymbal organ in the male of *C. punctiferalis* as the source of this acoustic communication. A similar mesothoracic structure, the anepisternal scale organ, has been reported by Clavijo Albertos (1990) in males of the genus *Diaphania* Hübner, 1818, although he assumed this organ to play a role in chemical communication due to its extensive scaling. It can be assumed that the organ reported by Nakano et al. (2012b) is homologous to the anepisternal scale organ of Clavijo Albertos (1990). This knowledge on the courtship acoustics of *Conogethes* is another potentially useful way of pest control. Furthermore, Nakano et al. (2014) reported that *C. punctiferalis* males use their ability of sound production to induce copulation and alternatively to mimic bat calls in order to disrupt the approach of other males towards copulation. This points to another way of influencing the mating success of *Conogethes*.

Li et al. (2010) report on infections with five different strains of the intracellular *Wolbachia* bacteria in four Chinese populations of *C. punctiferalis*, with an infection frequency of 2.0–8.0%. Three of the four populations carry a single *Wolbachia* strain, whereas the fourth population is superinfected with four strains. These different populations might be reproductively isolated from each other through *Wolbachia*-induced cytoplasmic incompatibility, a common consequence of *Wolbachia* infections among a wide range of arthropods (Werren et al. 2008).

## 1.2 Host Plant Use

The host plant range of *Conogethes* is wide, and the larvae are found to feed on Pinaceae as well as on a variety of mono- and dicotyledonous angiosperm plants, many of them of agricultural importance. Some species seem to have a rather narrow food spectrum, whereas especially *C. punctiferalis* is eminently polyphagous (Table 1.1). While the larvae of *C. pinicolalis* Inoue and Yamanaka 2006, feed on the needles of various conifers, the larvae of species feeding on angiosperms feed internally in the fruits, pods and stems of their host plants. This concealed feeding strategy hampers the detection of infestation and the application of methods for containment and control. Plants of the ginger family (Zingiberaceae) are utilised by larvae of at least three species, *C. evaxalis*, *C. pluto* and *C. sahyadriensis*. The phylogeny of Shashank et al. (2018), based on DNA barcode data, indicates that *C.*

**Table 1.1** Currently recognised *Conogethes* species and their synonyms (from Nuss et al. 2003–2017) as well as larval food plants, where known

| Species of *Conogethes* | Type locality | Known larval food plants |
|---|---|---|
| ***C. clioalis*** (Walker, 1859b) | Malaysia, Sarawak | *Dillenia* (Dilleniaceae), *Shorea* (Dipterocarpaceae) (Robinson et al. 2010) |
| ***C. diminutiva*** Warren, 1896 | India, Meghalaya, Khasi Hills | *Ipomoea* (Convolvulaceae) (Robinson et al. 2010) |
| ***C. ersealis*** (Walker, 1859c) | Australia, Moreton Bay | Unknown |
| ***C. evaxalis*** (Walker, 1859c) | India | *Curcuma* (Zingiberaceae), *Dillenia* (Dilleniaceae), *Dipterocarpus* (Dipterocarpaceae) (Robinson et al. 2010) |
| = *C. semistrigalis* Snellen, 1895 | Indonesia, Sumatra, Padang highlands; Java, Bogor | |
| ***C. haemactalis*** Snellen, 1890 | India, Sikkim | Unknown |
| = *C. nubifera* T. P. Lucas, 1892 | Australia, Brisbane, Burpengary | |
| ***C. mimastis*** Meyrick, 1897 | Indonesia, Sangir Island | Unknown |
| ***C. parvipunctalis*** Inoue and Yamanaka, 2006 | Japan, Ryukyu, Amami-ôshima, Hatsuno | Unknown |
| ***C. pinicolalis*** Inoue and Yamanaka, 2006 | Japan, Honshu, Saitama Pref., Iruma City, Bushi | *Pinus*, *Picea*, *Tsuga*, *Larix*, *Abies*, *Cedrus* (Pinaceae) (Inoue and Yamanaka 2006) |
| ***C. pluto*** (Butler, 1887) | Solomon Islands, Shortland Island | *Alpinia* (Zingiberaceae) (El-Sayed et al. 2012) |
| ***C. punctiferalis*** (Guenée, 1854) | India | On a large number of angiosperms, see, e.g. Inoue and Yamanaka (2006) |
| = *Astura guttatalis* Walker, 1866 | Indonesia, West Papua, Misool; North Maluku, Bacan Islands; Aru; Seram | |
| = *Botys nicippealis* Walker, 1859c | Indonesia, Maluku, Seram | |
| = *Conogethes punctiferalis* var. *jocata* T. P. Lucas, 1892 | Australia, Hamilton Scrub near Brisbane | |
| = *Deiopeia detracta* Walker, 1859a | Singapore | |

(continued)

**Table 1.1** (continued)

| Species of *Conogethes* | Type locality | Known larval food plants |
|---|---|---|
| ***C. sahyadriensis*** Shashank et al., 2018 | India, Chikmagalur, Mudigere, 12°25′11″N 75°43′48″N, 980 m | *Elettaria* (Zingiberaceae) (Doddabasappa et al. 2014) |
| ***C. semifascialis*** (Walker, 1866) | Australia, Moreton Bay | Unknown |
| = *C. jubata* T. P. Lucas, 1900 | Australia, Queensland, Brisbane | |
| = *Conogethes punctiferalis* var. *nigralis* Warren, 1896 | India, Meghalaya, Khasi Hills | |
| ***C. spirosticha*** Meyrick, 1935 | Indonesia, Java, Telawa | Unknown |
| ***C. tharsalea*** (Meyrick, 1887) | Australia | Unknown |
| ***C. umbrosa*** Meyrick, 1886 | New Guinea, Fly River | Unknown |

*evaxalis* and *C. pluto* + *C. sahyadriensis* are not closely related and that their Zingiberaceae host plant use has evolved independently.

## 1.3 Systematics and Delimitation

*Conogethes* was described by Meyrick, 1884, with *Astura punctiferalis* Guenée, 1854, as type species. Hampson (1896) synonymised *Conogethes* with *Dichocrocis* Lederer, 1863. This was followed by most authors, and *C. punctiferalis* was treated in *Dichocrocis* until the early 1980s. Honda and Mitsuhashi (1989) still consider *Conogethes* synonymous with *Dichocrocis*, but give priority to the former, younger name without explanation; they treat *punctiferalis* in *Conogethes*. *Conogethes punctiferalis* is redescribed by Inoue and Yamanaka (2006) in context of the description of two closely related species, *C. pinicolalis* and *C. parvipunctalis*.

*Conogethes* species (Fig. 1.1) resemble those of the African genus *Marwitzia* Gaede, 1917, and several Central- and South American species of *Polygrammodes* Guenée, 1854 (Fig. 1.2). Other species that are superficially similar to *Conogethes* are found in the genera *Dichocrocis*; *Notarcha* Meyrick, 1884; and *Trigonobela* Turner, 1915. The South American *Syllepte incomptalis* Hübner, 1823, the type species of the 'dustbin genus' *Syllepte* Hübner, 1823, also shows a wing pattern similar to that of *Conogethes*; the identity of *S. incomptalis*, however, could not be clarified so far as the type material is lost (Groll 2017).

## 1.4 Diagnosis

All adults of *Conogethes* exhibit a characteristic yellow wing ground colour and series of dark dots and dashes forming the ante- and postmedial as well as subterminal lines and the discal markings. In *C. haemactalis* Snellen, 1890, and *C.*

**Fig. 1.1** Adults of different *Conogethes* species, dorsal view. **(a)** *Conogethes clioalis* (Walker 1859a, b, c), Holotype (OUMNH); **(b)** *C. diminutiva* Warren, 1896 (ANIC); **(c)** *C. ersealis* (Walker 1859a, b, c) (ANIC); **(d)** *C. evaxalis* (Walker 1859a, b, c) (NHMUK); **(e)** *C. haemactalis* Snellen, 1890 (ANIC); **(f)** *C. mimastis* Meyrick, 1897, Cotype (NHMUK); **(g)** *C. parvipunctalis* Inoue and Yamanaka, 2006 (NHMUK); **(h)** *C. pinicolalis* Inoue and Yamanaka 2006 (ZMBN); **(i)** *C. pluto* (Butler, 1887) (ANIC); **(j)** *C. punctiferalis* (Guenée, 1854) (ANIC); **(k)** *C. sahyadriensis* Shashank et al. 2018, abdomen removed (ZMBN); **(l)** *C. semifascialis* (Walker 1866) (ANIC); **(m)** *C. spirosticha* Meyrick, 1934, Holotype (NHMUK); **(n)** *C. tharsalea* (Meyrick 1887) (ANIC). Specimens not to scale; OUMNH image (A) is © Copyright Oxford University Museum of Natural History, permalink: HYPERLINK "http://www.oum.ox.ac.uk/collections/irn/ca3241" www.oum.ox.ac.uk/collections/irn/ca3241; NHMUK images (D, F, G, M) are © Copyright Natural History Museum London; ANIC images (B, C, E, I, J, L, N) are © Copyright Len Willan and CSIRO Entomology, downloaded from http://www1.ala.org.au/gallery2/main.php?g2_itemId=11643&g2_page=3

**Fig. 1.2** Species that are superficially similar to Conogethes: **(a)** male Marwitzia dichocrocis (Hampson, 1913), a representative of the African genus Marwitzia Gaede, 1917; **(b)** male Polygrammodes cf. eleuata (Fabricius, 1777), a representative of the Neotropical P. eleuata species complex

*semifascialis* (Walker, 1866), the forewings' medial area is usually darker than the basal and postmedial areas. This dark-spotted wing pattern on yellow ground is also present in the *Polygrammodes eleuata* (Fabricius, 1777) complex and species of *Marwitzia* Gaede, 1917. The reasons for this striking similarity remain unknown, but features of the genitalia suggest that it might not be due to an evolutionary sister group relationship. Maes (1998) revised *Marwitzia* and its three species and distinguished it from *Conogethes* based on the labial palps and the distinct genitalia. From the genitalia of *Polygrammodes* species figured in Munroe (1958, 1960), it is evident that *Conogethes* is also distinct from this genus. *Conogethes* is distinguished from the *P. eleuata* complex and *Marwitzia* by the following characters:

The labial palps are upturned (also in *P. eleuata* s.l.), whereas *Marwitzia* species have porrect labial palps. In the male genitalia (Fig. 1.3), the uncus of *Conogethes* and *Marwitzia* is capitate with a bulbous ovate head on a tubular, curved neck; the uncus head is densely covered with deeply bifid chaetae on the dorsal side; the apex of the uncus head bears sparse simple chaetae (absent in *Marwitzia*); the uncus of the *P. eleuata* complex (and generally in *Polygrammodes*) is elongate conical and weakly sclerotised, its ventral side with sparse simple, hair-like chaetae. In *Conogethes* and *Polygrammodes*, the gnathos consists of a well-sclerotised band that is fused with the ventral end of the subscaphium; in *Marwitzia* a gnathos band is not evident. The saccus of the vinculum is broad V-shaped in *Conogethes*, while it is broad U-shaped in *Marwitzia* and the *P. eleuata* complex. *Conogethes* exhibits an elongate, narrow juxta of subulate shape, while *Marwitzia* and the *P. eleuata* complex have a shorter ovate to almost circular juxta. In *Conogethes*, the fibula emerges from the distal part of a sclerotised band traversing the inner valva surface from dorsal of the sacculus base to the valva apex, while in the *P. eleuata* complex and *Marwitzia*, the fibula emerges from the end of a short valva sclerotisation spanning from the costa base to the distal sacculus. The dorsal joint of valva with vinculum formed by an elongate, ventrally directed rodlike process emerging from the costa base (also in *P. eleuata* s.l.), while in *Marwitzia*, this costal process is much shorter. The phallus is narrow and long, corresponding to the long ductus bursae in the females, and the vesica contains a long needle-like cornutus stretching through almost the entire phallus length; the *P. eleuata* complex has a considerably shorter phallus which also contains a needle-like cornutus; the phallus of *Marwitzia* is weakly sclerotised and contains a small uni- or multidentate cornutus.

The female genitalia (Fig. 1.3) of *Conogethes* and the *P. eleuata* complex exhibit a completely membranous corpus bursae without sclerotised structures, while *Marwitzia* comprises a pair of ovate or elongate strip-like signa. The presence of a membranous appendix bursae emerging laterally from the corpus bursae in *Conogethes* is a feature not often seen in Spilomelinae, but it is a common feature among Pyraustinae, the putative sister group to Spilomelinae (Regier et al. 2012). Among Spilomelinae, a laterally attached appendix bursae is found in the genera *Cadarena* Moore, 1886; *Filodes* Guenée, 1854; *Hydriris* Meyrick, 1885; *Gonocausta* Lederer, 1863; and *Syllepis* Poey, 1832. The ductus bursae is longer and narrower than in *Marwitzia* and the *P. eleuata* complex. The size of the antrum varies, but in the *C. punctiferalis* complex, it is as wide as the ductus bursae and formed by a short and narrow, dorsally open sclerotised sheath around the copulatory duct (paralleled in *Marwitzia dichocrocis* Hampson, 1913), whereas in the *P. eleuata* complex and the other two *Marwitzia* species, the antrum is wider than the ductus bursae. The papillae anales are enlarged towards the dorsal end of the ovipositor as in *Marwitzia*, whereas in the *P. eleuata* complex, the papillae anales have a semicircular shape in lateral view.

Several morphological characters are useful for species distinction within *Conogethes*. Among the external morphology, these are colouration of the 2nd meron of the labial palps; maculation pattern and colouration of the wings, which is reddish-brown in the *C. haemactalis* complex, otherwise usually black; metathorax dorsally with two to three dark spots, in *C. tharsalea* only a transverse median streak; and colouration of legs and the presence of a hair tuft on the male hind leg tibia (in *C. evaxalis*). In the male genitalia, interspecific variation is found in the breadth and course of the sclerotised band that spans the inner side of the valva from sacculus base to apex, the shape and orientation of the fibula that emerges from this sclerotised valva band, the size and degree of sclerotisation of the transtilla, the shape of the tegumen roof (with a domed protrusion in most investigated species), the length of the phallus and the presence and length of a cornutus (usually needle-like). In the female genitalia, the size and degree of sclerotisation of the antrum, the length and degree of sclerotisation of the ductus bursae and the point of protrusion of the appendix bursae can vary between species. The shape of the tympanal organs and the venulae secundae could be an additional useful character to discriminate *Conogethes* species.

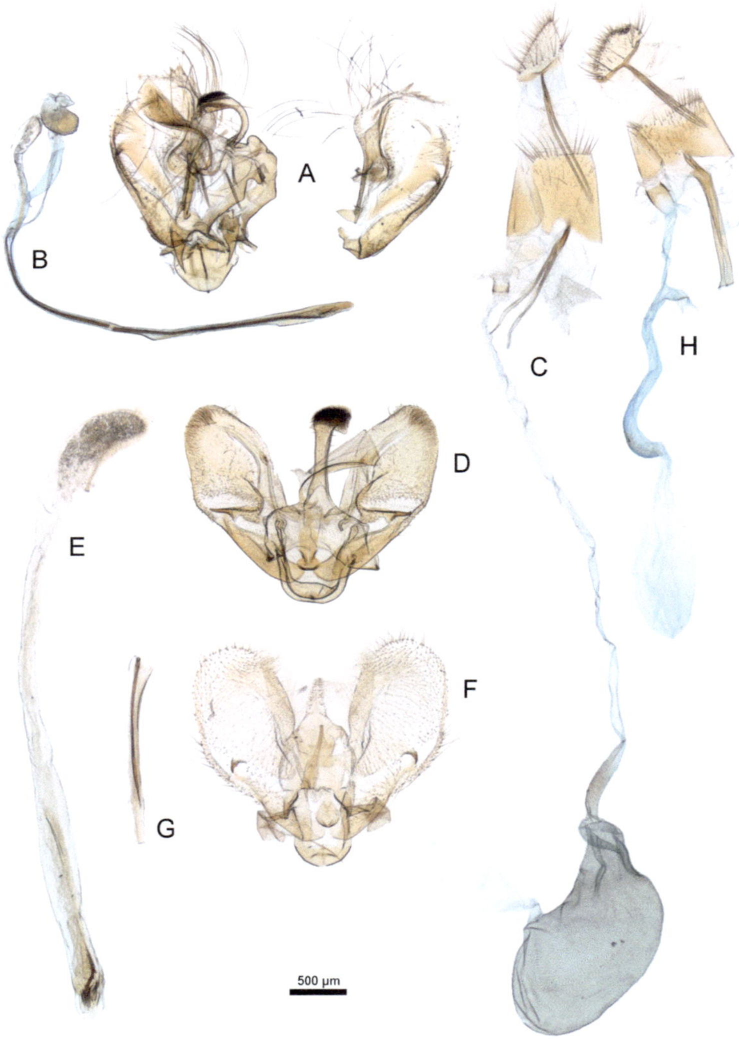


**Fig. 1.3** Genitalia of *Conogethes* and superficially similar species. (**a**–**c**) *Conogethes punctiferalis*; (**a**) male genital, right valva detached; (**b**) phallus; (**c**) female genital; (**d**–**e**) *Marwitzia dichocrocis* (Hampson, 1913); (**d**) male genital; (**e**) phallus; (**f**–**h**) *Polygrammodes* cf. *eleuata* (Fabricius, 1777); (**f**) male genital; (**g**) phallus; (**h**) female genital. All genitalia to scale

## 1.5 Phylogenetic Relationships

The phylogenetic relationships both within *Conogethes* and among Spilomelinae are poorly known. The only phylogenetic analysis that has been undertaken to infer relationships within *Conogethes* is that by Shashank et al. (2018), based on COI barcode data. Regarding the relationship of *Conogethes* to other genera, several morphological characters could be indicative of a close evolutionary relationship: despite the differences in the genitalia of *Conogethes*, *Marwitzia* and the *P. eleuata* complex as pointed out above, they appear more similar to each other than to those of many other Spilomelinae, indicating a potential closer relationship. The appendix bursae, emerging laterally from the corpus bursae, could be another phylogenetically useful character. Furthermore, Clavijo Albertos (1990) found the anepisternal scale organ – the putative tymbal organ of Nakano et al. (2012b) – in a number of other Spilomelinae, namely, *Anarmodia bistrialis* (Guenée, 1854), *Antigastra catalaunalis* (Duponchel, 1833), *Palpita flegia* (Cramer, 1777), *Sparagmia gonoptera* (Latreille, 1828) and species of *Omiodes* Guenée, 1854, and he did not find this organ in several other Spilomelinae and Pyraustinae. Considering the complexity of this organ, it is reasonable to assume that the species exhibiting this organ share a common ancestor in which this paired tymbal structure originally evolved.

Nakano et al. (2009) reported on three Spilomelinae species capable of ultrasound production, with a sonic click pattern similar to that observed later (Nakano et al. 2012a) in *C. punctiferalis*. These three Spilomelinae species are *Glyphodes pyloalis* Walker, 1989c, *Palpita nigropunctalis* (Bremer, 1864) and *Spoladea recurvalis* (Fabricius, 1775), and it is possible that they exhibit a tymbal organ similar to that of *C. punctiferalis* and of the species reported by Clavijo Albertos (1990).

Shashank et al. (2018) observe 17 species clades in their results, while currently only 15 species are recognised in *Conogethes*. This points to at least two undescribed species in the genus. Furthermore, Armstrong (2010) reports on another potential cryptic *Conogethes* species in New Zealand. These cases point out the need for a thorough systematic integrative revision of the genus in order to gain a deep understanding of the diversity within *Conogethes* and to establish a solid baseline on which future research can build. Future phylogenetic work should be based on a broad taxon and data sample and should address the relationships among *Conogethes* species, the relationship of *Conogethes* to the African genus *Marwitzia* and the Neotropical, *Polygrammodes eleuata* species group and the phylogenetic position of the genus within Spilomelinae.

**Acknowledgements** Study material was provided by Leif Aarvik (Natural History Museum Oslo), Bernard Landry (Natural History Museum Geneva), Scott E. Miller and M. Alma Solis (both Smithsonian Institution, Washington, D.C.) and Marja van der Straten (Dutch Food and Products Authority, Wageningen). James Hogan (Oxford University Museum of Natural History) and David Lees (Natural History Museum London) provided images of *Conogethes* imagines. Steffen Roth (University Museum of Bergen) provided helpful comments.

## References

Armstrong K (2010) DNA barcoding: a new module in New Zealand's plant security diagnostic toolbox. EPPO Bull 40:91–100

Bremer O (1864) Lepidopteren Ost-Sibiriens, insbesondere des Amur-Landes, gesammelt von den Herren G. Radde, R. Maack und P. Wulffius. Mémoires de l'Académie des Sciences de St-Pétersbourg. St. Petersburg (Ser 7) 8(1):1–104, pp 1–8

Butler AG (1887) Descriptions of new species of Heterocerous Lepidoptera (Pyralites) from the Solomon Islands. Ann Mag Nat Hist Incl Zool Bot Geol, Lond (ser. 5) 20:114–124

Clavijo Albertos JA (1990) Systematics of black and white species of the genus *Diaphania* Hübner, 1818 (Lepidoptera: Pyralidae: Pyraustinae). Dissertation, Department of Entomology, McGill University, Montreal, Canada, pp i–xi, 1–215, figs 1–159

Cramer P (1777) De uitlandsche kapellen, voorkomende in de drie waereld-deelen Asia, Africa en America [Papillons exotiques, des trois parties du monde l'Asie, l'Afrique et l'Amerique], vol 2. S. J. Baalde & B. Wild, Amsterdam/Utrecht, pp 1–151, pls 97–192

Doddabasappa D, Chakravarthy AK, Thyagaraj NE (2014) Comparative biology of shoot and capsule borer, *Conogethes punctiferalis* (Guenee), (Crambidae: Lepidoptera) on castor and cardamom. Curr Biotica 8(3):228–245

Duponchel PAJ (1833) Nocturnes 5 (2). Histoire naturelle des Lépidoptères ou Papillons de France. Paris 8(2):5–402, errata, pls 211–236

El-Sayed AM (2017) The pherobase: database of insect pheromones and semiochemicals. http://www.pherobase.com

El-Sayed AM, Gibb AR, Mitchell VJ, Manning LAM, Revell J, Thistleton B, Suckling DM (2012) Identification of the sex pheromone of *Conogethes pluto*: a pest of Alpinia. Chemoecology 23(2013):93–101

Fabricius JC (1775) Systema entomologicae, sistens insectorum classes, ordines, genera, species, adjectis synonymis, locis, descriptionibus, observationibus. Kortii, Flensburgi et Lipsiae. i–xxx, 1–832

Fabricius JC (1777) [imprint "1776"]: Genera insectorum eorumque charcteres naturales secundum numerum, figuram, situm et proportionem omnium partium oris adiecta mantissa specierum nuper detectarum. M. F. Bartschii, Chilonii. [i]–[xii], 1–310

Gaede M (1917) Neue Lepidopteren des Berliner Zoologischen Museums. I. Aethiopische Pyralididen. Mitteilungen aus dem Zoologischen Museum in Berlin 8(3):387–401

Groll EK (2017) Biographies of the entomologists of the world. Online database, version 8. Senckenberg Deutsches Entomologisches Institut, Müncheberg http://sdei.senckenberg.de/biographies/

Guenée MA (1854) Deltoïdes et Pyralites. In: de Boisduval JBAD, Guenée MA (eds) Histoire Naturelle des Insectes. Species Général des Lépidoptères 8 8. Roret, Paris, pp 1–448

Hampson GF (1896) Moths. The Fauna of British India, including Ceylon and Burma. London 4:i–xxviii, 1–594

Hampson GF (1913) Descriptions of new species of Pyralidae of the subfamily Pyraustinae. Ann Mag Nat Hist, Incl Zool Bot Geol Lond (Ser 8) 11(63, 66):322–342, 509–530

Honda H, Mitsuhashi W (1989) Morphological and morphometrical differences between the fruit- and Pinaceae-feeding type of yellow peach moth, *Conogethes punctiferalis* (Guenée) (Lepidoptera: Pyralidae). Appl Entomol Zool Tokyo 24(1):1–10

Hübner J (1808–1818) ["1818"] Zuträge zur Sammlung exotischer Schmettlinge [sic], bestehend in Bekundigung einzelner Fliegmuster neuer oder rarer nichteuropäischer Gattungen. Augsburg. [1]-[3]-4-6-[7]-8-32-[33]-[40], pls [1]-[35]

Hübner J (1819–1823) ["1823"] Zuträge zur Sammlung exotischer Schmettlinge [sic], bestehend in Bekundigung einzelner Fliegmuster neuer oder rarer nichteuropäischer Gattungen. Augsburg. [1]–[3]–4–6–[7]–8–32–[33]–[40], pls [36]–[69]

Inoue H, Yamanaka H (2006) Redescription of *Conogethes punctiferalis* (Guenée) and descriptions of two new closely allied species from Eastern Palaearctic and Oriental Regions (Pyralidae, Pyraustinae). Tinea, Tokyo 19(2):80–91

Kaneko JI (1978) Abdominal constriction in the copulated female of yellow peach moth, *Dichocrocis punctiferalis* Guenée (Lepidoptera: Pyralidae). Appl Entomol Zool Tokyo 13(2):131–133

Kimura T, Sakai J, Honda H (2002) Microstructure and pheromone producing function of male hair-pencils in the yellow peach moth, *Conogethes punctiferalis* (Lepidoptera: Pyralidae). Entomol Sci Tokyo 5(2):237–247

Konno Y, Honda H, Matsumoto Y (1980) Observations on the mating behavior and bioassay for the sex pheromone of the Yellow Peach Moth, *Dichocrocis punctiferalis* Guenée (Lepidoptera: Pyralidae). Appl Entomol Zool Tokyo 15(3):321–327

Konno Y, Arai K, Sekiguchi K, Matsumoto Y (1982) E-10-Hexadecenal, a sex pheromone component of the yellow peach moth, *Dichocrocis punctiferalis* Guenée (Lepidoptera: Pyralidae). Appl Entomol Zool Tokyo 17(2):207–217

Latreille (1828) In: Humboldt, A. von & A. Bonpland, Relation historique du voyage aux régions équinoxiales du nouveau continent, fait en 1799, p 138 pl. 43 figs 7–8. 1800, 1801, 1802, 1803, et 1804 par Al. de Humboldt et A. Bonpland. Smith & Gide, Paris

Lederer J (1863) Beitrag zur Kenntniss der Pyralidinen. Wiener Entomologische Monatschrift 7(8, 10–12):243–280, 331–504, pls 2–18

Li J, Zhang Y, Wang ZY, He KL (2010) *Wolbachia* infection in four geographic populations of yellow peach moth, *Conogethes punctiferalis* in China. J Environ Entomol 32(2):322–328

Lucas TP (1892) On twenty new species of Australian Lepidoptera. Proc Linnean Soc N S W, Sydn 7(2):249–266

Lucas TP (1900) New species of Queensland Lepidoptera. Proc R Soc Qld Brisbane 15:137–161

Maes KVN (1998) Revision of the genus *Marwitzia* Gaede, 1917 (Lepidoptera, Pyraloidea, Crambidae, Spilomelinae). Lambillionea, Bruxelles 98(3):365–371

Mally R, Hayden JE, Neinhuis C, Jordal BH, Nuss M (subm.) The phylogenetic systematics of Spilomelinae and Pyraustinae (Lepidoptera: Pyraloidea: Crambidae) inferred from DNA and morphology. Arthropod Syst Phylogeny

Meyrick E (1884) On the classification of the Australian Pyralidina. Trans Entomol Soc Lond 61–80:277–350

Meyrick E (1885) On the classification of the Australian pyralidina. Trans Entomol Soc Lond 33:421–456

Meyrick E (1886) On some Lepidoptera from the Fly River. Proc Linn Soc N S W, Sydn Ser 1(2):241–258

Meyrick E (1887) On Pyralidina from Australia and the South Pacific. Trans Entomol Soc Lond: 185–268

Meyrick E (1897) On Lepidoptera from the Malay Archipelago. Trans R Entomol Soc Lond: 69–92

Meyrick E (1930–1936) Exotic microlepidoptera. Taylor and Francis, London, pp 1–642

Moore F (1884–1887) The Lepidoptera of Ceylon. L. Reeve, London. i–xvi, 1–578, pls 144–214

Munroe EG (1958) Some species of *Polygrammodes* Guenée (Lepidoptera: Pyralidae). Can Entomol Ottawa 90:257–274

Munroe EG (1960) New species of *Polygrammodes* and a related new genus (Lepidoptera: Pyralidae). Can Entomol Ottawa 92(4):279–284

Munroe EG (1989) Changes in classification and names of Hawaiian Pyraloidea since the publication of Insects of Hawaii, Volume 8, by E. C. Zimmerman (1958) (Lepidoptera). Bish Mus Occas Pap Honol 29:199–212

Nakano R, Takanashi T, Fujii T, Skals N, Surlykke A, Ishikawa Y (2009) Moths are not silent, but whisper ultrasonic courtship songs. J Exp Biol 212:4072–4078

Nakano R, Takanashi T, Ihara F, Mishiro K, Toyama M, Ishikawa Y (2012a) Ultrasonic courtship song of the yellow peach moth, *Conogethes punctiferalis* (Lepidoptera: Crambidae). Appl Entomol Zool Tokyo 47:87–93

Nakano R, Ihara F, Mishiro K, Toyama M (2012b) Male courtship ultrasound produced by mesothoracic tymbal organs in the yellow peach moth *Conogethes punctiferalis* (Lepidoptera: Crambidae). Appl Entomol Zool Tokyo 47:129–135

Nakano R, Ihara F, Mishiro K, Toyama M, Toda S (2014) Double meaning of courtship song in a moth. Proc R Soc B: Biol Sci Lond 281:1–8

Nuss M, Landry B, Mally R, Vegliante F, Tränkner A, Bauer F, Hayden JE, Segerer A, Schouten R, Li H, Trofimova T, Solis MA, De Prins J, Speidel W (2003–2017) GlobIZ – Global Information System on Pyraloidea. www.pyraloidea.org

Poey F (1832–1833) Centurie de Lépidoptères de l'Île de Cuba. J. Albert Mercklein, Paris. 54 pp., 20 pls

Regier JC, Mitter C, Solis MA, Hayden JE, Landry B, Nuss M, Simonsen TJ, Yen SH, Zwick A, Cummings MP (2012) A molecular phylogeny for the pyraloid moths (Lepidoptera: Pyraloidea) and its implications for higher-level classification. Syst Entomol 37:635–656

Robinson GS, Ackery PR, Kitching IJ, Beccaloni GW, Hernández LM (2010) HOSTS – a database of the world's lepidopteran hostplants. Natural History Museum, London http://www.nhm.ac.uk/hosts. Accessed 16 June 2017

Shashank PR, Kammar V, Mally R, Chakravarthy AK (2018) A new Indian species of shoot and capsule borer of the genus *Conogethes* (Lepidoptera: Crambidae), feeding on cardamom. Zootaxa 4374(2):215–234

Snellen PCT (1890) A catalogue of the Pyralidina of Sikkim collected by Henry J. Elwes and the late Otto Möller, with notes by H. J. Elwes. Trans Entomol Soc Lond: 557–647, pls 19–20

Snellen PCT (1895) Aanteekeningen over Pyraliden met Beschrijving van nieuwe Genera en Soorten. Tijdschrift voor Entomologie, 's Gravenhage 38:103–161, pls 5–6

Truscott LAC (2007) *Conogethes punctiferalis* (Guenée, 1854) (Lepidoptera: Crambidae) at light in Cornwall, newly recorded for Europe. Entomol Gaz 58:203–204

Turner AJ (1915) Studies in Australian Lepidoptera. Proc R Soc Qld Brisbane 27:11–57

Walker F (1859a) Catalogue of the heterocerous Lepidoptera collected at Singapore by Mr. A. R. Wallace, with descriptions of new species. J Linn Soc Lond Zool 3:183–196

Walker F (1859b) Pyralides. List of the specimens of Lepidopterous insects in the collection of the British Museum, London 18:509–798

Walker F (1859c) Pyralides. List of the specimens of Lepidopterous insects in the collection of the British Museum, London 19:799–1036

Walker F (1866) ["1865"] Supplement 4. List of the specimens of Lepidopterous insects in the collection of the British Museum, London 34:1121–1533

Warren W (1896) New species of Pyralidae from the Khasia Hills. Ann Mag Nat Hist Incl Zool Bot Geol Lond (ser. 6) 18:107–119, 163–177, 214–232

Werren JH, Baldo L, Clark ME (2008) *Wolbachia*: master manipulators of invertebrate biology. Nat Rev Microbiol 6(10):741–751

# Molecular Status of *Conogethes* spp.: An Overview

# 2

Vasudev Kammar, P. R. Shashank, A. T. Rani, and V. Selvanarayanan

**Abstract**

*Conogethes* sp. is a large taxon that infests more than 120 wild and cultivated host plants across the world. So far, several scientists have studied this group of moth pests to appreciate the genetic variability in the genus; however, owing the complexity at genetic and ecological levels, it has become difficult to understand this group of moths. Till date 24 species have been deposited in the Barcode of Life Data (BOLD) system. In order to study the molecular diversity of *Conogethes*, multiple markers have been used world over. However, several researchers have revealed that combined analysis of molecular markers and morphological data with field observations may constitute powerful evidence for proper identification of pest species in this problematic taxa. Precise identification would then be utilized for developing realistic practices for the management of *Conogethes* sp. in diversified cultivated ecosystems.

**Keywords**

*Conogethes* sp. · Genetic variability · Molecular marker · Morphological data

V. Kammar (✉)
Department of Food and Public Distribution, Ministry of Consumer Affairs and Food and Public Distribution, Government of India, New Delhi, India

P. R. Shashank
Insect Taxonomy Laboratory, Division of Entomology, ICAR-Indian Agricultural Research Institute, New Delhi, India

A. T. Rani
Division of Crop Protection, ICAR-Indian Institute of Vegetable Research (IIVR), Varanasi, Uttar Pradesh, India

V. Selvanarayanan
Department of Entomology, Faculty of Agriculture, Annamalai University, Chidambaram, Tamil Nadu, India


A. K. Chakravarthy (ed.), *The Black spotted, Yellow Borer, Conogethes punctiferalis Guenée and Allied Species*,
https://doi.org/10.1007/978-981-13-0390-6_2

## 2.1 Introduction

The fruit and shoot borer *Conogethes* (*Dichocrocis*) *punctiferalis* is an important polyphagous pest as the larvae of this crambid moth typically attack more than 120 diversified and economically important plant species including wild and cultivated plants. *Conogethes* mainly occurs in tropical and subtropical countries (Pena et al. 2002) and is distributed in the eastern Palaearctic and Indo-Australian regions (Shaffer et al. 1996). At present, the genus *Conogethes* Meyrick 1884, includes 14 species (Shaffer et al. 1996; Inoue and Yamanaka 2006). The complex structure of male genitalia and a broad larval host range have made *Conogethes* taxonomically a difficult taxon. There are two different types of *C. punctiferalis*, i.e. the fruit-feeding type on angiosperms and the pinaceae-feeding type on pinaceae gymnosperms (Koizumi 1960). Inoue and Yamanaka (2006) redescribed *C. punctiferalis* along with two new species, i.e. the fruit-feeding-type *C. parvipunctalis* and the pinaceae-feeding-type *C. pinicolalis* (Inoue and Yamanaka 2006), from the eastern Palaearctic and Oriental regions. Wang et al. (2014) have confirmed *C. punctiferalis* and *C. pinicolalis* to be two different species by reconstructing the phylogenetic tree on the basis of the sequence data from the combined gene markers COI, COII and Cytb of the mitochondrial cytochrome. The type locality of *C. punctiferalis* is India; hence many closely allied species may be included; however, their taxonomic revision has been neglected for a long time. *C. punctiferalis* is in focus owing to the expanding host range, geographical occupancy and complexity involved in species identification. As the pest is an internal tissue borer, it is difficult to manage it in fruit orchards and plantations. Further, this insect group is undergoing speciation, genomic changes and is evolving actively.

The use of DNA sequences is a promising and effective tool for fast and accurate species identification (Hebert et al. 2003; Waugh 2007; Pereira et al. 2008). Molecular characterization and DNA barcoding is a taxonomic method that uses a short genetic marker in an insect DNA to identify a species, including the unknown. The DNA barcode method of identification includes, for example, identifying insect species from any developing stage and part; whereas, generally, morphological identification of insects depends on the adult stage and male genitalia (Jalali et al. 2015). Molecular identification technique possesses several advantages over the conventional ones. Molecular techniques have been successfully applied in vertebrate and invertebrate taxa for species delimitation and identification (Smith et al. 2005; Clare et al. 2006; Hubert et al. 2008; Smith and Fisher 2009; Zhou et al. 2009).

Accurate identification of insect species is one of the important aspects of entomological science. In most groups, traditional taxonomic research is based on morphological characters, and it is difficult to identify cryptic and polymorphic species through conventional taxonomy; however, there are hurdles for want of experts. In this connection, molecular methods have been found valuable in discriminating cryptic species (Jackson and Resh 1998; Pilgrim et al. 2002).

Cryptic species are defined as two or more distinct species which are classified as a single nominal species as they are morphologically indistinguishable (Bickford et al. 2007). One of the mechanisms thought to promote speciation in phytophagous insects is shifting to new hosts that lead to the establishment of new species by way of an intermediate step of host-race formation (Dres and Mallet 2002). The occurrence of insect host races reflects the recently evolved genetic differentiation with respect to host plant use (Dres and Mallet 2002).

## 2.2 Biosystematics

For the last several decades, the genus *Conogethes* was placed under Pyralidae. Maes (1998) demonstrated the differences in the structure of the tympanal organ, or ears, in Crambidae and Pyralidae which was further supported by Munroe and Solis (1999) and Kristensen (1999), and *Conogethes* was retained in the Crambidae family. *Conogethes* sp. is taxonomically and genetically a complex taxon belonging to the superfamily Pyraloidea, family Crambidae, subfamily Pyraustinae. The species *punctiferalis* was placed in the genus *Conogethes* by Meyrick (1884), although it was moved to *Dichocrocis* after that. It was reclassified as *Conogethes* by Munroe (1989), and Shaffer et al. (1996) placed it in *Conogethes* as a revised combination. Hampson (1896) in his "Fauna of British India" reported 20 species of *Dichocrocis* on the basis of wing venation and arrangements of black spots on wings. Subsequently, a few preliminary studies have been carried out on this genus by lepidopterists in India and abroad. Five species of *Dichocrocis* were identified, namely, *D. evaxalis* Walker, *D. punctiferalis* Walker, *D. nigrilnealis* Walker, *D. plutusalis* Walker and *D. surusalis* Walker from light trap collections in Kerala, India (Mathew and Menon 1984).

## 2.3 Molecular Identification

### 2.3.1 *Conogethes* Barcode

The family Crambidae, subfamily Pyraustinae, has 4585 species with barcodes. The genus *Conogethes* has 138 barcode sequences and about 19 species (Table 2.1) from six countries, namely, Australia, Papua New Guinea, Cambodia, China, Indonesia and Nepal, but these sequences are not available in the public domain (IBOL 2012).

Armstrong (2010) compared DNA barcoding of different populations of *Conogethes* and revealed that the Australian and Asian specimens form separate clades divergent by ~6%. The barcode data successfully distinguished *C. punctiferalis* and *C. pluto* but unexpectedly revealed divergence between the Asian and Australian populations. Morphologically these were determined to be the same species but distinct from other closely related species found on the east coast of Australia such as *C. haemactalis* Walker, *C. semifascialis* Walker or *C. tharsalea* Walker.

**Table 2.1** *Conogethes* species with records on barcode of life database

| Sl. No. | Species | Specimens | Sequences | Sequences >500 bp |
|---|---|---|---|---|
| 1. | *Conogethes diminutiva* | 2 | 2 | 1 |
| 2. | *Conogethes ersealis* | 3 | 3 | 3 |
| 3. | *Conogethes evaxalis* | 10 | 9 | 9 |
| 4. | *Conogethes haemactalis* | 5 | 5 | 5 |
| 5. | *Conogethes sp. nr. diminutiva* | 1 | 1 | 1 |
| 6. | *Conogethes sp. nr. haemactalis* | 3 | 3 | 3 |
| 7. | *Conogethes sp. nr. semifascialis* | 4 | 4 | 3 |
| 8. | *Conogethes parvipunctalis* | 1 | 1 | 1 |
| 9. | *Conogethes pinicotalis* | 1 | 1 | 1 |
| 10. | *Conogethes pluto* | 12 | 12 | 12 |
| 11. | *Conogethes punctiferalis* | 79 | 60 | 59 |
| 12. | *Conogethes semifascialis* | 17 | 14 | 11 |
| 13. | *Conogethes* sp. | 1 | 0 | 0 |
| 14. | *Conogethes* sp. ANIC1 | 2 | 0 | 0 |
| 15. | *Conogethes* sp. ANIC2 | 3 | 3 | 3 |
| 16. | *Conogethes sp.* ANIC3 | 1 | 1 | 0 |
| 17. | *Conogethes* sp. ANIC4 | 1 | 0 | 0 |
| 18. | *Conogethes* sp. complex | 4 | 3 | 3 |
| 19. | *Conogethes tharsalea* | 5 | 4 | 4 |

Source: http://www.boldsystems.org/views/speciessummary.php (Shashank 2012; Vasudev 2013; Bold 2012)

## 2.4 Japan

The biosystematics of Japanese *Conogethes* sp. was done by Honda (2013) with special reference to the host plant preference and reproductive isolation. In Japan, *C. punctiferalis* (CPU) and *C. pinicolalis* (CPI) are the most well-known pest species of agricultural and forest plants. The *C. punctiferalis* was called as the fruit-feeding type (FFT) in order to distinguish from the pinaceae-feeding type of *C. punctiferalis*, which was registered as *C. pinicolalis* in 2006.

## 2.5 India

Azam and Ali (1965) studied the morphology of larva of *D. punctiferalis* with special reference to chaetotaxy collected from the castor bean (*Ricinus communis* L.).

In the mid-1980s, Chakravarthy found differences in morphology of *Conogethes* moths reared on castor and cardamom (*Elettaria cardamomum* Maton) in the Western Ghats of Karnataka, South India. The male genitalia also differed between the two. The *Conogethes* larvae reared on castor bean and cardamom required two different mass-rearing techniques (Chakravarthy et al. 1991).

Nowadays, DNA barcoding is a major tool for species identification, and molecular taxonomy provides additional support for species identification through traditional taxonomy. Seventeen species from six countries, namely, Australia, Papua New Guinea, Cambodia, China, Indonesia and Nepal, were barcoded for genus *Conogethes* and deposited in BOLD to date (Shashank et al. 2015). Unfortunately, there is no single species from India that is barcoded, and the species listed in Hampson's Fauna are different from the ones that are barcoded. Recently, Shashank (2012) barcoded *Conogethes* moths reared on castor and cardamom from different geographical locations of India.

In India, larvae of *C. punctiferalis* that infest castor (*Ricinus communis* L.) and cardamom (*Elettaria cardamomum* Maton) have been identified as two different lineages based on different mass-rearing techniques (Chakravarthy et al. 1991, 2015). Subsequently, Shashank (2012) and Shashank et al. (2014a, b) accurately indicated differences in morphology and morphometry of male and female genitalia, larvae and pupae, in molecular data and in behaviour, like adult emergence pattern, calling and mating behaviour and effect of interbreeding on offsprings between the two types of *C. punctiferalis*, but they refrained from giving definite taxonomic action. Genetic analysis revealed significant genetic differentiations among the two sampled populations, reflecting the limited gene flow. Recently Shashank (In print) described the cardamom-feeding type as a new species based on morphological and molecular data as *Conogethes sahyadriensis* (Shashank et al. 2018).

Vasudev et al. (2016) conducted a study on the genetic diversity of *Conogethes* species infesting select host plants based on COI genes. The results showed that the pairwise genetic distance analysis between the individuals varied from 0.000 to 0.076, indicating a high genetic divergence. The nearest neighbour distance between *Conogethes* bred within the 15 populations was 5.32%, indicating wide genetic variability between two *Conogethes* populations. Sequence length showed significant variation from 477 bp (castor) to 726 bp (gingiberaceae) and per cent G + C content for COI showed low variations (0.17%) compared to the cardamom *Conogethes* species. In addition, topologies of neighbour-joining tree indicated that the *Conogethes* sp. breeding on castor, mango, pear, peach, plum, guava and sapota belongs to *C. punctiferalis* while those feeding on cardamom, turmeric and ginger are of a separate clade. Further genetic analysis revealed significant genetic differentiations among the two sampled populations, reflecting limited gene flow. The results of an analysis of molecular variance (AMOVA) indicated the existence of significant genetic variation among the examined host races, suggesting that the variations in *Conogethes* populations are genetically heterogeneous (Vasudev 2013).

Alagar et al. (2013) developed a DNA-based molecular identification system for the identification and confirmation of the close association of *C. punctiferalis* populations from cocoa with other related moth populations. They designed primers based on two nuclear genes, viz. ribosomal protein S5 (RPS5) gene and carbamoyl phosphate synthetase/aspartate transcarbamylase/dihydroorotase (CAD). PCR-amenable DNA was isolated from *C. punctiferalis* larva. The designed primers amplified single bands of the expected sizes using genomic DNA as template. The amplicons were purified, cloned and sequenced. Sequence analysis using BLASTn revealed close homology to the gene of interest from related moths. The phylograms constructed by BLASTn analysis showed the close association of *C. punctiferalis* from cocoa with *C. punctiferalis* from Hawaii moths of Spilomelinae subfamily and with other related moths. They also suggested that both RPS5 and CAD genes behave as a single copy in PCR reactions and a combination of these two genes could be useful in molecular systematics for specific amplification of *C. punctiferalis* DNA.

## 2.6 China

Zhang (2010) investigated the genetic diversity of 11 geographic populations of *Conogethes punctiferalis* in China using inter-simple sequence repeat (ISSR) markers. The results of the study showed that 209 bands were polymorphic, making up 99.05% of the total 211 amplified bands. The genetic distances between different *C. punctiferalis* populations were 0.0059–0.0237. The Nei's index, Shannon information index and coefficient of genetic (gene) differentiation among populations ($G_{st}$) were 0.1750, 0.2966 and 0.053, respectively, and the estimated value of gene flow from $G_{st}$ was 8.8724. These results suggested that *C. punctiferalis* populations in China kept a low level of population genetic differentiation due to a considerable gene flow. Recent study conducted by Wang et al. (2014) revealed that *C. punctiferalis* was originally considered as a single species with fruit-feeding type (FFT) and pinaceae-feeding type (PFT), but it has subsequently been divided into two different species of *C. punctiferalis* and *C. pinicolalis*. For further details on *Conogethes* sp. in China, refer to the other chapter in this book.

## 2.7 New Zealand

In New Zealand, *C. punctiferalis* are pests of quarantine status and were intercepted using DNA barcode diagnostic methods to improve discrimination of cryptic species within species complexes. Barcoding of *C.punctiferalis* distinguished it from *C. pluto*, a sympatric pest within the yellow peach moth complex (Armstrong 2010). *C. punctiferalis* has been categorized as a potentially high impact pest of stone fruit in New Zealand and is targeted for active surveillance using pheromone traps (Ganev and Braithwaite 2003; Stephenson et al. 2003). However, species in the complex have a very similar morphology, variable colour morphs and overlapping host range.

## 2.8 Other Molecular Works

The insect olfactory system enables them to detect and discriminate different odour molecules from their environment. The odourant and pheromone perception in insects is a complex series of processes that transform external stimulus from the environment into behavioural response. Diverse proteins constituting this signal transduction pathway include odourant binding proteins (OBPs), chemosensory proteins (CSPs), chemosensory receptor (CRs), odourant degrading enzymes (ODEs) and sensory neuron membrane proteins (SNMPs) (Fan et al. 2011; Leal 2013). The gene family encoding CRs in insects are divided mainly into three groups: olfactory receptors (ORs), gustatory receptors (GRs) and ionotropic receptors (IRs) (Nei et al. 2008; Cande et al. 2013). The insect ORs and the GRs were identified for the first time in the genomic analysis of *Drosophila melanogaster* (Robertson et al. 2003).

The chemosensory systems of *Conogethes* moth play an important role in detecting food, oviposition sites and mate attraction. Several antennal chemosensory receptors are involved in odour detection (Ge et al. 2016). Olfactory receptors have potential applications in behavioural studies of yellow peach moth. Identification of these chemosensory receptor genes provides control of this moth pest by semiochemical method. There are nearly 83 candidate chemosensory receptors, including 62 odorant receptors, 11 ionotropic receptors and 10 gustatory receptors that have been identified by transcriptomic analysis of male and female antennae (Xing et al. 2016).

Antennal transcriptomes of male and female yellow peach moths were sequenced and characterized by Xiao-Jian et al. (2016). In total, 15 putative odourant binding proteins (OBPs), 46 putative odourant receptors (ORs) and 7 putative ionotropic receptors (IRs) were annotated and identified as olfactory-related genes of *C. punctiferalis*. Further analysis of RT-qPCR revealed that all these olfactory genes are primarily or uniquely expressed in male and female antennae. Among which, 3 OBPs (*OBP4*, *OBP8* and *PBP2*) and 4 ORs (*OR22*, *OR26*, *OR44* and *OR46*) were especially expressed in male antennae, whereas 4 ORs (*OR5*, *OR16*, *OR25* and *OR42*) were primarily expressed in female antennae. The predicted protein sequences were compared with homologs in other lepidopteran species and model insects, which showed high sequence homologies between *C. punctiferalis* and *Ostrinia furnacalis* (Guenée).

**Acknowledgement** Authors are thankful to the BOLD website for the contents of Table 2.1 and to the authorities of the University of Agricultural Sciences, GKVK, Bengaluru, for encouragement and facilities.

## References

Alagar M, Rachana KE, Keshava Bhat S, Rahman S, Rajesh MK (2013) Biology, damage potential and molecular identification of *Conogethes punctiferalis* Guenee in cocoa (*Theobroma cacao* Linn.). J Plant Crop 41(3):350–356

Armstrong K (2010) DNA barcoding: a new module in New Zealand's plant bio-security diagnostic toolbox. Bull OEPP/EPPO 40:91–100

Azam KM, Ali MH (1965) Morphology of larva of *Dichocrocis punctiferalis* Guen., the castor shoot and capsule borer with special reference to chaetotaxy (Lepidoptera: Pyralidae). Indian J Entomol 27(4):423–431

Bickford D, Lohman DJ, Sodhi NS, Pkl NG, Meier R, Winker K, Ingram KK, Das I (2007) Cryptic species as a window on diversity and conservation. Trends Ecol Evol 22:148–155

Bold (2012) International barcode of life, Lepidoptera barcode of life. [Cited 11 May 2012] Available from URL: http://www.lepbarcoding.org/

Cande J, Prud'homme B, Gompel N (2013) Smells like evolution: the role of chemoreceptor evolution in behavioural change. Curr Opin Neurobiol 23:152–158

Chakravarthy AK, Honda H, Thyagaraj NE (1991) Comparison of containers for larval rearing in stalk and fruit feeding type of *Conogethes punctiferalis* (Guen.) Lepidoptera: Pyralidae. Placrosym IX:127–131

Chakravarthy AK, Vasudevkammar, Kumar KP (2015) Monograph on shoot and fruit borer, *Conogethes punctiferails* and allied species. Consortia Research Platform on Borers, ICAR-Indian Institute of Horticultural Research (IIHR), Bengaluru

Clare EL, Lim BK, Engstrom MD, Eger JL, Hebert PDN (2006) DNA bar-coding of neotropical bats: species identification and discovery within Guyana. Mol Ecol Notes 7:184–190

Dres M, Mallet J (2002) Host races in plant-feeding insects and their importance in sympatric speciation. Philos Trans R Soc Lond Ser B Biol Sci 357:471–492

Fan J, Francis F, Liu Y, Chen JL, Cheng DF (2011) An overview of odorant-binding protein functions in insect peripheral olfactory reception. Genet Mol Res 10:3056–3069

Ganev S, Braithwaite M (2003) Resource requirements for national active surveillance programmes of high impact exotic pests in New Zealand. N Z Plant Protect 56:10–15

Ge X, Zhang T, Wang Z, He K, Bai S (2016) Identification of putative chemosensory receptor genes from yellow peach moth *Conogethes punctiferalis* (Guenée) antennae transcriptome. Sci Rep:1–13. www.nature.com/scientificreports

Hampson GF (1896) The fauna of British India including Ceylon and Burma (Moths). Taylor and Francis, London, p 594

Hebert PDN, Cywinska A, Ball SL, Dewaardjr (2003) Biological identifications through DNA barcodes. Proc R Soc Lond B 270:313–321

Honda H (2013) Bio systematics of Japanese *Conogethes* sp. With special reference to host plant preference and reproductive isolation. In: Chakravarthy AK, Ashok Kumar CT, Varghese A, Thyagaraj NE (eds) New horizons in insect science. Abstracts ICIS-2013. Springer, Bangalore/New Delhi, p 136

Hubert N, Hanner R, Holm E, Mandrak NE, Taylor E, Burridge M, Watkinson D, Dumont P, Curry A, Bentzen P, Zhang J (2008) Identifying Canadian freshwater fishes through DNA barcodes. PLoS One 3:e2490

Inoue H, Yamanaka H (2006) Redescription of *Conogethes punctiferalis* (Guenée) and description of two new closely allied species from Eastern Palaearctic and oriental regions (Pyralidae: Pyraustinae). Tinea 19(2):80–91

Jackson JK, Resh VH (1998) Morphologically cryptic species confound ecological studies of the caddisfly genus *Gumaga* (Trichoptera: Sericostomatidae) in Northern California. Aquat Insects 20:69–84

Jalali SK, Ojha R, Venkatesan T (2015) DNA barcoding for identification of agriculturally important insects. In: Chakravarthy AK (ed) New horizons in insect science: towards sustainable pest management. Springer, New Delhi, pp 13–22

Jia X-J et al (2016) Antennal transcriptome and differential expression of olfactory genes in the yellow peach moth, Conogethes punctiferalis (Lepidoptera: Crambidae). Sci Rep 6:29067. https://doi.org/10.1038/srep29067

Koizumi K (1960) Two forms of *Dichocrocis punctiferalis* (Guenee) presumably representing separate species. The main purport of a lecture the 20th annual meeting of the Entomological Society of Japan, pp 8–9

Kristensen NP (1999) Lepidoptera, moths and butterflies. In: Volume 1: lEvolution, systematics, and biogeography (Handbuch der Zoologie [founded by Willy Kukenthal; ed.: Fischer, M.] Band 4, Teilband 35). Walter de Gruyter, Berlin, pp 217–232

Leal WS (2013) Odorant reception in insects: roles of receptors, binding proteins, and degrading enzymes. Annu Rev Entomol 58:373–391

Maes KVN (1998) Pyraloidea: crambidae, pyralidae pp. 87–98, 426–435. In: Polaszek A (ed) African cereal stem borers: economic importance, taxonomy, natural enemies and control. CAB International, Oxon, p 74

Mathew G, Menon MGR (1984) The pyralid fauna (Lepidoptera: Pyralidae) of Kerala (India). J Entomol Res 8(1):5–13

Meyrick E (1884) On the classification of the Australian Pyralidina. Trans R Entomol Soc Lond 32(3):277–350

Munroe E (1989) Changes in classification and names of Hawaiian Pyralidae since the publication of Insects of Hawaii, vol 8, by Zimmerman EC (1958). Bishop Mus Occas Pap 29:199–212

Munroe E, Solis MA (1999) Pyraloidea, pp 233–256. In: Kristensen N (ed) Lepidoptera, moths and butterflies, Vol. 1, Arthropoda, Insect, Vol. 4, Part 35. Handbook of Zoology. Walter de Gruyter & Co., Berlin, 491 p

Nei M, Niimura Y, Nozawa M (2008) The evolution of animal chemosensory receptor gene repertoires: roles of chance and necessity. Nat Rev Genet 9:951–963

Pena JE, Nadel H, Pereira MB, Smith D (2002) Pollinators and pests of *Annona* Sp. In: Pena J, Sharp J, Wysoki (eds) Tropical fruit pests and pollinators: biology, economic importance, natural enemies and control. CABI Publication, Wallingford/New York, p 208

Pereira F, Carneiro J, Amorim A (2008) Identification of species with DNA-based technology: current progress and challenges. Recent Pat DNA Gene Seq 2:187–200

Pilgrim EM, Roush SA, Krane DE (2002) Combining DNA sequences and morphology in systematics: testing the validity of the dragonfly species *Cordulegaster bilineata*. Heredity 89:184–190

Robertson HM, Warr CG, Carlson JR (2003) Molecular evolution of the insect chemoreceptor gene superfamily in *Drosophila melanogaster*. Proc Natl Acad Sci U S A 100:14537–14542

Shaffer M, Nielsen ES, Horak M (1996) Pyraloidea. In: Nielsen ES, Edwards ED, Rangsi TV, Checklist of the Lepidoptera of Australia. Monograph of Australia Lepidoptera 4:164–199

Shashank PR (2012) Biosystematics and pheromone components of *Conogethes punctiferalis* (Guénee) (Lepidoptera: Crambidae) with special reference to populations infesting Castor (*Ricinus communis* L.) and cardamom (*Elettaria cardamomum* Maton), Ph.D thesis, UAS, GKVK, Bangalore-65

Shashank PR, Chakravarthy AK, Raju BR, Bhanu KRM (2014a) DNA barcoding reveals the occurrence of cryptic species in host-associated population of *Conogethes punctiferalis* (Lepidoptera: Crambidae). Appl Entomol Zool 49:283–295

Shashank PR, Chakravarthy AK, Chandrashekharaiah R, KRM B (2014b) Behavioural studies on shoot and fruit borer, *Conogethes punctiferalis*, Guenée, (Crambidae: Lepidoptera) host-associated populations reveal occurrence of cryptic species. Entomol Generalis 35(2):103–115

Shashank PR, Doddabasappa B, Kammar V, Chakravarthy AK, Honda H (2015) Molecular characterization and management of shoot and fruit borer *Conogethes punctiferalis* Guenee (Crambidae: Lepidoptera) populations infesting cardamom, castor and other hosts. In: Chakravarthy AK (ed) New horizons in insect science: towards sustainable pest management. Springer, New Delhi, pp 207–227

Shashank PR, Kammar V, Mally R, Chakravarthy AK (2018) A new Indian species of shoot and capsule borer of the genus *Conogethes* (Lepidoptera: Crambidae), feeding on cardamom. Zootaxa 4374(2):215–234

Smith MA, Fisher BL (2009) Invasion, DNA barcodes, and rapid biodiversity assessment using the ants of Mauritius. Front Zool 6:31

Smith MA, Fisher BL, Hebert PDN (2005) DNA barcoding for effective biodiversity assessment of a hyper diverse arthropod group: the ants of Madagascar. Philos Trans R Soc B Biol Sci 360:1825–1834

Stephenson BP, Gill GSC, Randall JL, Wilson JA (2003) Biosecurity approaches to surveillance and response for new plant pest species. N Z Plant Protect 56:5–9

Vasudev K (2013) Molecular diversity of *Conogethes* sp. Guenèe, (Lepidoptera: Crambidae) infesting castor and cardamom. M.Sc. dissertation, University of Agricultural Sciences, GKVK, Bangalore, pp 55–66

Vasudev K, Chakravarthy AK, Ashokan R (2016) Molecular evidence for divergence in the shoot and fruit borer, *Conogethes sp.* (Lepidoptera: Crambidae) infesting select host-plants in India. Mysore J Agric Sci 50(2):245–249

Wang J, Zhang T, Wang Z, HE K, Liu Y, Li J (2014) Molecular taxonomy of *Conogethes punctiferalis* and *Conogethes pinicolalis* (Lepidoptera: Crambidae) based on mitochondrial DNA sequences. J Integr Agric 13(9):1982–1989

Waugh J (2007) DNA bar-coding in animal species: progress, potential and pitfalls. BioEssays 29:188–197

Xing Ge, Tiantao Zhang, Zhenying Wang, Kanglai He & Shuxiong Bai, 2016, Identification of putative chemosensory receptor genes from yellow peach moth Conogethes punctiferalis (Guenée) antennae transcriptome. Scientificreports, pp.1–13. www.nature.com/scientificreports.

Zhang Y (2010) Genetic diversity of *Conogethes punctiferalis* (Guenée) (Lepidoptera: Crambidae) populations from different geographic regions in China. Acta Entomol Sin 53(9):1022–1029

Zhou X, Adamowicz SJ, Jacobus LM, DeWalt RE, Hebert PDN (2009) Towards a comprehensive barcode library for arctic life – Ephemeroptera, Plecoptera, and Trichoptera of Churchill, Manitoba, Canada. Front Zool 6:30

# *Conogethes sahyadriensis*: A New Borer on Zingiberaceous Crop Plants from India

3

A. K. Chakravarthy, P. R. Shashank, K. P. Kumar, and Vasudev Kammar

**Abstract**

A new borer, *Conogethes sahyadriensis* (Shashank PR, Kammar V, Mally R, Chakravarthy AK, Zootaxa 4374:215–234, 2018), has been reported on cardamom from South India. This borer is a sister species of *Conogethes pluto* that is widely distributed in Australia and Thailand. Hitherto, *C. sahyadriensis* was classified under *Conogethes punctiferalis*. The new borer, *C. sahyadriensis*, is an economically important species as it causes on an average yield losses of more than 20% on cardamom, turmeric and ginger in different parts of India. Phenotypical, biosystematic, genetic and phylogenetic differences have been found among *C. punctiferalis*, *C. sahyadriensis* and *C. pluto*.

**Keywords**

*C. sahyadriensis* · Zingiberaceous · Distribution · Management practices

A. K. Chakravarthy (✉)
Division of Entomology and Nematology, ICAR-Indian Institute of Horticultural Research, Bengaluru, Karnataka, India

P. R. Shashank
Insect Taxonomy Laboratory, Division of Entomology, ICAR-Indian Agricultural Research Institute, New Delhi, India

K. P. Kumar
Department of Entomology, University of Agricultural Sciences, Gandhi Krishi Vignan Kendra (GKVK), Bengaluru, Karnataka, India

V. Kammar
Department of Food and Public Distribution, Ministry of Consumer Affairs and Food and Public Distribution, Government of India, New Delhi, India



A. K. Chakravarthy (ed.), *The Black spotted, Yellow Borer, Conogethes punctiferalis Guenée and Allied Species*,
https://doi.org/10.1007/978-981-13-0390-6_3

## 3.1 Introduction

In September–October 1983, *Conogethes* borer moths emerging from cardamom (*Elettaria cardamomum* Maton) plants at Mudigere (12° 25′ 11″ 75° 43′ 48″N), Chikmagalur, Karnataka, South India, were found to be morphologically different from the moths emerging from castor (*Ricinus communis* L.) plants. Moths reared on cardamom were larger and bright yellow compared to the smaller, dull-yellow ones reared on castor. Further, female *Conogethes* moths reared on cardamom were characterized by a tuft of black hairs at the abdominal tip, a feature absent in those reared on castor. Dissection of adult moths revealed differences in male genitalia of *Conogethes* moths reared on castor and cardamom. Chakravarthy et al. (1991) demonstrated that the rearing methods and containers required by *Conogethes* larvae reared on castor were different from those required by them when raised on cardamom. Concomitantly, workers in Japan and China categorized *Conogethes punctiferalis* populations into two feeding forms, namely, fruit-feeding and stalk-feeding types. Fruit-feeding type of *C. punctiferalis* is a polyphagous population, while the stalk-feeding one is oligophagous (Honda and Mitsuhashi 1989). Parallelly, in Australia, two forms of *C. punctiferalis* populations were identified and designated as southern and northern populations (Shaffer et al. 1996). Inoue and Yamanaka (2006) redescribed *C. punctiferalis* along with two closely related new species, *C. parvipunctalis* and *C. pinicolalis* from eastern Palaearctic and oriental regions. These observations and revelations prompted us to further investigate in detail into the *Conogethes* populations infesting cardamom and castor in India. Shashank et al. (2018) described the new species, *C. sahyadriensis* on cardamom from India, and this chapter surmises comprehensively the select aspects of this new species.

## 3.2 Biosystematics

Visually *C. sahyadriensis* is indistinguishable from *C. pluto* in characteristics with respect to size and patterns of punctuations. The salient differences between the two species are tabulated below (Table 3.1).

*C. sahyadriensis* also closely resembles *C. punctiferalis* and *C. evaxalis*, although their host plant use might have evolved independently. The select differences among the above species are as given in Table 3.2.

A detailed comparative picture of the morphological and anatomical characters of the six *Conogethes* species has been brought out by Shashank et al. (2018). Systematics, delimitation and diagnosis of adult *Conogethes* have been dealt with in

**Table 3.1** Select morphological differences between *C. sahyadriensis* and *C. pluto*

| Character | *C. sahyadriensis* | *C. Pluto* |
|---|---|---|
| Male genitalia: dorsal tegmen roof | Not so bulged | Prominently bulged |
| Clasper base | Narrower, rectangular | Broader, trapezoidal |
| Vulva towards the apex | Blunt, right angled | Arched, not angled |

**Table 3.2** Differences among select species of *Conogethes* to *C. sahyadriensis*

| Character | *C. sahyadriensis* | *C. punctiferalis* | *C. evaxalis* |
|---|---|---|---|
| Metathorax | Two dark spots | Three black spots | Three dark spots |
| Distal tarsus of male hind leg | Slight tuft of hairs | Black yellow hairs | Tuft of dark hairs |
| Hindwing | No large dark spot in anal area | A large black spot was present on discocellular | Large dark black spot in hindwing anal area |
| Hair pencils | Slender, phylliform hair scales | Broad ovate | Tuft of dark hairs |
| Male genitalia: Transtilla | Not enlarged and not highly sclerotized | Completely sclerotized and lateral arms are narrow | Enlarged and strongly sclerotized |
| Female genitalia: Antrum | Smaller, not sclerotized ductusbursae | Shorter, corpus bursae smaller, ovate | Larger ductusbursae sclerotized |
| Labial palp: Second segment | Broadly tinted with black fuscous | Narrow without black fuscous | Broad brown band |

another chapter (see Richard Mally, this volume). Armstrong (2010), Shashank et al. (2018) and Richard Mally (2018, this volume) undertook a comprehensive examination of the specimens of *Conogethes* and related genera along with the species within *Conogethes* and contended that a thorough systematic and integrated revision of the genus is required. Area-wide data and specimen samples across locations and a robust analysis will further explain species diversity within *Conogethes*, relationships among the species and the resemblance *Conogethes* genus bears with African genus *Marwitzia* and the Neotropical *Polygrammodes eleuata* species group.

## 3.3 Geographical Distribution and Host Range

All the specimens of *C. pluto* are from Austral-Asian region (Australia, New Guinea), and it is possible that the species is restricted to this region (Shashank et al. 2018). To date *C. sahyadriensis* is known to be from South India and Sri Lanka, and on the basis of the current knowledge, it can be summarized that the distribution range of *C. sahyadriensis* may be different. Geographical distribution of three related *Conogethes* spp., namely, *C. punctiferalis*, *C. pluto* and *C. sahyadriensis*, is depicted in Fig. 3.1. While *C. punctiferalis* is polyphagous feeding on more than 24 cultivated crops in India (Chakravarthy et al. 2015b), *C. sahyadriensis* is oligophagous feeding on zingiberaceous plants like cardamom, turmeric and ginger. Some of the plants were tested in the laboratory against *C. sahyadriensis* at Mudigere, Chikmagalur, South India, and it was found that cardamom was the most preferred plant followed by *Hedycium* sp., *Alpinia* sp. and *Amomum* sp. (Thyagaraj 2003). Further, *C. sahyadriensis* has been found feeding on *Curcuma nilgiriensis* in the field and *Amomum subulatum* under lab conditions; however, in Northeast Indian regions like Sikkim, *C. sahyadriensis* has not been found feeding on large

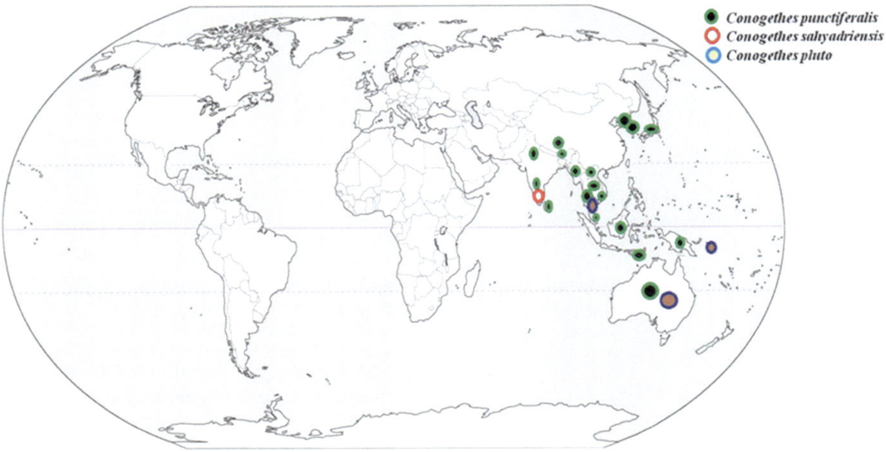


**Fig. 3.1** Geographical distribution of three closely related *Conogethes* species

cardamom, *Amomum subulatum* Roxb. The new borer species, *C. sahyadriensis*, was found infesting spice crops, such as turmeric and ginger in the Northeast India but not large cardamom which is cultivated in the sub-Himalayan region of Northeast India, Nepal and Bhutan under the shade of trees at 800–2000 amsl. In Andaman and Nicobar islands too, several species of zingiberaceous plants were found in the wild but were not infested by *C. sahyadriensis*. These plants were found fed by another lepidopteran pest, *Glyphipterix* sp., and large cardamom was found infested with a specific borer larvae that caused dead heart (Bhowmick 1962; Azad Thakur 1982). This could be on account of differing climatogeographic conditions and plant phenotype with different biochemical profiles. Recently *C. sahyadriensis* has been found in Borneo (Pers comm. Richard Mally, 2018) and possibly in South Vietnam (Pers comm. Loc, 2018).

Hybridization experiments were conducted with *Conogethes* moths reared on castor and cardamom under laboratory conditions. The mating experiments were conducted with single (1 pair/cage) and multiple pairs (4 pairs/cage). Higher mating success was achieved in cages with multiple pairs and no mating occurred when the moths were reared on 'shifted' plants although they attempted to mate several times (Shashank 2012). Thus, the moths on the two plants were reproductively isolated.

When *Conogethes* reared on cardamom were implanted on castor, the larvae suffered almost cent per cent mortality. Besides, few larvae that pupated emerged as deformed adults and completed the life cycle early probably because of the physiological stress caused by host plant shift (11–12 days). When *Conogethes* reared on castor were implanted on cardamom, about 5% larvae emerged as deformed adults in 14–16 days. Further, when cardamom larvae were implanted on cardamom, they took 20–22 days to complete the life cycle. Field and laboratory observations for the past two decades have revealed that the composition of natural enemies on *C. punctiferalis* and *C. sahyadriensis* larvae and pupae is different although several species of the biocontrol agents were common.

## 3.4 Molecular Characterization

Phylogenetically *C. sahyadriensis* is most similar to *C. pluto*. COI data revealed that *C. sahyadriensis* belonged to a distinct clade that is sister to *C. pluto*. Shashank (2012) barcoded *Conogethes* moths reared on castor and cardamom from different geographic locations in India and found that the moths belonged to two clades.

Mitochondrial genes are often chosen for evolutionary and barcoding studies as they have a number of positive characteristics. The protein coding genes are the most frequently sequenced mitochondrial genes for evolutionary studies and phylogenetic analysis. The CBOL initiated the "All-Leps Barcodes of Life" project as Lepidoptera is the second most diverse order of the insects. There are about 1,80,000 known species in this order, and it is likely that there are another 30,000 species awaiting description. The initiative involves campaigns on three geographic scales: global (Geometridae, Saturniidae and Sphingidae), continental (North America and Australia) and regional (Great Smokey Mountains National Park, USA, and Area de Conservation, Guanacaste) (Bravo et al. 2008). Until now about 6000 lepidopteran species are with barcodes, of which 4443 are Crambidae (International Barcode of Life 2012). Herbert et al. (2003) studied the morphological and DNA barcoding of *Astraptes fulgerator* Walch, a widely distributed Neotropical skipper butterfly (Lepidoptera: Hesperiidae) in north western Costa Rica with museum specimens. It was found that *A. fulgerator* is a complex of at least 10 species in Costa Rica. Largely sympatric, these taxa have mostly different caterpillar food plants, largely distinct caterpillars and different ecosystem preferences, but only subtly differing adults with no genitalic divergence. It is likely that *Conogethes* genus represents an almost similar situation, but it needs to be deciphered.

The DNA barcodes were appended to an existing dataset from BOLD for *Conogethes* species. The entire dataset of *Conogethes* included 115 DNA signatures categorized into four clades and which have been further classified into more than 50 clusters. *Conogethes* on castor and cardamom belong to two distinct clades and the moths were distinguished based on the host plants. The specimen was matching with the signature corresponding to *C. punctiferalis* up to 91% of the standard signature in BOLD. The *Conogethes* specimens on cardamom had a new DNA signature which did not match with any signature deposited in BOLD earlier. Therefore, this indicated that the moths reared on cardamom belong to a new species which has now been confirmed (Shashank et al. 2018) and identified as *Conogethes sahyadriensis*. The mean divergence within species for *C. punctiferalis* was 2.102%. However, the pairwise divergence between *C. punctiferalis* and the moths reared on cardamom (*Conogethes sahyadriensis*) was >5% (Shashank 2012) (Fig. 3.2).

The family Crambidae, subfamily Pyraustinae, has 4585 species with barcode sequences, and 17 species are designated under *Conogethes*. Armstrong (2010) compared DNA barcoding of different populations of *Conogethes* and revealed that Australian and Asian specimens form separate clades divergent by ~6%. The barcode data successfully distinguished *C. punctiferalis* and *C. pluto* but unexpectedly revealed divergence between the Asian and Australian populations. Morphologically these were determined to be the same species but distinct from other closely related

**Fig. 3.2** Divergence in barcodes within *Conogethes* species (Shashank 2012)

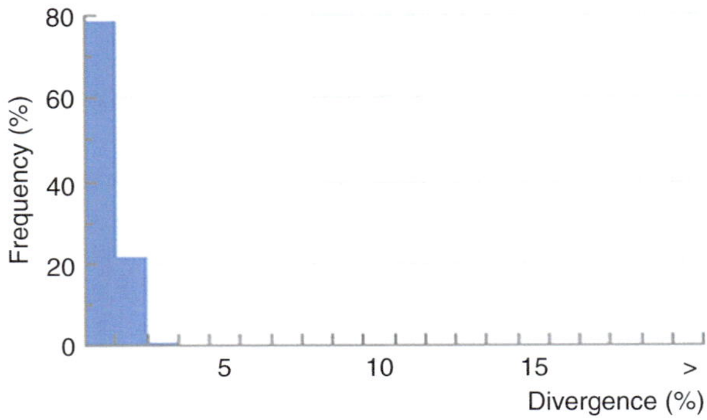


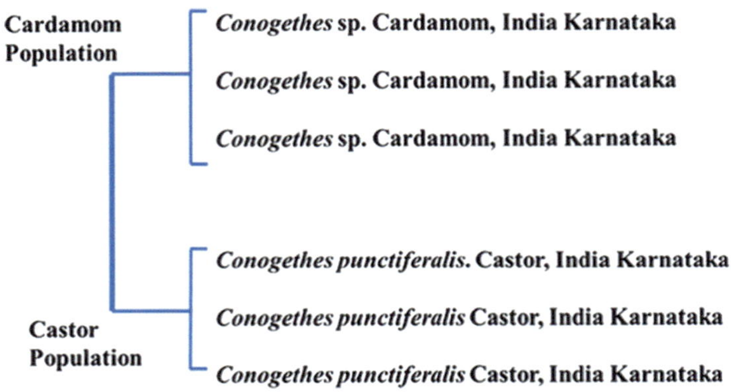


**Fig. 3.3** Neighbour-joining analysis of COI DNA barcode sequences of *Conogethes* species breeding on castor and cardamom (Shashank 2012)

species found on the east coast of Australia such as *Conogethes haemactalis* Walker, *Conogethes semifascialis* Walker or *Conogethes tharsalea* Walker.

The neighbour-joining tree (NJ tree) was constructed based on all the 33 DNA barcodes using BOLD analysis tool. On the basis of the NJ tree, two clades were recognized: Those clades which include all the *Conogethes* individuals breeding on cardamom were named as *Conogethes* sp., while those with individuals breeding on castor were named as *C. punctiferalis* (Figs. 3.2 and 3.3).

Differences between the two species of moths were recorded in the size of egg, larvae, pupae and adults, and duration of life cycle (in days) and the details have been furnished in the other chapters of this book. *C. sahyadriensis* required longer time on cardamom to complete a life cycle compared to *C. punctiferalis* on castor. On an average *C. punctiferalis* required 34.5 days on castor compared to 39 days required by *C. sahyadriensis* on cardamom (Doddabasappa 2012); select differences in morphology and bioecology between the two species are tabulated (Tables 3.3 and 3.4)

**Table 3.3** Morphological differences in the life stages of *Conogethes* between *C. sahyadriensis* and *C. punctiferalis*

| Life stage | *C. sahyadriensis* | *C. punctiferalis* |
|---|---|---|
| Egg | White, flattened | Oval, yellow |
| Chorion surface | Network of threads | Smooth |
| Larva | Pinkish with light brown head | Grey, dark brown head |
| Distance between dorsal setae D, on 8th and 9th abdominal segment on larva (mm) | 0.70 ± 0.017 mm | 0.63 ± 0.020 mm |
| Pupa (size) | Larger | Smaller |
| Adult (labial palp on 2nd segment) | Black band | Absent |
| Metathorax (adult) | Two black dots | Three black dots |
| Male genitalia (length of aedeagus) | Shorter, not strongly curved at tip | Lengthier, curved sharply at tip |
| Female genitalia (ductusbursae) | Shorter | Longer |

**Table 3.4** Differences in the bioecology of *Conogethes* between *C. sahyadriensis* and *C. punctiferalis*

| Stages | *C. sahyadriensis* | *C. punctiferalis* |
|---|---|---|
| Egg | Laid singly between sheaths on pseudostem of cardamom or unopened leaf of ginger | Singly or in groups of 1–6 on inflorescences between the warts or on the ovary |
| Larva | Neonate larvae bore into pseudostem, base of leaf axils and capsule. The pseudostem is plugged with excreta, yellow when fresh. Pupation is completed in pseudostem and at times just above soil especially on ginger and turmeric | Neonate larvae bore into tender shoot/capsule; such affected shoots are plugged with brown excreta/frass. Pupation is completed in the capsule or shoot |

## 3.5 Seasonal Incidence and Crop Loss

*C. sahyadriensis* occurs throughout the year on cardamom, ginger and turmeric. Usually, two peaks in the population were observed annually in Karnataka, South India, i.e. April–May and November–December (Thyagaraj 2003). The peak populations coincided with no rainfall period, i.e. during pre-monsoon and post-monsoon periods. Least numbers of the moths were recorded during summer, i.e. February–March, when temperatures are generally above 35° C with low relative humidity of 60–70% and when the cardamom plant does not put forth fresh/new reproductive plant parts.

It is difficult to assess accurately the crop loss caused by *C. punctiferalis* on castor and *C. sahyadriensis* on cardamom, turmeric and ginger. This is because it occurs in conjunction with other pests which also affect the quality and quantity of capsule yield. In general, *C. punctiferalis* causes a yield loss of 63% on castor in Gujarat (Kapadia 1996), while *C. sahyadriensis* causes a loss of more than 20% in cardamom (Chakravarthy et al. 2015a). Even on castor, *C. punctiferalis* completes the total life cycle of 21–25 days with two annual population peaks during June–July and November–December (Yathish 2012).

## 3.6 Mating and Feeding Behaviour

Males of *C. punctiferalis* sexually communicate with females by emitting loud ultrasound (103 dB at frequency 82 kHz) before attempting copulation. The male ultrasound consists of consecutive clicks in the early phase of the second train and consecutive pulses (bursts) in the late phase. When females were deafened by puncturing the abdominal tympanic membranes, copulation never occurred (Nakano et al. 2012, 2014). The male ultrasounds consist of pulses and bursts and owing to the resolution of the temporal coding in the moth auditory processing; *Conogethes* would recognize a series of consecutive clicks as a single continuous pulse. This pulse was analogous to the time structure of the pulse emitted by a horseshoe bat, *Rhinolophus* sp., and in order to avoid predation, hearing moths have a freezing response upon exposure to a bat call (Nakano et al. 2010). There are subtle differences in the mating repertoire of the *C. punctiferalis* and *C. sahyadriensis*. Larvae implanted from castor plant did not feed on the cardamom plant and vice versa (Doddabasappa 2012), and differences were observed in the feeding behaviour of the two species on host plants.

Gravid females of *C. punctiferalis* reared on castor emerged 4 h (17.78%) after lights off (ALO), whereas those reared on cardamom emerged an hour (23.46%) ALO. The calling frequency was more pronounced in female *Conogethes* reared on castor compared to that reared on cardamom. *C. punctiferalis* (now recognized as a different species) moths reared on cardamom showed peak mating activity between 4 and 6 h ALO, while it was between 6 and 9 ALO hours in those reared on castor. Failure of hybridization between *C. punctiferalis* reared on castor and cardamom suggests that the two *Conogethes* populations segregated into two species (Shashank et al. 2014).

## 3.7 Sex Pheromone Components

Analysis of the sex pheromone gland of female *C. pluto* by gas chromatography/electroantennogram detector revealed the presence of seven candidate pheromone compounds that elicited electroantennogram responses (El-Sayed et al. 2013). Using gas chromatography/mass spectrometry analysis and micro-derivatization reactions, six compounds were identified as (E)-10-hexadecenal, as the main pheromone compound, and (Z)-10-hexadecenal, hexadecanal, (E)-10-hexadecen-1-ol, (10E,12E)-hexadeca-10,12-dienal and (3Z,6Z,9Z)-tricosa-3,6,9-triene as minor pheromone compounds. In two-field trapping experiments, *C. pluto* responded to the six-component blend and three of six compounds, i.e. (E)-10-hexadecenal, (3Z,6Z,9Z)-tricosa-3,6,9-triene and (10E,12E)-hexadeca-10,12-dienal were shown to be necessary for attraction. In a subsequent experiment, three doses (i.e. 0.01, 0.1 and 1 mg) of the six-component blend were tested; the largest number of males was captured in traps baited with a lure loading of 1 mg. The availability of the sex pheromone of *C. pluto* will provide basis for additional control options for this pest. The proven pheromone composition of *C. punctiferalis* embrace E-10-hexadecenal (E10–16-Ald), Z-10-hexadecenal (Z10–16: Ald) and hexadecenal (16:Ald) at 100:8:16 (3 mg/septa). The functional pheromone components of *C. sahyadriensis* are not yet determined despite attempts to elucidate them.

**Management Practices** The management practices were also different for both the species. Patel and Patel (2009) revealed that the lowest infestation (34.18%) of capsule borer was recorded on castor intercropped with cowpea (1:2) and maximum castor seed yield was obtained from green gram (1:1) intercrop, but it did not differ from sesame (1:1) or (1:2) intercrop. Overall results revealed that green gram and sesame were profitable by resulting higher seed yield and reducing the capsule borer infestation, but intercropping is often not practicable in cardamom ecosystem.

Removal of infested stems/shoots, fruits and crop residues can be adopted against both *C. punctiferalis* on castor and *C. sahyadriensis* on cardamom. Again, balanced application of NPK fertilizers is essential for both the crops which act against both the *Conogethes* borer species. Healthy capsules in pairs can be separated by placing sticks or pieces of cardboard in between; however, this cannot be practised in cardamom. Conserving natural enemies of castor capsule borer is important as they actively suppress this pest, so that pesticides can be avoided.

Collection and destruction of borer-infested shoots by burning and removal of alternate hosts around the field can be adopted against *C. sahyadriensis* on spice crops. Rodrigo (1940) suggested that the use of natural enemies such as *Phanerotoma hendecasiella* and *Xanthopimpla* sp. was found effective against shoot and fruit borer in turmeric. The infestation by early stages of larva of this pest in emerging panicle, immature capsule and leaf bud can be controlled effectively by insecticide applications. The use of biopesticides, Bioasp (0.25%, 0.50% and 0.75%) and DiPel (*B. thuringiensis kurstaki*) (0.1%, 0.2% and 0.3%), proved effective in ginger at Kerala (Devasahayam 2010). Neem gold (0.03%), neem seed cake (0.5 kg/plant), NSKE (4.0%), neem oil (0.03%) and econeem plus (0.03%) were found effective on small cardamom (cv. −2) at Mudigere, Karnataka, India (Naik et al. 2006).

Fish oil insecticidal soap (Na-based) 2.5%, FOIS (K-based) 2.5%, FOIS (K-based) + tobacco extract (2.5%) or FOIS (Na-based) + tobacco extract (2.5%) or nimbecidine (0.2%) or garlic extract (2.5%) + nimbecidine (0.2%) and quinalphos (0.05%) were found effective against cardamom capsule and shoot borer (Rajkumar et al. 2003). It is crucially important that nature-friendly methods are adopted against cardamom borer as the crop is cultivated in evergreen tropical forest tract which is characterized by endemism, high biodiversity and genetic variability and where pollinators are indispensable for crop production. Need-based insecticides + spraying lambda cyhalothrin at 20 ppm ai/ha was found effective against the cardamom borer (Sureshkumar et al. 2004) (Fig. 3.4).

*C. sahyadriensis* sp. nov *C. punctiferalis* *C. pluto*

**Fig. 3.4** The three closely related *Conogethes* species

**Acknowledgement** Authors are thankful to the authorities of the Indian Council of Agricultural Research (ICAR), New Delhi; Director, Indian Agricultural Research Institute, New Delhi; and University of Agricultural Sciences, GKVK, Bengaluru, for facilities and encouragement, and to Dr. Richard Mally for facilitating identification of *Conogethes* specimens.

## References

Armstrong K (2010) DNA barcoding: a new module in New Zealand's plant bio-security diagnostic toolbox. Bull OEPP/EPPO 40:91–100

Azad Thakur NS (1982) Seasonal incidence of insect pests of large cardamom (*Amomum subulatum* Roxb.) in Sikkim. Res Bull ICAR 21:1–21

Bhowmick TP (1962) Insect pests of large cardamom and their control in West Bengal. J Ent 24:283–286

Bravo JP, Silva JLC, Mumhoz REF, Fernandez MA (2008) DNA barcode information for the sugar cane moth borer *Diatraea saccharalis*. Genet Mol Res 7(3):741–748

Chakravarthy AK, Honda H, Thyagaraj NE (1991) Comparison of containers for larval rearing in stalk and fruit feeding type of *Conogethes punctiferalis* (Guen.) Lepidoptera: Pyralidae. Placrosym 9:127–131

Chakravarthy AK, Shashank PR, Doddabasappa B, Kandakoor SB, Chandrashekaraiah (2015a) Biosystematics, molecular characterization and management of shoot and fruit borer *Conogethes* spp. (Lepidoptera: Crambidae). In: Singh B, Arora R, Gosal SS (eds) Biological and molecular approaches in pest management. Scientific Publishers, Jodhpur, pp 329–343

Chakravarthy AK, Vasudevkammar, Kumar KP (2015b) Monograph on shoot and fruit borer, *Conogethes punctiferails* and allied species. Consortia research platform on borers. ICAR-Indian Institute of Horticultural Research (IIHR), Bengaluru

Devasahayam S, Jacob TK, Abdulla Koya KM, Sasikumar B (2010) Screening of ginger (Zingiber officinale) germplasm for resistance to shoot borer (*Conogethes punctiferalis*). J Med Aromat Pl Sci 32:137–138

Doddabasappa B (2012) Factors associated with resistance mechanism with shoot and fruit borer *Conogethes punctiferalis* Guenee (Lepidoptera: Crambidae) population infesting castor and cardamom. PhD dissertation, UAS, GKVK, Bengaluru

El-Sayed AM, Andrew RG, Vanessa JM, Lee-Anne MM, Revell J, Thistleton B, David MS (2013) Identification of the sex pheromone of *Conogethes pluto:* a pest of Alpinia. Chemoecology 23:93–101. https://doi.org/10.1007/s00049-012-01239

Herbert PDN, Cywinska A, Ball SL, Dewaardjr (2003) Biological identifications through DNA barcodes. Proc R Soc Lond B 270:313–321

Honda H, Mitsuhashi W (1989) Morphological and morphometrical differences between the fruit and pinaceae feeding type of yellow peach moth, *Conogethes punctiferalis* (Guenee) (Lepidoptera:Pyralidae). J Appl Entomol Zool 24:1–10

Inoue H, Yamanaka H (2006) Redescription of *Conogethes punctiferalis* (Guenée) and description of two new closely allied species from Eastern Palaearctic and oriental regions (Pyralidae, Pyraustinae). Tinea 19:80–91

International Barcode of Life (2012) Lepidoptera barcode of life. [Cited 11 May 2012] Available from http://www.lepbarcoding.org/

Kapadia MN (1996) Estimation of losses due to pod borer in oil seed crops. J Oilseeds Res 13(1):139–140

Nakano R, Takanashi T, Skals N, Surlykke A, Ishikawa Y (2010) Ultrasonic courtship songs of male Asian corn borer moths assist copulation attempts by making the females motionless. Physiol Entomol 35:76–81

Nakano R, Takanashi T, Ihara F, Mishiro K, Toyama M, Ishikawa Y (2012) Ultrasonic courtship song of the yellow peach moth, *Conogethes punctiferalis* (Lepidoptera: Crambidae). Appl Entomol Zool 47:87–93. https://doi.org/10.1007/s13355-012-0092-z

Nakano R, Ihara F, Mishiro K, Toyama M, Toda S (2014) Double meaning of courtship song in a moth. Proc R Soc B 281:20140840. https://doi.org/10.1098/rspb.2014.0840

Patel BS, Patel IS (2009) Management of shoot and capsule borer, *Conogethes punctiferalis* Guenee in castor by intercropping. Trends Biosci 2:66–67

Rajkumar AJ, Kurian PS, Backiyarani S, Murugan M (2003) Evaluation of biorationals against thrips (Sciothrips cardamomi Ramk.) and shoot and capsule borer, *Conogethes punctiferalis* Guen. in cardamom. J Spices Aromat Crops 11:132–134

Richard Naik DJ, Belavadi VV, Thippesha D, Kumar MD, Madaiah D (2006) Field efficacy of neem products against thrips and capsule borer of small cardamom. Karnataka J Agril Sci 19:144–145

Rodrigo E (1940) Administration report of the acting director of agriculture for 1940. RAE 30:338

Shaffer M, Nielsen ES, Horak M (1996) Pyraloidea. In: Nielsen ES, Edwards ED, Rangsi TV (eds) Checklist of the Lepidoptera of Australia. Monograph of Australia Lepidop 4:164–199

Shashank PR (2012) Bio-systematics and pheromone components of *Conogethes punctiferalis* (Guénee) (Lepidoptera: Crambidae) with special reference to populations infesting castor (*Ricinus communis* L.) and cardamom (*Elettaria cardamomum* Maton). Ph.D thesis, UAS, GKVK, Bangalore-65

Shashank PR, Chakravarthy AK, Raju BR, Bhanu KRM (2014) DNA barcoding reveals the occurrence of cryptic species in host-associated population of *Conogethes punctiferalis* (Lepidoptera: Crambidae). Appl Entomol Zoo 49:283–295

Shashank PR, Kammar V, Mally R, Chakravarthy AK (2018) A new Indian species of shoot and capsule borer of the genus *Conogethes* (Lepidoptera: Crambidae), feeding on cardamom. Zootaxa 4374(2):215–234

Suresh Kumar RS, Rajabaskar D, Chozhan K, Regupathy A (2004) Evaluation of a new insecticide Lambda cyhalothrin (Kerate®) )5EC for management of shoot and capsule borer, *Conogethes punctiferalis* (Guen.) on cardamom. Pestology 28:430-433

Thyagaraj NE (2003) Integrated management of important cardamom pests of hill region of Karnataka, South India. PhD thesis, Dr. B.R. Ambedkar University, Uttar Pradesh, India

Yathish KR (2012) Life table studies of castor shoot and capsule borer, *Conogethes punctiferalis* Guenee. M Sc (agri.) thesis, Dr. Panjabrao Deshmukh Krishi Vidyapeeth, Krishinagar, Akola, Maharashtra India

# Status of Shoot and Fruit Borer, *Conogethes* spp. (Crambidae: Lepidoptera) in Asia: Central, South, and the Southeast

4

K. P. Kumar, Naveen Kumar, and A. K. Chakravarthy

**Abstract**

An attempt has been made in this chapter to determine status of *Conogethes* moths in Afghanistan, Bhutan, Bangladesh, Maldives, Myanmar, Nepal, and Island countries. The landscape of these countries includes species from the Oriental and Palearctic regions. This region is rich in moth including *Conogethes* diversity as revealed by entomological experditions. This region serves as an important natural reservoir for not only host plants of *Conogethes* moths but their species itself. Further, the status of species under *Conogethes* and related genera in this region has remained largely uncertain. The chain of Islands that constitute Andaman and Nicobar, Maldives, and Lakshadweep may be a pathway for *Conogethes* species distribution both inside and outside the region. This region needs expeditions for gathering information on crambids including *Conogethes* and native parasites, parasitoids, and predators. Growers in this region need to adopt newer and environmental friendly methods against *Conogethes* spp.

**Keywords**

*Conogethes* spp. · Fruit crops · Castor · Zingiber plants · Asia

K. P. Kumar (✉)
Department of Entomology, University of Agricultural Sciences, Gandhi Krishi Vignan Kendra (GKVK), Bengaluru, Karnataka, India

N. Kumar
Denthottu House, Ballamanja, Manchina, Belthangady, Mangaluru, Karnataka, India

A. K. Chakravarthy
Division of Entomology and Nematology, ICAR-Indian Institute of Horticultural Research, Bengaluru, Karnataka, India





A. K. Chakravarthy (ed.), *The Black spotted, Yellow Borer, Conogethes punctiferalis Guenée and Allied Species*,
https://doi.org/10.1007/978-981-13-0390-6_4

## 4.1 Introduction

Asia is the world's largest and most diverse continent. It has the world's most extremes of climatic conditions and holds incredible forms of plants and animals. In this book, the status of the group of moths belonging to the genus *Conogethes* in India, China, Sri Lanka, Vietnam, Malaysia, and neighboring Southeast Asian countries has been dealt in different chapters. However, a vast majority of the countries in Central, South, and Southeast Asia have not been addressed. Therefore, the species richness and the status of *Conogethes* spp. in select countries, viz., Afghanistan, Bhutan, Bangladesh, Myanmar, Maldives, and Nepal with Andaman, Nicobar, and Lakshadweep islands (Fig. 4.1), have been looked into. Documented information on *Conogethes* spp. from this region is meager and that too is mostly in local languages/dialects and not easily accessible to nationals outside countries in the region; virtually there are no data on pest identification, yield losses, and management

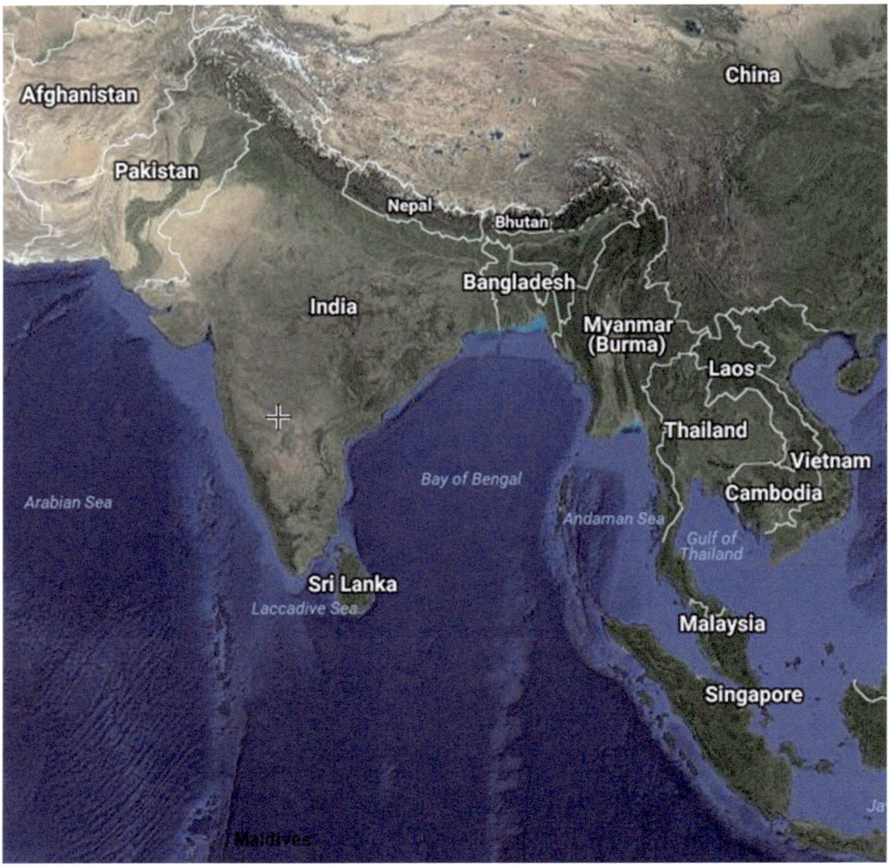


**Fig. 4.1** The map depicts countries in South Asia where *Conogethes* spp. of moths occur and are economically important. (Source: Google maps)

practices. The above countries have a diverse geography embracing forested hills, fertile planes, tropical, subtropical, temperate zones, river deltas, and arid and semi-arid regions (Chandra 1996). The landscapes comprising wild and cultivated ecosystems are frequently subjected to natural calamities like floods, earthquakes, landslides, drought, avalanches, and extreme climatic conditions. The peoples' communities are struck by poverty and difficult livelihood patterns. The horticultural and agricultural sectors are not mechanized and are of subsistence nature. Only about 4–5% farmers resort to application of insecticides. The cultivated ecosystems are usually characterized by poor crop husbandry practices and low crop productivity.

The subfamily Spilomelinae is the largest in the family Crambidae and in superfamily Pyraloidea consisting of 3300 species in more than 300 genera having worldwide distribution (Ullah et al. 2017; Munroe and Solis 1999). Integrative taxonomic study is required to uncover cryptic species complex of *C. punctiferalis*. Nine species of *Conogethes* have been recorded from this region.

## 4.2 Afghanistan

Afghanistan is a land-locked mountainous country located in South and Central Asia. Land mass is characterized by arid to semiarid habitats experiencing cold winters and hot summers. Cultivated areas are mostly plateaus with deserts, range lands, and fertile plane lands. The boundaries are porous. Therefore, it has the potential for introduction of exotic pests and pathogens into or from Pakistan, Iran, China, and India. Guava fruits imported from Pakistan, for instance, have been intercepted and found to contain *Conogethes* eggs and larvae. Almonds, olives, pistachios, apricot, peaches, grapes, guava, loquat, figs, cherries, pomegranate, etc. are mainly cultivated. Most of these fruits are infested with shoot and fruit borer, *Conogethes punctiferalis*, and farmers deploy mostly physical and cultural practices like fruit bagging, timely fruit harvests, and destruction of borer affected fruits (Lim 2016).

## 4.3 Bhutan

Bhutan is a rugged land of steep mountains and deep valleys. Geographically, it is divided in to three areas: southern border with India, lower Himalayas, and central northern borders with China. Bhutan lies on the border between Oriental and Palearctic regions and hence is considered very rich in lepidopteran diversity. Altitude varies from 200 m to >7000 m. The low lying parts are rich in Oriental elements, while the high altitude is so in Palearctic elements. The information on Bhutan lepidopetra is meager; few works are available in "The Fauna of British India Series" by Hampson from 1892 to 1899. For instance, from Bhutan 75 species of Pyraloidea has been reported. Robinson et al. (1995) and Yamanaka (1995, 1998, 2000) have reported 558 species of Pyraloidea from Nepal through the specimen collections of the Natural History Museum (NHM), London. Few species of *Conogethes* recorded in this region are "iceberg species" as very little is known about them. So any inference on these species will be a matter of conjunctures. Jatishwor Singh Irungham (2012 to 2016) with other workers conducted surveys on crambid moths in Bhutan and made a taxonomic review of superfamily Pyraloidea. Investigations of Irungham on lepidopteran fauna of central and southern Bhutan revealed a list of 182 moth species belonging to Pyralidae and Crambidae. Ninety-two species were recorded as new records for Bhutan. This investigation was a part of the Invertebrate Documentation Project of Bhutan initiated by the National Biodiversity Centre, Thimphu, funded by the Bhutan Trust Fund for Environmental Conservation, Thimphu.

The following are the species of *Conogethes* listed by Irungbam et al. (2016): *Conogethes haemactalis* (Snellen 1890), *Conogethes punctiferalis* (Guenée 1854), *Dichocrocis bistrigalis* (Walker 1866), *Dichocrocis definita* (Butler 1889), *Dichocrocis evaxalis* (Walker 1859), *Dichocrocis zebralis* (Moore 1867), and *Dichocrocis rigidalis* (Snellen 1890).

Interestingly *Botyodes caldusalis* (Walker 1859) which look morphologically similar to a *Conogethes* sp. moth has been identified and placed under *Botyodes* genus and subfamily Spilomelinae (Fig. 4.2). *Dichocrocis definata*, *D. rigidalis*, and *D. zebralis* moths do not appear morphologically similar to *Conogethes* (Fig. 4.2) but were conventionally identified and placed under genus *Dichocrocis*. Richard Mally (2018) (this volume) do not mention the above three species as recognized *Conogethes* species. Richard Mally lists nine *Conogethes* species from this region and other neighboring countries in Southeast Asia, of the total 15 *Conogethes* species recognized today. In the forewings of *Polygrammodes eleuta* (Fabricius 1777) complex and species of *Marwitzia gaede* (1917), medial areas are darker than basal and post medial areas. In addition to *Marwitzia* and *Polygrammodes* species complex, *Botyodes caldusalis* species complex should also be considered as similar to *Conogethes*. This species may not share evolutionary sister's species relationship, but in-depth studies are needed in this regard. Rose (2002) surveyed Jatinga area of Assam in the Northeast India and recorded and identified 180 species of moths belonging to four subfamilies and eight species of *Conogethes/Dichocrocis*. Hampson (1896) and Nuss et al. (2003–2014) mention that *Pycnarmon alboflavalis* (Moore 1888) has been recorded in India (Darjeeling, Sikkim, Andaman, Arunachal Pradesh) and Bhutan (Fig. 4.2). This species has been synonymized with *Conogethes alboflavalis*. However, Richard Mally (this volume, Chapter 1) does not mention this species in his list of recognized species. So collection, examination, and molecular identification of the specimens of this moth are required.

Turmeric and ginger in Bhutan have been known to be infested with *C. punctiferalis*. But again here, the species needs to be identified as Shashank et al. (2018) have identified *C. sahyadriensis* on turmeric, ginger, and cardamom from South India. The management practices against the *Conogethes* group of moths infesting castor, fruit, and spice crops adopted in India are also followed in Bhutan (see other chapters in the volume).

*Botyodes caldusalis* Walker 1859

*Dichocrocis definita* (Butler, 1889)

*Dichocrosis rigidalis* (Snellen, 1890)

*Dichocrocis zebralis* (Moore, 1867)

*Conogethes punctiferalis* Guenée, 1854.

*Pycnarmon alboflavalis* (Moore, 1888)
Syn: *Conogethes alboflavalis*

**Fig. 4.2** Species of *Conogethes* recorded from Bhutan and adjacent area. (Source: Hampson 1896, Irungbam et al. 2016, and Nuss et al. 2003–2014)

## 4.4 Bangladesh

Bangladesh is in South Asia, bordering Bay of Bengal between India and China. It represents a broad deltaic plane. It also has a small hilly region where cardamom is being cultivated and *Conogethes* species is a pest. In Lalmonirhat and hill region, *Conogethes* species is recorded in Bangladesh, and a combination of chemical and mechanical methods forms crop protection measures. On castor and fruit crops, *C. punctiferalis* is recorded, and the methods suggested in India are practiced (see other chapters in this volume).

## 4.5 Myanmar

Myanmar is the northwestern most country of Southeast Asia, limited by steep, rugged islands in the north located between Bangladesh and Thailand with India and China to the north. Ginger and turmeric are cultivated in Rakihine and Shan states. Small cardamom, *Elettaria cardamomum*, castor, durian, mangosteen, rambutan, mango, pears, grapes, and other fruits are cultivated in Myanmar. Chemical methods are adopted by only 2–3% farmers. Management practices include cultural, mechanical, and traditional methods. Unfortunately, endophytic behavior of larvae makes this insect difficult to control with traditional methods like conventional insecticides and cultural practices. Thus there is a need to develop new methods to monitor and suppress *C. punctiferalis* populations. Mori et al. (1990) and Xiao-Jian-Jia et al. (2016) made attempts to synergize sex attractant pheromones with usable host plant volatiles in *C. punctiferalis*. Three odorant-binding proteins (OBP) and four putative odorant receptors (ORS) were expressed in male moths, whereas four ORS were expressed in female antennae. Further functional studies are worthwhile on pheromone and odorant detection genes which are promising for managing the pest. Wei-Xiao et al. (2016) worked on chemoreception genes which may benefit control of the pest.

## 4.6 Nepal

Nepal is in South Asia located in the Himalayas. It borders China in the north and India in the south, east, and west. Bhutan is separated from Nepal by Sikkim. Bangladesh is within 27 kilometer in southeastern boundary. Nepal has a diverse geography. Robinson et al. (1995) compiled a checklist and bibliography of Lepidoptera from Nepal. In China, *C. punctiferalis* was originally considered as one species but subsequently two species, viz., *C. punctiferalis* and *C. pinicolalis*. Wang-Jing et al. (2014) reported from a combined analysis of mitochondrial DNA sequences from three genes and morphological data (Fig. 4.3) that *C. punctiferalis* and *C. pinicolalis* are significantly different. Nepal has also regions where pines are present, and it is likely that *C. pinicolalis* occurs. Boundaries among these countries are porous, and exchange of agricultural and horticultural products frequently occurs, and this has the potential for dispersion of *Conogethes* eggs and larvae (Kirti

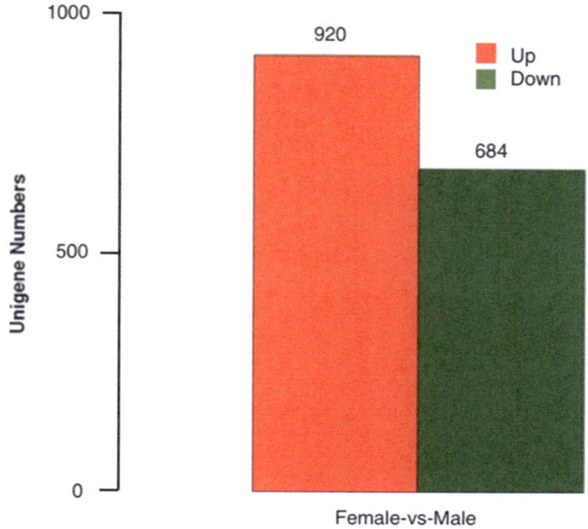


**Fig. 4.3** Statistics of differently expressed genes between male and female antennae transcriptomes. (Source: Wei Xiao et al. (2016))

and Sodhi 2001; Kirti et al. 2016). Fruit crops like apple, pear, plum, peach, mango, banana, and guava are cultivated. Again in Nepal, crop protection practices generally advocated in India are adopted (Government of Nepal 1913) (Fig. 4.4).

## 4.7 Maldives and Other Islands

Maldives is a country of South Asia in Indian Ocean, southwest of India. Andaman and Nicobar are in the Bay of Bengal. Lakshadweep islands are in Arabian Sea, near to Kerala, the southernmost state of India in the south. Fruits like mango, grapes, papaya, strawberry, etc. are cultivated in these islands together with ginger and turmeric. Traditional methods of crop protection practices are adopted. This chain of islands may serve as a pathway for exchange of horticultural and agricultural goods together with weeds, pests, and pathogens.

## 4.8 Networking

*Conogethes* is a pest whose populations cannot be effectively suppressed at individual or field level. Community-based participatory approach with suitable integration of methods is required. At the borders of the each state/country, quarantine stations with trained staff on phytosanitary measures/protocols are needed. *Conogethes* moths attack fruit, spice, and plantation crops that are being cultivated from the time immemorial. The research and development trade and marketing practices for above crops are still largely traditional. The infrastructure, transfer of technology, and awareness on phytosanitation are lacking. Need-based training programs on *Conogethes* and other pest and diseases are essential. Small and marginal

**Fig. 4.4** Microscope photograph of different male legs between pinaceae-feeding type (PFT) and fruit-feeding type (FFT) of yellow peach moth. As the arrows shows, for PFT, the male adults develop black hair tufts on distal part of the hind tibia (**a**), whereas FFT male moths have no such structure (**b**) (Source: Wang-Jing et al. 2014)

farmers should be able to adopt technologies for higher yields including novel crop protection methods. Intensive campaigns for use of biocontrol agents are needed. Organized and corporate sector approach, cooperation, and coordination for functional networking for the management of *Conogethes* among countries in Asia are urgently required. Mass multiplications of parasites and predators and mass productions of biopesticides and microbial agents against *Conogethes* will substantially contribute to suppress pest populations.

## 4.9 Future Thrusts

Expeditions for collections and field observations on *Conogethes* species, integrative taxonomic study, synergistic approach for utilizing plant odorant and pheromone lures, mass production and release of effective bioagents, practical effective tools to monitor populations of *Conogethes* in different habitats, and strict quarantine and phytosanitary measure are essential to effectively suppress the pest. For example, diafenthiuron is a thio urea compound with novel mode of action; this compound inhibits respiration, mitochondrial action, and energy metabolism and molting (Ishaaya et al. 1993). It can serve as an effective tool for managing *Conogethes* populations. Diafenthiuron at 800 ga.i./ha can be recommended for effective suppression of *C. punctiferalis* in cultivated fields (Aravind et al. 2017).

**Acknowledgment** Authors are thankful to the authorities of the University of Agricultural Sciences, GKVK, Bengaluru; Indian Council of Agricultural Research (ICAR), New Delhi; Director, IIHR, Bengaluru; and the websites of the countries, publishers from which select contents have been taken and dealt within this chapter, and Google resources for the map.

## References

Aravind J, Samiayyan K, Kuttalam S (2017) A novel insecticide Diafenthiuron 50WP against cardamom shoot and fruit borer, *C. punctiferalis* Guenee. Int J Curr Microbiol App Sci 6(10):4995–5004

Butler AG (1889) Illustrations of typical specimens of Lepidoptera Heterocera in the collection of the British Museum. The Trustees of the British Museum (Natural History), London

Chandra K (1996) Moths of great Nicobar Biosphere Reserve. India Malay Nat J 50:109–116

Government of Nepal (1913) Statistical information on Nepalese agriculture. Ministry of Agriculture Development, Kathmandu, p 50

Guenée MA (1854) Deltoïdes et Pyralites. In: de Boisduval JBAD, Guenée MA (eds) Histoire Naturelle des Insectes. Species Général des Lépidoptères 8 8. Roret, Paris, pp 1–448

Hampson GF (1892) The Fauna of British India including Ceylon and Myanmar. Moths Vol (I). Taylor and Francis, London

Hampson GF (1896) The Fauna of British India including Ceylon and Myanmar. Moths Vol (IV). Taylor & Francis, London Retrieved February 1, 2018

Hampson GF (1899) A revision of the moths of the subfamily Pyraustinae and family Pyralidae. In: Part I: Proceedings of the general meetings for scientific business of the Zoological Society of London, vol 1898. Zoological Society of London, London, pp 590–761

Irungbam JS, Chib MS, Wangdi K (2016) Taxonomic review of the superfamily Pyraloidea in Bhutan (Lepidoptera). J Asia-Pac Biodivers 9:355–382

Ishaaya I, Mendelson Z, Horowitz AR (1993) Toxicity and growth suppression exerted by Diafenthiuron in the sweetpotato whitefly, *Bemisia tabaci*. Phytoparasitica 21:199–204

Jia XJ, Wang HX, Yan ZG, Zhang MZ, Wei CH, Qin XC, Ji WR, Falabella P, Du YL (2016) Antennal transcriptome and differential expression of olfactory genes in the yellow peach moth, *Conogethes* punctiferalis (Lepidoptera: Crambidae). Nature Publishing Group, London, p 29067

Kirti JS, Sodhi JS (2001) A systematic list of Pyraustinae of North-Eastern India (Pyralidae: Lepidoptera). Zoo's Print J 16:607–614

Kirti JS, Singh N, Singh H (2016) Inventory of subfamily Pyraustinae (Crambidae: Lepidoptera) from Sikkim. J Entomol Zool Stud 4:700–705

Lim TK (2016) Edible medicinal and non-medicinal plants. Springer, Dordrecht, p 650

Moore F (1867) On the Lepidopterous insects of Bengal. In: Proceedings of the general meetings for Scientific Business of the Zoological Society of London 1867:4498

Moore F (1888) Descriptions of new Indian lepidopterous insects from the collection of the late Mr. WS Atkinson. Heterocera (continued) (Pyralidae, Crambidae, Geometridae, Tortricidae, Tineidae). In: Hewitson WC, Moore F (eds) Descriptions of new Indian lepidopterous insects from the collection of the late Mr. WS Atkinson 3. The Asiatic Society of Bengal/Taylor & Francis, Calcutta/London, pp 199–299

Mori K, Watanabe H, Fujiwhara M, Kuwahara S (1990) (E)- and (Z)-tetradecenyl formate, potent sex pheromone mimics against the yellow peach moth. Pheromone Synth 122:1257–1259

Munroe E, Solis MA (1999) The Pyraloidea, Lepidoptera, moths and butterflies. Volume 1: Evolution, systematics, and biogeography. Walter de Gruyter, Berlin, pp 233–256

Nuss M, Landry B, Mally R, Vegliante F, Tränkner A, Bauer F, Hayden JE, Segerer A, Schouten R, Li H, Trofimova T, Solis MA, De Prins J, Speidel W (2003–2014) GlobIZ – Global Information System on Pyraloidea. www.pyraloidea.org

Robinson GS, Sattler K, Shaffer M, KR TK, Allen MG (1995) Microlepidoptera and Pyraloidea of Nepal – a checklist and bibliography. In: Haruta T (ed) Moths of Nepal, Part 4, Tinea, vol 14 (Supplement 2). The Japanese Heterocerists' Society, Tokyo, pp 150–181

Rose HS (2002) An inventory of the moth fauna (Lepidoptera) of Jatinga, Assam, India. Zoos' Print J 17:707–721

Shashank PR, Vasudev K, Mally R, Chakravarthy AK (2018) A new Indian species of shoot and capsule borer of the genus Conogethes (Lepidoptera: Crambidae) feeding on Cardamom. Zootaxa 4372(2):215–234

Snellen PCT (1890) A catalogue of the Pyralidina of Sikkim collected by Henry J. Elwes and the late Otto Möller, with notes by H. J. Elwes. Trans Entomol Soc Lond 1890:557–647, pls 19–20

Ullah M, Dong Y, Qiao P, Zhang Y, Yang Z (2017) Delineating closely related species of Tylostega Meyrick (Lepidoptera: Crambidae: Spilomelinae) from mainland China using DNA barcodes. Mitochondrial DNA Part A. Taylor & Francis, pp 1–7

Walker F (1859) Pyralides. List of the specimens of Lepidopterous insects in the collection of the British Museum, London 18:509–798

Wang J, Zhang T-t, Wang Z-y, He K-l, Liu Y, Li J (2014) Molecular taxonomy of *Conogethes* punctiferalis and *Conogethes* pinicolalis (Lepidoptera: Crambidae) based on mitochondrial DNA sequences. J Integr Agric 13:1982–1989 Elsevier Publisher

Xiao W, Yang L, Xu Z, He L (2016) Transcriptomics and identification of candidate chemosensory genes in antennae of *Conogethes* punctiferalis (Lepidoptera: Crambidae). J Asia-Pac Entomol 19:911–920 Elsevier

Yamanaka H (1995) Pyralidae of Nepal (I). In: Haruta T (ed) Moths of Nepal, Part 4. Tinea. Japan Heterocerist's Society, Tokyo, pp 182–193

Yamanaka H (1998) Pyralidae of Nepal (II). In: Haruta T (ed) Moths of Nepal, Part 5. Tinea. Japan Heterocerist's Society, Tokyo, pp 99–116

Yamanaka H (2000) Pyralidae of Nepal (III). In: Haruta T (ed) Moths of Nepal, Part 6. Tinea. Japan Heterocerist's Society, Tokyo, pp 45–62

# 5 Research Progress of *Conogethes punctiferalis* (Lepidoptera: Crambidae) in China

Yan-Li Du, Jing Li, and Zhen-Ying Wang

**Abstract**

The yellow peach moth, *Conogethes punctiferalis* (Guenée), is a major insect pest feeding on various crops and fruit trees in China. In the present chapter, research progress in several aspects of *Conogethes* moths is reviewed, including distribution and host range, morphology, damage, host-plant interactions, artificial diet, diapause, *Wolbachia* infection, genetic diversity, gene flow, molecular taxonomy, and management in China. This chapter provides insights into studies on *C. punctiferalis*, particularly in China.

**Keywords**

*Conogethes punctiferalis* in China · Research progress · *Wolbachia* infection

## 5.1 Introduction

The yellow peach moth, *Conogethes punctiferalis* (Guenée) (Lepidoptera: Crambidae), is an insect widely distributed in South and East Asia, Australia, and Papua New Guinea (CAB International 2011). The larva of *C. punctiferalis* is a

Y.-L. Du
Plant Science and Technology College, Beijing University of Agirculture, Beijing, China

J. Li
School of Biological and Environmental Engineering, Xi'an University, Xi'an, Shaanxi Province, China

Z.-Y. Wang (✉)
State Key Laboratory for Biology of Plant Diseases and Insect Pests, Institute of Plant Protection, Chinese Academy of Agricultural Sciences, Beijing, China
e-mail: zywang@ippcaas.cn



A. K. Chakravarthy (ed.), *The Black spotted, Yellow Borer, Conogethes punctiferalis Guenée and Allied Species*,
https://doi.org/10.1007/978-981-13-0390-6_5

typical generalist and can feed on a broad range of hosts, including peach, chestnut, apple, pear, plum, apricot, durian, citrus, papaya, cardamom, ginger, eggplant, and maize (Sekiguchi 1974; Waterhouse 1993). Because of its broad host range and the potential spread to long distances, many basic studies on the morphology, biology, ecology, physiology, population genetics, and management of *C. punctiferalis* have been conducted in China in recent years. A summary of research work on *C. punctiferalis* in China will be introduced in this chapter.

## 5.2 Distribution and Host Range

In China, *C. punctiferalis* expanded from north to south all over the country, including Liaoning, Shaanxi, Shanxi, Hebei, Beijing, Tianjin, Henan, Shandong, Anhui, Jiangsu, Jiangxi, Zhejiang, Fujian, Taiwan, Guangdong, Hainan, Guangxi, Hunan, Hubei, Sichuan, Yunnan, and Tibet. There has been a distribution record of *C. punctiferalis* in Chayuxituo, Tibet, at 2200 m (Lu et al. 2010).

As to the host plants, *C. punctiferalis* was reported to damage more than 100 species of fruit trees and vegetables in China (Lu et al. 2010). The larvae bore into the fruits of peach (*Prunus persica*), chestnut (*Castanea mollissima*), plum (*Prunus domestica*), apple (*Malus pumila*), avocado (*Persea americana*), pear (*Pyrus serotina*), pomegranate (*Punica granatum*), hawthorn (*Crataegus pinnatifida*), longan (*Dimocarpus longan*), litchi (*Litchi chinensis*), loquat (*Eriobotrya japonica*), mango (*Mangifera indica*), cherry (*Cerasus pseudocerasus*), fig (*Ficus carica*), walnut (*Juglans regia*), etc. but also feed on many field crops, including maize (*Zea mays*), sorghum (*Sorghum bicolor*), castor (*Ricinus communis*), sunflower (*Helianthus annuus*), soybean (*Glycine max*), haricot bean (*Lablab purpureus*), sugarcane (*Saccharum officinarum*), and others (Wang and Cai 1997; Ni 1998; Luo et al. 2000; Wang et al. 2004, 2006; Wang 2009; Zhao et al. 2004; Chen et al. 2008; Lu et al. 2010).

## 5.3 Morphology

The morphological characters of eggs and one to five instar larvae, as well as pupae of two sexes and adults of *C. punctiferalis*, especially the ultrastructures of abdominal legs and spiracles of larvae, were described and illustrated under a stereoscopic microscope and scanning electron microscope by Ai et al. (2014). The eggs are milk-white at the beginning and then turn yellow and red gradually and successively (Fig. 5.1). The length of newly hatched larva is about 2.05 mm, gray white and somewhat reddish; along with the instars increasing, the color is darkened gradually (Fig. 5.2). Average length of the 2nd instar, 3rd instar, 4th instar, and last stage larva were 5.13 mm, 9.28 mm, 13.49 mm, and 20.67 mm, respectively (Table 5.1). Abdominal legs were with biordinal, penellipse crochets (Fig. 5.3). Average length of pupa is about 10.50 mm, is soft and yellowish at its early stage, and then turns orange and dark brown near eclosion (Fig. 5.4). Adult is yellow, forewing length 10.00~11.50 mm, dotted with 25~28 leopard black spots, the ninth abdomen with

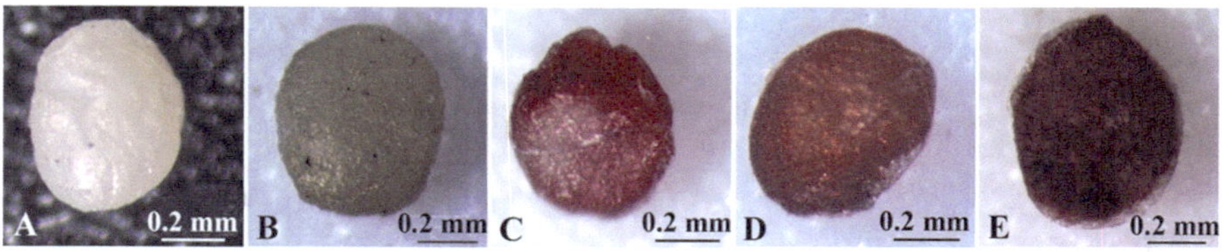


**Fig. 5.1** Color trends during egg stage of *Conogethes punctiferalis*. (**a**) 1st day; (**b**) 2nd day; (**c**) 3rd day; (**d**) 4th day; (**e**) 5th day. (Source: Ai et al. 2014)

**Fig. 5.2** 1~5 instars larvae of *Conogethes punctiferalis*. (**a**) 1st instar; (**b**) 2nd instar; (**c**) 3rd instar; (**d**) 4th instar; (**e**) 5th instar. (Source: Ai et al. 2014)

black corema (Fig. 5.5). It is distinguished from closely related species, *C. pinicolalis*, by the following characters: the second segment of labial palp is almost black; hind tibia and hind tarsus have large tufts of fuscous scales (Inoue and Yamanaka 2006).

Recently, scanning electron microscope (SEM) was also used to observe the sensilla of *C. punctiferalis*. The results indicated that most of the sensilla are located on the ventral and latero-ventral side of antennae. Among clavola, scape, and pedicel, only clavola have reticulate structure. Seven types of sensilla were identified in both sexes, including sensilla trichodea (type I and type II), sensilla chatica, sensilla auricillica, sensilla campaniform, sensilla basiconica, and sensilla coeloconica (type I and type II) (Li et al. 2014).

**Table 5.1** Body length and head capsule width of different instars larvae of *Conogethes punctiferalis*

| Instars | Body length (mm) | Length range (mm) | Head capsule width (mm) | Width range (mm) |
|---|---|---|---|---|
| 1st instar | 2.05 ± 0.09 A | 1.20~2.98 | 0.34 ± 0.01 A | 0.23~0.40 |
| 2nd instar | 5.13 ± 0.18 B | 3.54~7.66 | 0.70 ± 0.01 B | 0.62~0.76 |
| 3rd instar | 9.28 ± 0.29 C | 6.24~11.94 | 1.24 ± 0.02 C | 1.10~1.38 |
| 4th instar | 13.49 ± 0.30 D | 10.20~16.88 | 1.57 ± 0.01 D | 1.50~1.64 |
| 5th instar | 20.67 ± 0.34 E | 17.14~25.50 | 1.78 ± 0.02 E | 1.69~1.94 |

Data in the table are means ± *SE*; data with different letters in the same column mean significant difference ($P < 0.01$, Duncan's multiple range test)
Source: Ai et al. (2014)

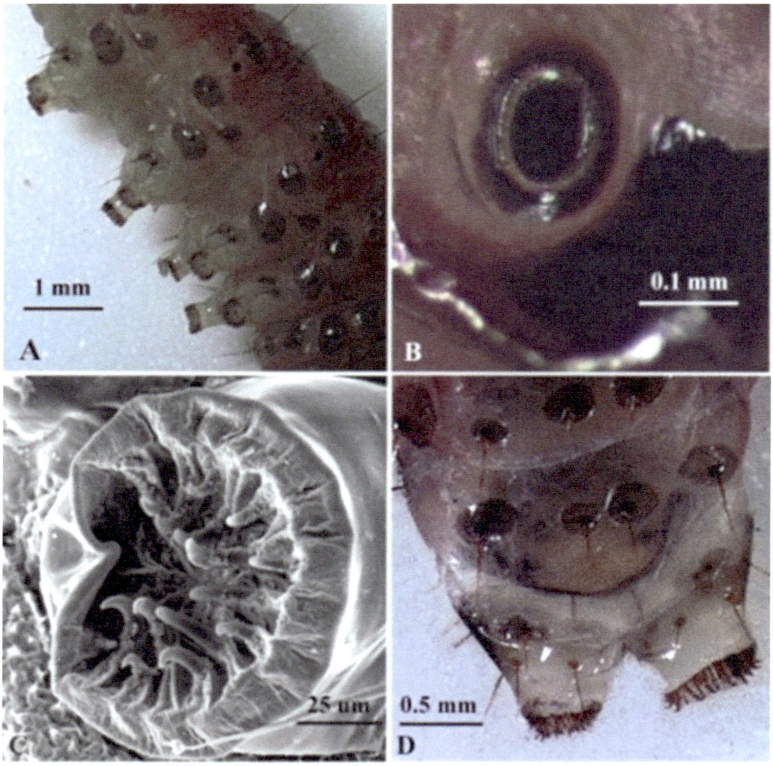


**Fig. 5.3** Abdominal structures of *Conogethes punctiferalis* larvae. (**a**) Proleg; (**b**) spiracle; (**c**) crochet; (**d**) caudal proleg. (Source: Ai et al. 2014)

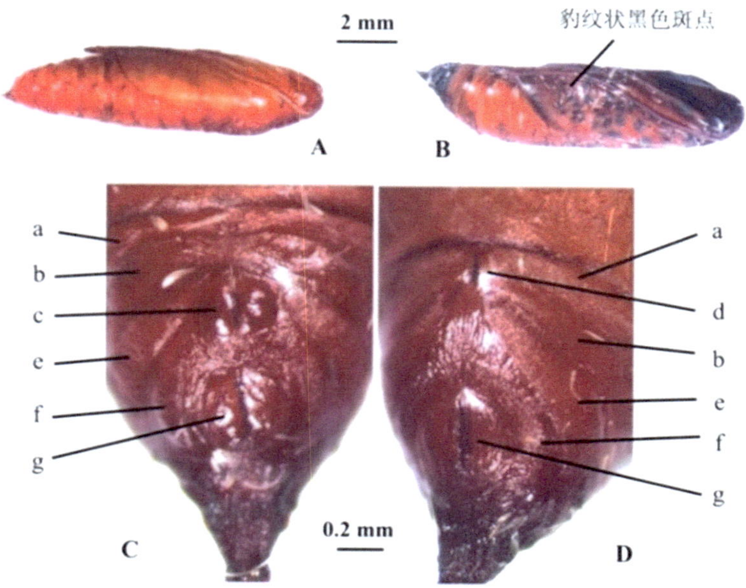


**Fig. 5.4** Male and female pupae and abdominal ends of *Conogethes punctiferalis*. (**a**) Early period of pupa; (**b**) later period of pupa; (**c**) abdominal end of male pupa; (**d**) abdominal end of female pupa; *a*. 7th abdominal segment; *b*. 8th abdominal segment; *c*. genital pore; *d*. genital pore, oviposition aperture; *e*. 9th abdominal segment; *f*. 10th abdominal segment; *g*. excretory pore. (Source: Ai et al. 2014)

**Fig. 5.5** Male and female adults of *Conogethes punctiferalis*. (Source: Ai et al. 2014)

## 5.4 Occurrence and Damage

Since the developmental duration of egg, larva, pupa, and preoviposition were affected by temperature, humidity, photoperiod, host plants, and other factors, the life cycle of *C. punctiferalis* was different due to different environmental conditions. Generally, *C. punctiferalis* completed one life cycle in 25–40 days with 2–4 generations a year in North China whereas 3–5 generations a year in South China (Lu et al. 2010).

The development and fecundity of *C. punctiferalis* have been investigated at five temperatures (15, 19, 23, 27, and 31 °C) with chestnut serving as food (Du et al. 2012). The results showed that temperature had significant effects on the developmental duration, survival rate, pupal weight, and reproduction of *C. punctiferalis.* The developmental duration of every stage of *C. punctiferalis* reduced with increasing temperatures from 15 to 27 °C; a positive correlation between developmental rate and temperature was present. Further increase of temperature to 31 °C, however, is unfavorable to the development of larva. The generation survival rate of the yellow peach moth at different temperatures decreased in the order of 23 °C > 27 °C > 19 °C > 31 °C. The number of eggs of *C. punctiferalis* was the maximum at 23 °C (average 55.00 eggs per female), followed by 19 °C and 27 °C (43.30 and 39.70 eggs average per female, respectively), and the least was 20.90 eggs per female at 31 °C. Based on the direct optimal method, the developmental threshold temperatures for egg stage, larval stage, pupal stage, preoviposition stage, and the whole generation were 10.37, 10.06, 14.27, 7.47, and 11.85 °C, respectively, and the corresponding effective accumulated temperatures were 70.84, 287.71, 118.42, 58.33, and 509.06 degree-days, respectively (Table 5.2). These results can form the basis for forecasting the occurrence of the yellow peach moth, and hence, in pest management.

## 5.5 Host-Plant Interactions

Although the larvae of *C. punctiferalis* feed on a broad range of plants, different host plants have very different impacts on the growth and development of *C. punctiferalis*. Honda et al. (1979) reported significant differences among peach, persimmon, and chestnut in sustaining larval growth and survival of *C. punctiferalis*, and an artificial diet based on host-plant materials was proposed for laboratory mass rearing. Kadoi and Kaneda (1990) compared larval development and survival of *C. punctiferalis* on apple and fresh maize. Choi et al. (2006) also fed *C. punctiferalis* with chestnut, peach, and cypress and found significant effects of different host plants on the survival and development of the insect. In order to determine the effects of different host plants on the fitness and performance of the yellow peach moth, experiments were carried out to test the developmental duration and reproduction of *C. punctiferalis* by feeding larvae with chestnut (*Castanea mollissima*), maize (*Zea mays*), plum (*Prunus salicina)*, apple (*Malus pumila*), pear (*Pyrus sorotina*), and peach (*Prunus persica*) (Li et al. 2015). The results showed that there

**Table 5.2** Developmental threshold temperature and effective accumulated temperature in each stage of *Conogethes punctiferalis*

| Developmental stage | Linear regression method | | | Direct optimal method | | |
|---|---|---|---|---|---|---|
| | Developmental threshold temperature (°C) | Effective accumulated temperature (degree-day) | Coefficient of variance (CV, %) | Developmental threshold temperature (°C) | Effective accumulated temperature (degree-day) | Coefficient of variance (CV, %) |
| Egg | 12.72 ± 1.75 | 55.02 ± 8.26 | 28.42 | 10.37 | 70.84 | 17.05 |
| Larva | 11.20 ± 4.83 | 262.49 ± 88.15 | 13.96 | 10.06 | 287.71 | 12.42 |
| 1st instar | 9.10 ± 3.54 | 56.52 ± 13.47 | 17.87 | 10.41 | 52.14 | 18.02 |
| 2nd instar | 9.78 ± 3.26 | 45.51 ± 10.42 | 15.73 | 10.46 | 44.41 | 16.12 |
| 3rd instar | 9.00 ± 3.32 | 55.62 ± 12.34 | 11.14 | 8.68 | 61.63 | 10.74 |
| 4th instar | 11.33 ± 2.71 | 48.21 ± 10.17 | 21.50 | 9.28 | 59.47 | 14.58 |
| 5th instar | 15.75 ± 2.79 | 41.17 ± 11.30 | 27.82 | 14.59 | 49.77 | 21.25 |
| Pupa | 15.17 ± 0.85 | 101.16 ± 8.00 | 13.74 | 14.27 | 118.42 | 10.94 |
| Preoviposition stage | 6.17 ± 4.05 | 61.63 ± 12.92 | 7.85 | 7.47 | 58.33 | 8.09 |
| Generation | 12.22 ± 2.58 | 490.18 ± 78.57 | 9.95 | 11.85 | 509.06 | 9.53 |

Data in the table are mean ± *SE*, and those in the same row followed by different letters are significantly different ($P < 0.05$, Ducan's multiple range test)
Source: Du et al. (2012)

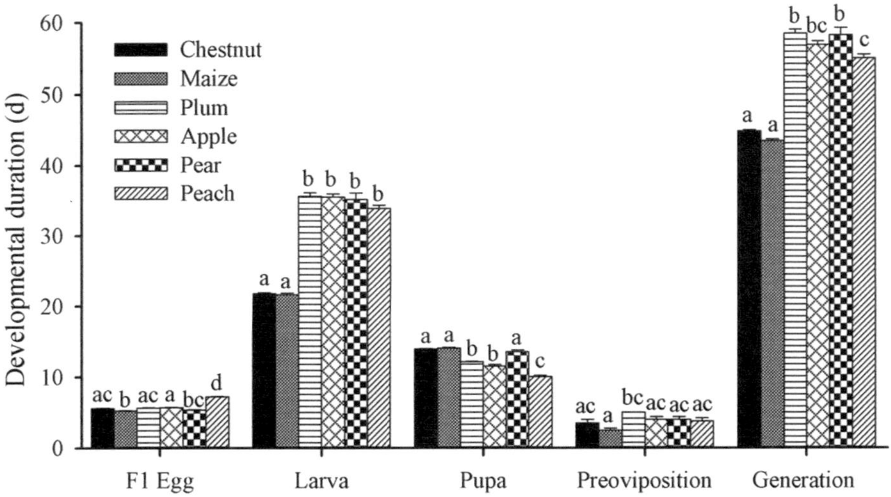


**Fig. 5.6** Developmental duration of *Conogethes punctiferalis* fed on different host plants. Bars represent means ± SE; significant differences among six host plants are indicated by different letters on each bar (Bonferroni correction, $P < 0.05$). (Source: Li et al. 2015)

were significant differences in developmental duration and reproduction among *C. punctiferalis* groups fed on different host plants (Figs. 5.6 and 5.7). The longest larval duration was observed for groups fed on plum and apple (35.63 and 35.55 days, respectively). In contrast, the shortest was for larvae fed on maize and chestnut (21.61 and 21.77 days, respectively). Larval survival of *C. punctiferalis* was lowest on plum (24.1%) and highest on maize (80.0%). Adult females developed from the larvae fed on chestnut and maize laid significantly more eggs (averagely 141.8 and 133.5 eggs per female, respectively) than those fed on the other four host plants. Analysis of life-table parameters indicated that *C. punctiferalis* larvae fed on chestnut and maize had better performance than those fed on plum, apple, pear, and peach. These findings may aid in understanding of the population dynamics of *C. punctiferalis* on different host plants. Recently Gong-Min Chen et al. (2018) demonstrated demography and population growth of *C. punctiferalis* reared on five host plants. Chestnut and maize were the most suitable host plants for *C. punctiferalis*. However, as *C. punctiferalis* can occur on a broad range of host plants, it seems necessary to take more host plants to determine their effects on the fitness and performance of *C. punctiferalis* populations.

In lepidopteran insects, many polyphagous species have been found to show better fitness on certain host plants than on others (Kirsten and Topp 1991). However, the fitness of lepidopteran insects is largely dependent on the host-finding abilities of the female adults because neonate larvae are often relatively immobile, and their proximity to feeding sites is mostly determined by the judicious choice of the female adults. Female lepidopteran moths have been shown to preferentially select certain host-plant species for oviposition (Renwick and Chew 1994; Mayhew 1997). For most insect herbivores, olfactory cues are usually used to orientate toward a specific host plant within a plant patch (Bruce and Pickett 2011). The most important

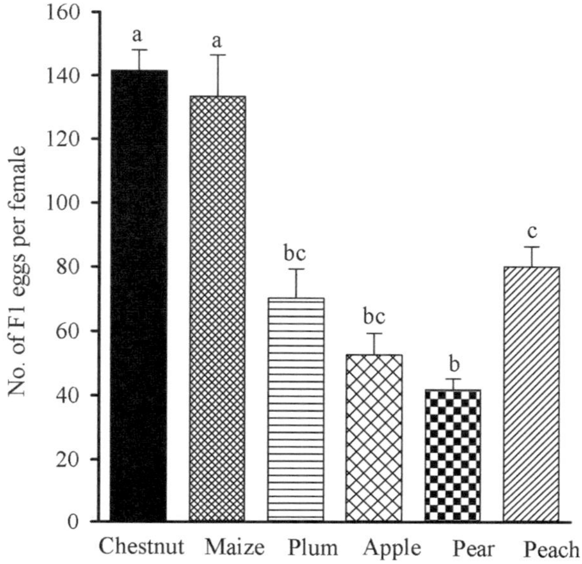


**Fig. 5.7** Effects of host plants on the reproduction of *Conogethes punctiferalis*. Bars represent means ± SE; significant differences among six host plants are indicated by different letters on each bar (Bonferroni test, $P < 0.05$). (Source: Li et al. 2015)

semiochemicals used by herbivores are plant volatiles (Kirsten and Topp 1991; Bruce and Pickett 2011; Ehrlich and Raven 1964; Becerra 1997; Hare 2011; Keesey and Barrett 2012). In some species, the concentrations of particular chemical compounds determine the relative preference of females for potential host plants, whereas in other species, the relative proportions of compounds determine the female's response (Thompson and Pellmyr 1991). As it has been reported that one of the potential mechanisms underlying the cultivar discrimination of *C. punctiferalis* is that the adult females can discriminate among volatiles emitted by different chestnut cultivars (Dong et al. 2012), it is essential to understand which plant volatiles mediate the insect's foraging and oviposition behavior. In order to identify plant-originated attractants for forecasting and controlling *C. punctiferalis* in chestnut orchards, volatiles emitted from involucres of four chestnut cultivars (Huaihuang, Huaijiu, Yanhong, and Shisheng), widely planted in Beijing area, were collected and analyzed using headspace trapping and GC-MS techniques, in young fruit period and mature fruit period, respectively (Du et al. 2014a). The results showed that ten volatile compounds were emitted from the four chestnut cultivars in young fruit period, and five out of the ten volatiles were shared in the four cultivars, including $\alpha$-pinene, camphene, $\beta$-thujene, $\beta$-pinene, and (*E*)-2-butenoic acid, 2-(methylenecyclopropyl)prop-2-yl ester. Cedrol and 3-carene were only detected in Yanhong cultivar, and *D*-limonene was only emitted by Huaijiu, Yanhong, and Shisheng cultivars. $\beta$-ocimene and farnesol were specifically released by Huaihuang and Huaijiu cultivars. In addition, the kinds and quantity of volatiles from mature fruit period were both significantly smaller than those in young fruit period. The results further indicated that the volatiles from involucres of chestnut differed by different growth periods, and few volatiles were shared among different chestnut cultivars, while others were cultivar-specific.

Recently, electroantennogram (EAG) was used to test the antennae response to five kinds of corn silk volatiles. As a result, antennal EAG response was positively correlated with the concentration of the volatiles tested. Female antennal strongly responded to 1-heptanol, $\beta$-pinene, and $\beta$-ionone, especially to 1-hexanol. Although male antennae also responded to 1-hexanol and 1-heptanol, the responses were much weaker than female antennae (Li et al. 2014).

## 5.6 Artificial Diet

In the past several decades, artificial diets have been widely used for mass rearing insects in commercial insectaries and scientific laboratories (Cohen 2004; Vanderzant 1974). Artificial diets for mass laboratory rearing of *C. punctiferalis* had been proposed early by Honda et al. (1979) and Utsumi et al. (1990). Recently, a study was undertaken to develop and assess four meridic diets for rearing *C. punctiferalis* (Du et al. 2015). As a result, the diet content containing 30 g chestnut, 70 g corn, and 70 g soybean per 700 ml diet yielded a larval survival rate of 94.5%, a generation developmental time of 42.4 days, mean pupal weights of 73.6 mg for males and 77.3 mg for females, and adult fecundity of 97.9 eggs/female. A separate chapter on mass rearing of *Conogethes* is included in this book.

## 5.7 Overwinter and Diapause

The shoot and fruit borer is a multivoltine species that overwinters as diapausing larvae. Effect of photoperiod and temperature on larval diapause was examined under empirical laboratory conditions (Xu et al. 2014). Short-day treatments caused larval diapause at 25 °C, and the critical photoperiod was between 12 and 13 h light per day. No sensitive instar was identified for diapause induction under alternated short- (L:D = 11:13 h) and long-day (L:D = 14:10 h) treatments at different larval stages (Fig. 5.8). However, accumulative treatment of three instars and 10 days under short-day treatment was required for the induction of 50% larval diapause (Fig. 5.9). All larvae entered diapause at 20 °C irrespective of the long- or short-day treatment. Under the short-day treatment, more than 90% of larvae went into diapause with ≤25 °C, but less than 17% did so at 28 °C. In contrast, under the long-day treatment, less than 19% of larvae went into diapause with ≥23 °C. The forward shift (5 °C) of critical temperature under the long-day regime demonstrated the compensatory effect of temperature and photoperiod on diapause induction (Fig. 5.10). In conclusion, *C. punctiferalis* had a temperature-dependent type I photoperiodic diapause response; there was no sensitive instar for diapause determination, but the photoperiodic accumulation time countermeasures both of the short-day cycles and the number of instars exposed. The photoperiodic diapause response was a temperature-compensated phenomenon (Xu et al. 2014).

The relationship between the cold hardiness of overwintering larvae feeding on three different crops – maize, sorghum, and sunflower – and larval water; lipid and sugar content was investigated (Xu et al. 2012). Larvae feeding on maize developed

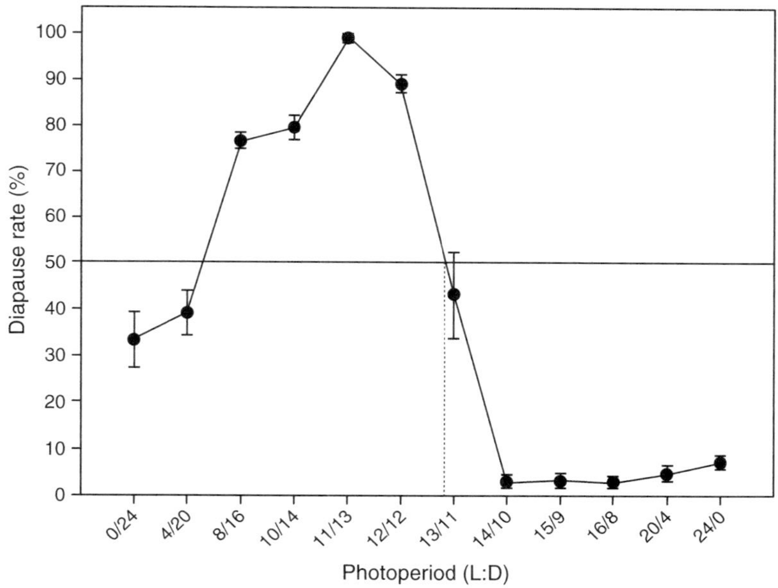


**Fig. 5.8** Photoperiodic diapause response of *Conogethes punctiferalis* at 25 °C ($n = 85–198$). The critical photoperiod (to induce 50% diapausing larvae) is between 12 and 13 h (or 12 h 51 min) as indicated by the dotted vertical line. (Source: Xu et al. 2014)

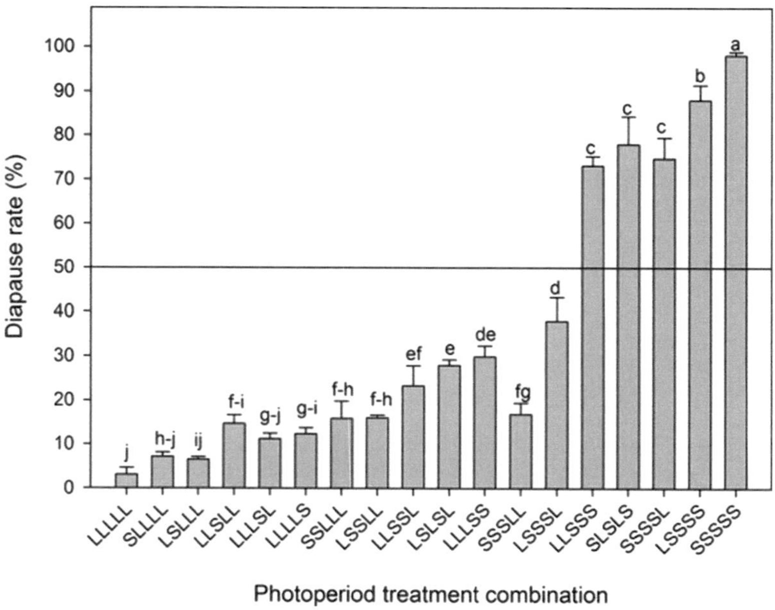


**Fig. 5.9** Impact of 18 empirical treatments on diapause induction during the 5 instars of *Conogethes punctiferalis* larval development at 25 °C ($n = 86–246$). (Source: Xu et al. 2014)

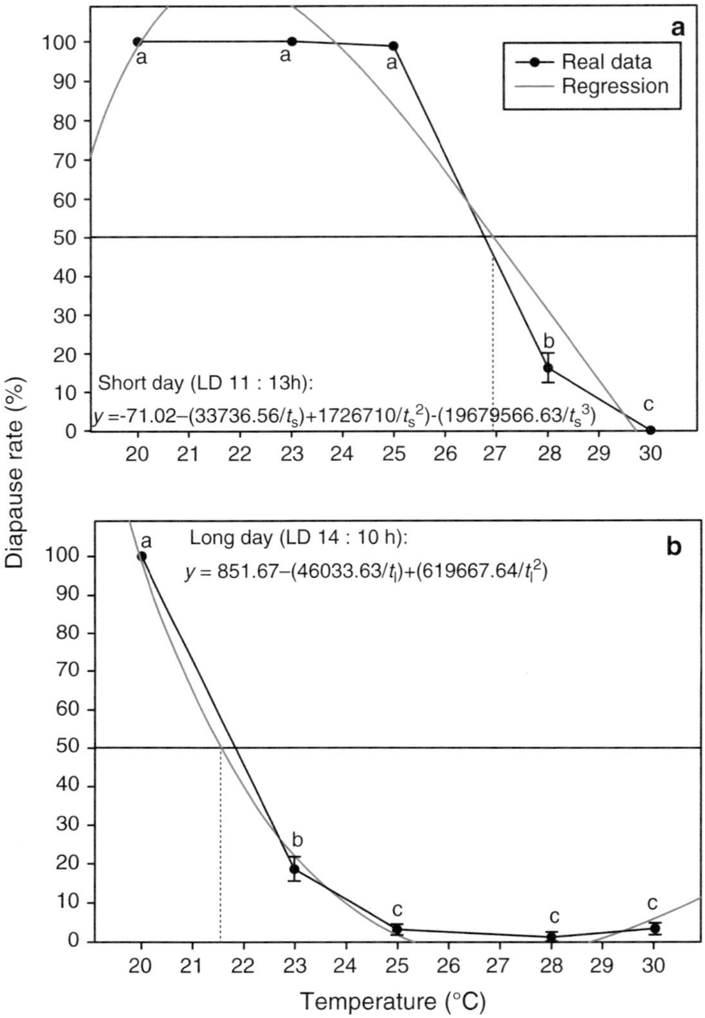


**Fig. 5.10** Modulation of temperature on photoperiodic diapauses responses of *Conogethes punctiferalis* larvae: (**a**) short-day treatment ($n$ = 81–187); (**b**) long-day treatment ($n$ = 68–196). The vertical dotted lines indicate the critical temperatures (to induce 50% diapausing larvae) (short day, $t_{s50}$ = 26.9, and long day, $t_{l50}$ = 21.5 °c). The scattered line plots showed the effect of temperature on larval diapause incidence under (**a**) short- and (**b**) long-day treatments. The blue curved lines in (**a**) and (**b**) were generated based on the regression model using the data shown in the scattered line plot in each graph. (Source: Xu et al. 2014)

better, were heavier, and had higher lipid content than those feeding on sorghum and sunflower. The reduction of larval water content occurred earlier in larvae feeding on maize than those feeding on sorghum and sunflower; larvae feeding on maize contained just 74.0% water in early December, whereas those feeding on sorghum and sunflower contained 82.3% and 84.6% water, respectively. In addition, larvae feeding on maize had the highest cold tolerance capability and the lowest super cooling point (−17.74 °C), followed by those feeding on sorghum (−14.62 °C) and sunflower (−11.68 °C). There was no significant difference in the super cooling capability of larvae from the three host-plant species in early winter and early spring. These results suggest that larvae developed better on maize than on sorghum and sunflower. The major sugars contained in the overwintering larvae were trehalose, glycerol, myoinositol, glucose, sorbitol, dulcitol, and mannitol. The glycerol content of larvae significantly increased from early to deep winter; however, glucose content significantly decreased over the same period, suggesting that glycerol plays an important role in increasing the cold tolerance of larvae (Xu et al. 2012).

## 5.8 *Wolbachia* Infection

*Wolbachia* are maternally inherited endosymbiotic bacteria infecting a wide range of arthropods. They can induce several alterations in reproduction in hosts, such as thelytokous parthenogenesis, cytoplasmic incompatibility (CI), feminization, and male killing. Prevalence and diversity of *Wolbachia* infection in *C. punctiferalis* was investigated in four populations collected in Hebei, Henan, Zhejiang, and Sichuan provinces, respectively, in China (Li et al. 2010). The results showed that the *Wolbachia* infection frequency ranged from 2.0% to 8.0% (average rate = 4.5%) in the sampled populations. Five strains of *Wolbachia* (*w*Pun1–*w*Pun5) were identified based on *wsp* sequences, indicating high diversity of *Wolbachia* infection in *C. punctiferalis* in China. Phylogenetic analysis suggested that these *Wolbachia* variants belong to subgroup Uni, Dro in Group A, and subgroup Div, Con in Group B (Fig. 5.11). In addition, superinfection was found in one population collected in Hebei province (Li et al. 2010).

## 5.9 Genetic Diversity and Gene Flow

Zhang et al. (2010) investigated genetic diversity among 11 geographic populations of *C. punctiferalis* collected in 6 provinces in China using inter-simple sequence repeat (ISSR) markers. The results showed the genetic distances between *C. punctiferalis* populations were 0.0059–0.0237, coefficient of genetic differentiation among populations (Gst) was 0.053, and the estimated value of gene flow was 8.87, indicating low level of genetic differentiation and high level of gene flow among geographic populations of *C. punctiferalis* in China (Zhang et al. 2010). Wang et al. (2014) analyzed mitochondrial cytochrome oxidase II (COII) sequences of 24 geographical populations collected from 10 provinces in China, to investigate the genetic differentiation and phylogeny evolution of *C. punctiferalis* in China. Fifty three haplotypes were detected in the total number of 622 individuals. The average gene flow (Nm) was 1.09 and the overall fixation index (Fst) was 0.585. The results showed Fst value of south region was higher than north, east, and southwest regions, which was in accordance with gene flow results. It revealed that there was different

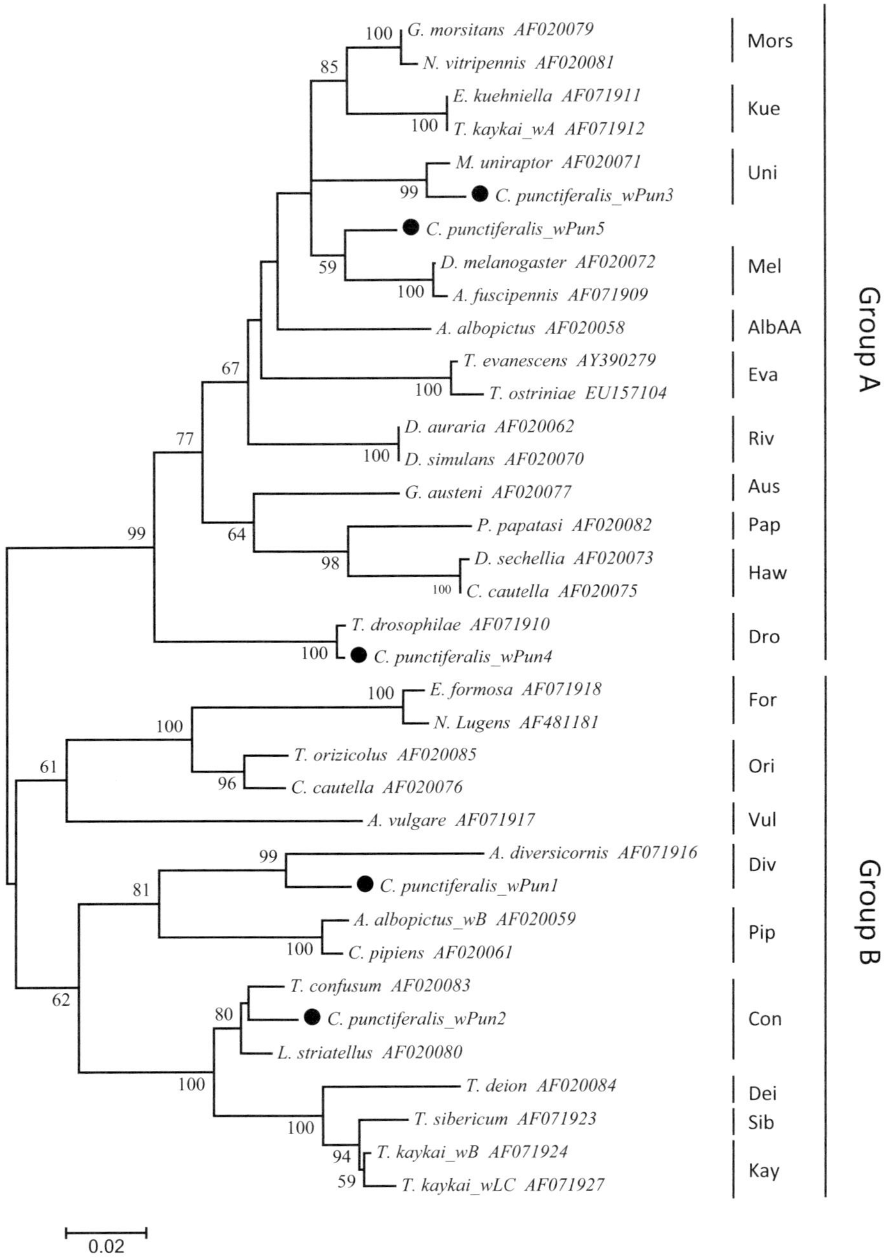


**Fig. 5.11** Phylogenetic tree of *Wolbachia* strains in *C. punctiferalis* based on *wsp* gene sequence. (Source: Li et al. 2010)

level of genetic differentiation among *C. punctiferalis* populations in China. The nucleotide mismatch distribution and neutrality test suggested the possibility of population expansion of *C. punctiferalis* in history, and the expansion time was estimated at around 46,700~116,800 years ago (Wang et al. 2012).

## 5.10 Molecular Taxonomy

*Conogethes punctiferalis* (Guenée) (Lepidoptera: Crambidae) was originally considered as one species with fruit-feeding type (FFT) and pinaceae-feeding type (PFT). However, it has subsequently been divided into two different species of *C. punctiferalis* and *C. pinicolalis* (Inoue and Yamanaka 2006). Three mitochondrial genes including cytochrome c oxidase subunits I(COI), II(COII), and cytochrome b(Cytb) were used to detect genetic differentiation and phylogenetic relationship between *C. punctiferalis* and *C. pinicolalis*. The mtDNA sequences data suggested that haplotypes of COI, COII, and Cytb genes in *C. pinicolalis* populations differed distinctly from *C. punctiferalis* populations. Kxy and Fst values indicated that there was a strong genetic structure between *C. punctiferalis* and *C. pinicolalis*. Analyses of molecular variance (AMOVA) of combined three genes also indicated significant genetic structure between groups, considering FFT and PFT as separate groups respective with 94.04% of the total genetic variance explained by variations between the two types of yellow peach moths based on the three genes. In addition, the migration rate (Nm) suggested little gene flow between the two species. The phylogenetic tree and network showed that conspecific sequences were clustering together despite intraspecific variability, indicating far genetic distance between FFT and PFT (Figs. 5.12 and 5.13). The molecular data provided the powerful molecular evidence that *C. pinicolalis* and *C. punctiferalis* are two different species (Wang et al. 2014).

## 5.11 Integrated Management

*C. punctiferalis* is multivoltine and polyphagous insect pest. The female adults feed, oviposit, and develop primarily in buds and fruits of all kinds of host plants (Li et al. 2015). After hatching, the larvae remain within the reproductive structures and use them as a food source and a protected habitat to complete their life circle. The endophytic behavior of the larvae has made this insect difficult to control with conventional insecticides and cultural practices. Thus, many approaches have been applied to control this insect pest. For example, the pesticide effect on *C. punctiferalis* of 2% thiacloprid DP was tested in chestnut orchard in July 2010 in China (Yuan et al. 2011). The results showed that orchards sprayed with 2% thiacloprid DP achieved significant results to kill the borer. The killing efficiency became stronger as the

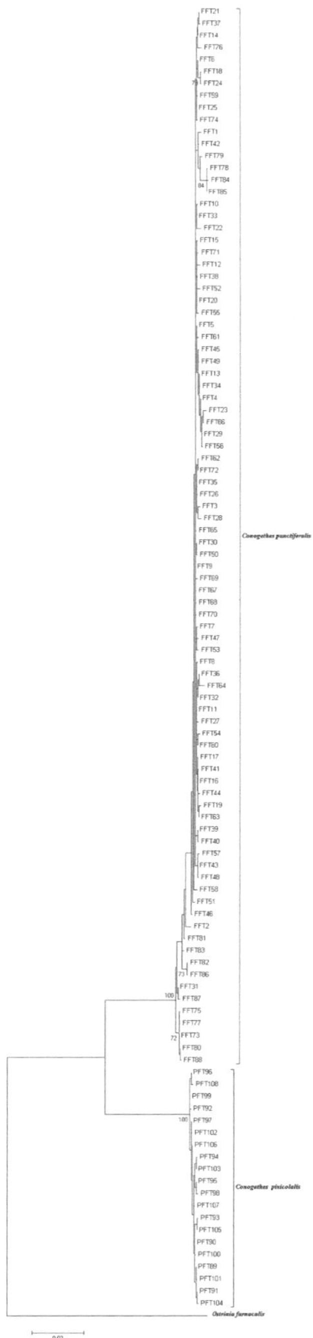

**Fig. 5.12** Phylogenetic maximum likelihood (ML) tree of a concatenated three-gene haplotypes. Branch lengths indicate phylogenetic distance. Value on the nodes indicates bootstrap confidence levels. *Ostrinia furnacalis* was utilized as an out-group. The trees were generated by MEGA based on 1000 bootstraps. (Source: Wang et al. 2014)

**Fig. 5.13** Network profile of the concatenation of three-gene haplotypes of the yellow peach moth. Each haplotype is represented by a circle. The size of the circle relates to the number of individuals sampled. Different colors represent different populations. (Source: Wang et al. 2014)

dose increased. Some predators and parasites were also identified and can be exploited in the management of this pest (Huang et al. 2000). For example, *Apanteles* sp., *Brachymeria lasus*, *Temelucha* sp., and *Trathala flavoorbitalis* were reported as parasites of *C. punctiferalis* (Huang et al. 2000; Li et al. 2005). Recently, *Trichogramma dendrolimi* was released to parasitize the eggs of *C. punctiferalis* in chestnut orchards and a higher parasitism rate (Li 2007; Zhang 2007) was realized. It was also reported that the parasitism rate of *Trichogramma chilonis* on *C. punctiferalis* was 28.8% (He et al. 2008). In addition, some microorganisms, including pathogenic nematode, *Bacillus thuringiensis*, and *Beauveria bassiana* were also used to control *C. punctiferalis* (Choo et al. 1995; Devasahayam 2000; Chen 2004; Zu and Qin 2009). *Bacillus thuringiensis* formulation was recorded to have a control efficiency of 75%, significantly higher than chemical pesticides in chestnut orchards (Xv et al. 2002).

In insects, chemosensation serves to detect and react to environmental chemical cues, in virtually every aspect of their life cycle (Field et al. 2000; Pitts et al. 2011). Olfaction, as a kind of chemosensation, is critical to food source identification, predator avoidance, oviposition site selection, kin recognition, mate choice, and toxic compound avoidance. Up to the present, several olfactory-based strategies have been developed to control moth populations, such as mass trapping and mating disruption (Witzgall et al. 2010). During the recent years, the olfactory communication systems of the yellow peach moth have received considerable attention because of their potential in population outbreak monitoring and pest controlling. For example, sex pheromone composites of *C. punctiferalis* have been analyzed, synthetized, and made into lure to attract male moths and disrupt their mating in fields (Konno et al. 1982; Liu et al. 1994; Xiao et al. 2012; Du et al. 2014b). Among which, (*E*)-10-hexadecenal (E-16:Ald) was considered as principal component of pheromones, whereas (*Z*)-10-hexadecenal (Z-16:Ald) as secondary component, and the field trapping effect was best when the ratio of these two components was 9:1 at the dose of 300.0 ug/lure (Konno et al. 1982). About 10 years later, hexadecenal (16:Ald) was identified as another secondary component, and the field trapping effect was improved when the ratio of 16:Ald, Z-16:Ald, and E-16:Ald was 16:8:100 at the dose of 250.0 ug/lure. Song et al. (2008) synthesized *Z*/*E*-10-hexadecenal by Wittig reaction, and the field experiments showed that the sex pheromone had the best effect when the dose was 50.0 μg/bait at the ratio of *Z*/*E* 20/80. Recently, 16 formulations of sex pheromones and corresponding lures were developed with different compositions, ratios, and dosages of (*E*)-10-hexadecenal (E-16:Ald), (*Z*)-10-hexadecenal (Z-16:Ald), and hexadecanal (16:Ald), and the field trapping tests were conducted to investigate their efficiencies. The results showed that the number of moths trapped by formulation D400-1:4 ((Z-16:Ald): (E-16:Ald) = 1:4 and the dose used was 400.0 μg) was the highest among the 16 tested formulations (Du et al. 2014b). A comparative study was conducted on the efficiency of different doses, lure color, number and types, as well as hanging height of traps on capture of *C. punctiferalis*. The results showed that water barrel trap with a lure of Zhongjie company and hung it in the middle of crown could reach to the best trapping effect (Ren and Guo 2015).

At the same time, better knowledge on the molecular mechanisms by which an odor generates a neuronal signal could lead to the identification of targets for the development of new safe control strategies. To study the function of olfactory receptors (ORs) in sex pheromone and general odorant detection of *C. punctiferalis*, the *Orco* gene was cloned from adults of *C. punctiferalis* and named as *CpunOrco* (Ge et al. 2013). Real-time quantitative PCR results indicated that *CpunOrco* was mainly expressed in the antennae and maxillary palpi of *C. punctiferalis* while expressed in other tissues at a relatively low level. The expression level of *CpunOrco* was higher in antennae of male adults than in antennae of female adults. This study makes clear the expression level of *CpunOrco* in different tissues of *C. punctiferalis* adults and provides a basis for further functional study of this gene (Ge et al. 2013). Recently, in order to study the function and ligand-binding characteristics of pheromone-binding proteins (PBPs), a novel PBP cDNA named as *CpunPBP*1 was obtained from *C. punctiferalis* by Jia et al. (2015). Real-time quantitative PCR results indicated that *CpunPBP*1 was dominantly expressed in adult antennae whereas scarcely expressed in egg stage and not expressed in larval and pupal stages. In addition, the binding affinity of CpunPBP1 to 2 sex pheromones and 14 plant volatiles showed that CpunPBP1 had a strong capability of binding 2 pheromone odorants (*cis*-10-hexadecenal and hexadecanoyl). Especially, CpunPBP1 could also bind with eight plant volatiles, with the highest binding specificity to camphene. These results indicated that CpunPBP1 play important roles in the process of discriminating odorants of pheromones and host-plant volatiles of *C. punctiferalis*. Phylogenetic tree of PBPs from ten insect species in Crambidae showed that CpunPBP1 was closely related to DindPBP (Fig. 5.14).

However, any individual methods are partially effective in keeping this pest below the economic threshold level; an integrated approach to monitor the population outbreaks is essential to control *C. punctiferalis*.

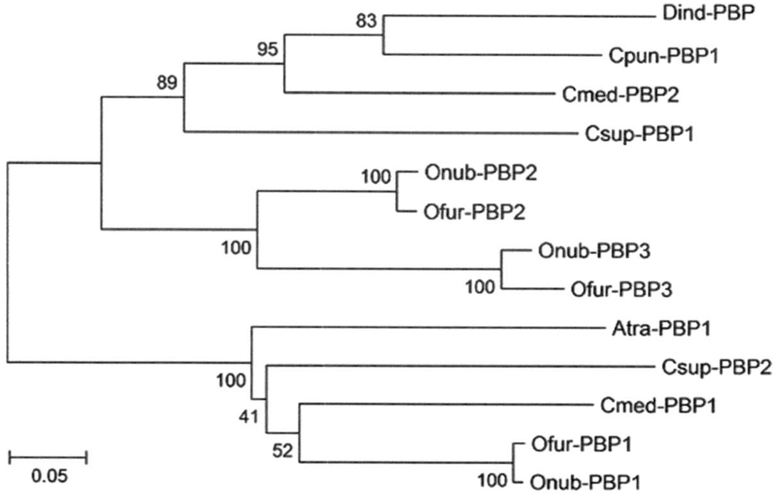


**Fig. 5.14** Phylogenetic tree of PBPs from *Conogethes punctiferalis* and other insects in the family of Crambidae based on amino acid sequences (neighbor-joining method). Numerals in the figure are bootstrap values. The scale bar indicates the genetic distance. Cpun, *Conogethes punctiferalis*; Dind, *Diaphania indica*; Csup, *Chilo suppressalis*; Atra, *Amyelois transitella*; Cmed, *Cnaphalocrocis medinalis*; Cmed, *Cnaphalocrocis medinalis*; Onur, *Ostrinia nubilalis*; Ofur, *Ostrinia furnacalis*. (Source: Jia et al. 2015)

**Acknowledgment** Authors sincerely thank all the scientists and publishers for the source material provided in the tables and figures.

## References

Ai PP, Yang R, Zhang MZ, Sun SL, Du YL (2014) A preliminary observation of the morphological characters of different stages of *Conogethes punctiferalis* (Guenée). J Beijing Univ Agric 29(3):53–55

Becerra JX (1997) Insects on plants: macroevolutionary chemical trends in host use. Science 276:253–256

Bruce TJA, Pickett JA (2011) Perception of plant volatile blends by herbivorous insects-finding the right mix. Phytochemisty 72:1605–1611

CAB International (2011) *Conogethes punctiferalis* datasheet. In: Crop protection compendium. CAB International, Wallingford

Chen WJ (2004) Study on occurrence and integrated management of *Dichocrocis punctiferalis* on *Punica granatum*. J Hennan Univ Sci Technol 24(2):50–52

Chen BX, Dong YZ, Lu H (2008) Studies on the spatial distribution pattern of *Dichocrocis punctiferalis* larvae in chestnut trees. J Environ Entomol 30(4):301–304

Choi KS, Han KS, Park IK, Hong JI, Kim CS, Chung YJ, Chin SC (2006) Difference of developmental time, survival rate and sex ratio of *Dichocrocis punctiferalis* (Lepidoptera: Pyralidae) on three hosts. J Korean For Soc 95:174–176

Choo HY, Lee SM, Chung BK, Park YD, Kim HH (1995) Pathogenicity of Korean entomopathogenic 消nematodes (Steinernematidae and Heterorhabditidae) against local agricultural and forest insect pests. Korean J Appl Entomol 34(4):314–320

Cohen AC (2004) Insect diets: science and technology. CRC Press, Boca Raton, p 312

Devasahayam S (2000) Evaluation of biopesticides for the management of shoot borer (*Conogethes punctiferalis* Guenée) on ginger (*Zingiber officinale* Rosc.). Spices and aromatic plants: challenges and opportunities in the new century. Contributory papers. Centennial conference on spices and aromatic plants, Calicut, Kerala, India, pp 20–23

Dong YZ, Chen BX, Xu S, Li PY, Zhang XW (2012) Analysis of volatiles from fruit and leaf of insect-resistant and insect-susceptible chestnut cultivars with GC-MS. J Fruit Sci 29:1052–1056

Du YL, Guo HM, Sun SL, Zhang MZ, Zhang AH, Wang JB, Qin L (2012) Effects of temperature on the development and reproduction of yellow peach moth, *Conogethes punctiferalis* (Guenée). Acta Entomol Sin 55:561–569

Du YL, Zhang JX, Yang MM, Sun SL, Qin L, Wang JB, Zhang ZY (2014a) Volatiles from involucres of chestnut and their potential applications in control of *Conogethes punctiferalis*. Chin J Ecol 33(8):2096–2100

Du YL, Zhang MZ, Ma YQ, Sun SL, Wang JB, Liu JL, Qin L (2014b) Formulation screening of sex pheromones and field trapping tests for the yellow peach moth, *Conogethes punctiferalis* (Guenée) (Lepidoptera: Pyralidae). Acta Phytophyl Sin 41:187–191

Du YL, Ai PP, Lang LX, Zhang MZ, Sun SL (2015) Development of an artificial diet for rearing *Conogethes punctiferalis* (Lepidoptera: Crambidae). J Entomol Sci 50:89–98

Ehrlich PR, Raven P (1964) Butterflies and plants: a study in coevolution. Evolution 18:586–608

Field LM, Pickett JA, Wadhams LJ (2000) Molecular studies in insect olfaction. Insect Mol Biol 9:545–551

Ge X, Zhang TT, He KL, Wang QY, Li YL, Wang ZY (2013) Cloning and tissue expression profiling of the olfactory receptor coreceptor gene in adults of *Conogethes punctiferalis* (Lepidoptera: Crambidae). Acta Entomol Sin 56(3):243–250

Gong-Min Chen, Hsin Chi, Rong-Cheng Wang, Yun-Peng Wang, Yong-Yu Xu, Xiong-Dong Li, Ping Yin, Fang-Qiang Zheng (2018) Demography and uncertainty of population growth of *Conogethes punctiferalis* (Lepidoptera: Crambidae) reared on five host plants with discussion on some life history statistics, J Econ Entomol, toy202. https://doi.org/10.1093/jee/toy202

Hare JD (2011) Ecological role of volatiles produced by plants in response to damage by herbivorous insects. Annu Rev Entomol 56:161–180

He JX, Guo LF, Zhou H, Wang XG, Yuan H (2008) Report on using *Trichogramma dendrolimi* and *Trichogramma chilonis* to control *Conogethes punctiferalis* in litchi orchards. China Fruits 5:50–51

Honda H, Kaneko J, Konno Y, Matsumoto Y (1979) A simple method for mass-rearing of the yellow peach moth, *Dichocrocis punctiferalis* Guenée (Lepidoptera: Pyralidae), on an artificial diet. Appl Entomol Zool 14:464–468

Huang YQ, Zhang XJ, Wei H, Hu QY, Zhan ZX (2000) Studies on the *Dichocrocis punctiferalis* Guenée and its enemies. Acta Agric Univ Jiangxiensis 22(4):523–525

Inoue H, Yamanaka H (2006) Redescription of *Conogethes punctiferalis* (Guenee) and description of two new closely allied species from Eastern Palaearctic and oriental regions (Pyralidae, Pyraustinae). Tinea 19(2):80–91

Jia XJ, Hao SD, Du YL, Zhang MZ, Wang JZ, Wang HX, Ji WR (2015) cDNA cloning, expression profiling and binding affinity assay of the pheromone binding protein Cpun-PBP1 in the yellow peach moth, *Conogethes punctiferalis* (Lepidoptera: Crambidae). Acta Entomol Sin 58(11):1167–1176

Kadoi M, Kaneda M (1990) Development of yellow peach moth, *Conogethes punctiferalis* (Guenée) on apple fruit. Res Bull Plant Protect Serv 26:61–63

Keesey IA, Barrett BA (2012) Behavioral and electroantennographic responses of the lesser chestnut weevil, *Curculio sayi* (Coleoptera: Curculionidae), to odors emanating from different chestnut plant tissues. J Kansas Entomol Soc 85:145–154

Kirsten K, Topp W (1991) Acceptance of willow-species for the development of the winter moth, *Operophtera brumata* (Lepidoptera: Geometridae). J Appl Entomol 111:457–468
Konno Y, Arai K, Sekiguchi K et al (1982) (E)-10-hexadecenal, a sex pheromone component of the yellow peach moth, *Dichocrocis punctiferalis* Guenée (Lepidoptera: Pyralidae). Appl Entomol Zool 17(2):207–217
Li X (2007) Almost ten million parasites of *Trichogramma dendrolimi* are protecting the growth of chestnut in Huairou District, Beijing. http://bj.news.163.com/07/0727/02/3KCFDRO70038006Lhtml
Li HJ, Lei HD, Liu HQ, Yao TS, Tian WH, Qian KM (2005) Occurrence and management of *Dichocrocis punctiferalis* in citrus orchard. China South Fruit Tree 34(6):22–23
Li J, Zhang Y, Wang ZY, He KL (2010) *Wolbachia* infection in four geographic populations of yellow peach moth, *Conogethes punctiferalis* in China. J Environ Entomol 32(3):322–328
Li Q, Zhang TT, Bai SX, He KL, Wang QY, Li YL, Wang ZY (2014) Ultrastructure observation and electroantennogram responses of *Conogethes punctiferalis* antennae to corn silk volatiles. Plant Prot 40(1):70–76
Li DY, Ai PP, Du YL, Sun SL, Zhang MZ (2015) Effects of different host plants on the development and reproduction of yellow peach moth, *Conogethes punctiferalis* (Guenée) (Lepidoptera; Crambidae). Aust Entomol 54:149–153
Liu MY, Tian Y, Li X (1994) Identification of minor components of the sex pheromone of yellow peach moth, *Dichocrocis punctiferalis* Guenée, and field trials. Entomol Sin 1(2):150–155
Lu JQ, Wang ZY, He KL, Liu Y (2010) Research history, progresses and prospects in the yellow peach moth, *Conogethes punctiferalis*. Plant Prot 36:31–38
Luo ZJ, Zhao SP, Zeng J, Chang J, Zhang DY (2000) Life history, habitus and management of *Conogethes punctiferalis* in chestnut orchards. J Hubei For Sci Technol 113:22–23
Mayhew PJ (1997) Adaptive patterns of host-plant selection by phytophagous insects. Oikos 79:417–428
Ni DW (1998) Main pests and diseases of Castame amollissinia in Zhejiang and their control technique. J Zhejiang For Sci Technol 6:68–71
Pitts RJ, Rinker DC, Jones PL, Rokas A, Zwiebel LZ (2011) Transcriptome profiling of chemosensory appendages in the malaria vector *Anopheles gambiae* reveals tissue and sex-specific signatures of odor coding. BMC Genomics 12:271
Ren S, Guo SJ (2015) Study on the applied technology of *Conogethes punctiferalis*' sex pheromone in chestnut orchard. J Environ Entomol 37(1):96–101
Renwick JAA, Chew FS (1994) Oviposition behavior in Lepidoptera. Annu Rev Entmol 39:377–400
Sekiguchi K (1974) Morphology, biology and control of the yellow peach moth, *Dichocrocis punctiferalis* Guenée (Lepidoptera: Pyralidae). Ibaraki-ken Horticultural Experiment Station Press, Ibaraki-ken
Song W, Tang GH, Feng JT (2008) Synthesis of the sex pheromone of *Dichocrocis puncticralis* Guenee (Z/ E-10-Hexadecenal) by Wittig reaction. J Northwest Agric For Uni (Nat Sci Ed) 36(1):179–182
Thompson JN, Pellmyr O (1991) Evolution of oviposition behavior and host preference in Lepidoptera. Annu Rev Entomol 36:65–89
Utsumi K, Aihara E, Watanabe N (1990) Artificial diet of the yellow peach moth, *Conogethes punctiferalis* (Guenée) using corn seeds (Lepidoptera: Pyralidae). Res Bull Plant Protect Serv 26:65–67
Vanderzant ES (1974) Development, significance, and application of artificial diets for insects. Annu Rev Entomol 19:139–160
Wang CC (2009) Study on the harm of *Dichocrocis punctiferalis* Guenée to *Ricinus communis* and its prevention. J Anhui Agric Sci 37(4):1631–1632
Wang YX, Cai RT (1997) The emergence and control of *Dichocrocis punctiferalis* for Younai plum variety. S China Fruits 26(3):45
Wang OF, Wang JG, Hu HR (2004) The reason of serious occurrence and management of *Dichocrocis punctiferalis* in Dongyang. China Plant Protect 24(2):23–24

Wang ZY, He KL, Shi J, Ma SY (2006) Analysis of the heavily occurrence trend of the yellow peach borer in corn and management strategy for the insect pest. Plant Prot 32(2):67–69

Wang J, Li J, Wang ZY, He KL, Liu Y, Bai SX (2012) Reseach on phylogeny in different geographical populations of the *Conogethes punctiferalis* (Guenée) in China estimated by mitochondrial cytochrome oxidase II gene (COII) sequences. J Agric Biotechnol 20(10):1106–1116

Wang J, Zhang TT, Wang ZY, He KL, Liu Y, Li J (2014) Molecular taxonomy of *Conogethes punctiferalis* and *Conogethes pinicolalis* (Lepidoptera: Crambidae) based on mitochondrial DNA sequences. J Integr Agric 13(9):1982–1989

Waterhouse DF (1993) The major arthropod pests and weeds of agriculture in Southeast Asia: distribution, importance and Origin. ACIAR monograph no. 21, Canberra, Australia

Witzgall P, Kirsch P, Cork A (2010) Sex pheromones and their impact on pest management. J Chem Ecol 36:80–100

Xiao W, Matsuyama S, Ando T, Millar JG, Honda H (2012) Unsaturated cuticular hydrocarbons synergize responses to sex attractant pheromone in the yellow peach moth, *Conogethes punctiferalis*. J Chem Ecol 38:1143–1150

Xu LR, He KL, Wang ZY (2012) Studies on variation in cold hardiness in relation to the in vivo water,lipid,and sugar content of *Conogethes punctiferalis* (Guenée) larvae living on three different host plant species. Chin J Appl Entomol 49(1):197–204

Xu LR, Ni XZ, Wang ZY, He KL (2014) Effects of photoperiod and temperature on diapause induction in *Conogethes punctiferalis* (Lepidoptera: Pyralidae). Insect Sci 21:556–563

Xv JP, Wu LP, Hong Q, Lang XJ, Xv YL, Mao QG (2002) Experiment about *Bacillus thuringiensis* control of *Dichocrocis punctiferalis* Guenee on chestnut. J Jiangsu For Sci Technol 29(5):31–32

Yuan Z, Wang MY, Lu JM, Li QL (2011) Efficacy of 2% thiacloprid DP to control *Dichocrocis punctiferalis* in chestnut orchard. J Jiangsu For Sci Technol 4:21–23

Zhang N (2007) It is the first time to use *Trichogramma dendrolimi* control *Conogethes punctiferalis* in chestnut orchards in Miyun District, Beijing. http://student.gov.cn/kjxx/bjxx/2007/9/2/131620.shtml

Zhang Y, Li J, Wang ZY, He KL (2010) Genetic diversity of *Conogethes punctiferalis* (Guenée) (Lepidoptera: Crambidae) populations from different geographic regions in China. Acta Entomol Sin 53(9):1022–1029

Zhao HT, Zhang DZ, Wang ZL, Pu HY (2004) Occurrence regularity and management of *Dichocrocis punctiferalis* on *Punica granatum*. Hebei Fruits 4:6–8

Zu WF, Qin LIZ (2009) A review on *Dichocrocis punctiferalis* Guenee. J Hebei Agric Sci 13(1):18–20

# Status of *Conogethes punctiferalis* (Guenée) in South of Vietnam

# 6

H. T. Loc, K. P. Kumar, and A. K. Chakravarthy

**Abstract**

Currently over a dozen species of *Conogethes* have been recognized. Of which, *Conogethes punctiferalis* is prevalent on several plants and well distributed in South of Vietnam. Species of *Conogethes* infesting ginger and other Zingiberaceae belong to another species. Durian, soursop, longan, and rambutan are the major fruit crops infested by *C. punctiferalis* in Vietnam. Integration of cultural, mechanical, biological, and chemical methods has proven effective against this borer pest. Augmentation and supplement releases of parasites and encouragement and conservation of predators and natural enemies is necessary for suppressing *Conogethes* borer populations in fruit orchards and plantations.

**Keywords**

Durian · Ginger · Integrated pest management · Vietnam

H. T. Loc (✉)
Plant Protection Division, Southern Horticultural Research Institute (SOFRI), My Tho, Tien Giang, Vietnam

K. P. Kumar
Department of Entomology, University of Agricultural Sciences, Gandhi Krishi Vignan Kendra (GKVK), Bengaluru, Karnataka, India

A. K. Chakravarthy
Division of Entomology and Nematology, ICAR-Indian Institute of Horticultural Research, Bengaluru, Karnataka, India





A. K. Chakravarthy (ed.), *The Black spotted, Yellow Borer, Conogethes punctiferalis Guenée and Allied Species*,
https://doi.org/10.1007/978-981-13-0390-6_6

## 6.1 Introduction

Fruit borer, *Conogethes punctiferalis* (Guenée), is a highly polyphagous pest; the larvae bore into fruits, seeds, and stems of plants of many different families. It has been reported from various parts of the world, mainly because larvae are imported alongside fruits. Records include Hawaii, Great Britain, and the Netherlands. This is a species complex and incomplete information on their separation has been published to date. *C. punctiferalis* is found in Asia (from India eastward) and Australasia. *C. punctiferalis* species complex has been recorded from Australia, Brunei, Darussalam, Burma, Cambodia, China, India, Indonesia, Japan, Laos, Malaysia, North Korea, Papua New Guinea, the Philippines, South Korea, Sri Lanka, Taiwan, Thailand, and Vietnam (CABI 2011).

## 6.2 Characteristics of *C. punctiferalis* (Guenée) in South of Vietnam

### 6.2.1 Species of Genus *Conogethes* in South of Vietnam

Among 12 species of genus *Conogethes, Conogethes clioalis* (Walker, 1859), *C. diminutiva* (Warren, 1896), *C. ersealis* (Walker, 1859), *C. haemactalis* (Snellen, 1890), *C. minimastis* (Meyrick, 1897), *C. parvipunctalis* (Inoue & Yamanaka, 2006), *C. pinicolalis* (Inoue & Yamanaka, 2006), *C. pluto* (Butler, 1887), *C. punctiferalis* (Guenée, 1854), *C. semifascialis* (Walker, 1866), *C. tharsalea* (Meyrick, 1887), and *C. umbrosa* (Meyrick, 1886). There is one species present in South of Vietnam, i.e., *C. punctiferalis* (Guenée, 1854).

### 6.2.2 Synonyms

*Astura punctiferalis* (Guenée), *Dichocrocis punctiferalis* (Guenée), *Deiopeia detracta* (Walker), *Botys nicippealis* (Walker), and *Astura guttatalis* (Walker)

### 6.2.3 Host Range

Major host range: Durian (*Durio zibethinus*), guava (*Psidium guajava*), longan (*Dimocarpus longan*), rambutan (*Nephelium lappaceum*), and soursop (*Annona muricata*)

Minor host range: Ginger (*Zingiber officinale*), macadamia (*Macadamia ternifolia*), white mulberry (*Morus alba*), cotton (*Gossypium* spp.), starfruit (*Averrhoa carambola*), corn (*Zea mays*), and sunflower (*Helianthus annuus*). The species of *Conogethes* infesting Zingiberaceae is not *C. punctiferalis.* It has now been identified as *C. sahyadriensis* (Shashank et al. 2018).

### 6.2.4 Morphological Characteristics

The size of the larvae and adults, the numbers of black dots, as well as the distribution of the number of black dots on the wings are depending on the food as well as the host plant. The largest size of *C. punctiferalis* is obtained when reared on guava. The smallest size is seen when it is reared on soursop. Eggs of *C. punctiferalis* are oval shaped, 2–2.5 mm in length. Freshly laid eggs are creamy white and then become yellowish.

Fully grown larva is 22–28 mm in length. The head of the larva is brown, the body is white blush, prothoracic and mesothoracic segments and two ends of abdominal segments are usually pinkish white, and another one segment is pink. There are four light brown dots with spin hair on every segment of the body, two top dots are large, and two lower dots are narrow and long in shape. Each body segment has small brown dots on the side of the body, beside the black spiracle. The ventral side of the body has light brown dots on every segment with small spin hair (Fig. 6.2).

Adult wingspan is 25–30 mm in length; the body is 12–14 mm in length. The body and wings are yellow with black spots scattered (Fig. 6.1). Pupa is brownish yellow turning brown when ready to emerge, 13–15 mm in length and 4–5 mm in width.

### 6.2.5 Biological Characteristics

The larvae have five instars. Larvae need 12–13 days to fully develop and then go to pupal stage. The pupal stage lasts for 7–9 days.

Adults are nocturnal. They are active during 20–22 pm until 5 am. During daytime moths hide in the leaves of the trees. Both male and female often live on nectar of host plants in the orchard. After emerging, females often release sex pheromone to attract males (CABI 2005). Two to three days after mating, the females start laying eggs. Each female lays 20–30 eggs. The eggs hatch in the morning. The incubation period is 4–6 days. The life cycle of *C. punctiferalis* completes in 29–32 days. Surprisingly, the number of eggs laid/female is less.

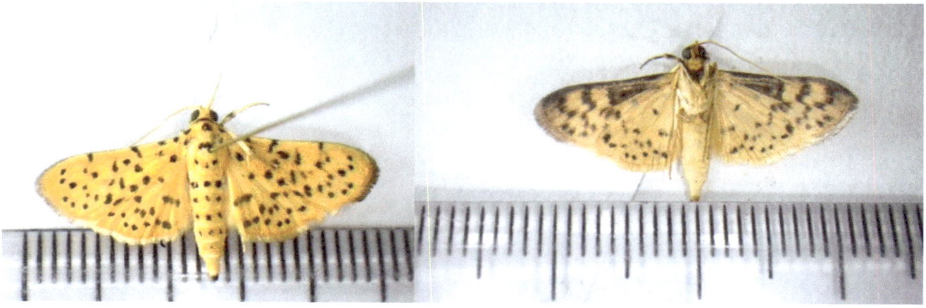

**Fig. 6.1** Adults of *Conogethes punctiferalis*

**Fig. 6.2** Larvae of *Conogethes punctiferalis* (**a**) first instar, (**b**) second instar, (**c**) third instar, (**d**) fourth instar, (**e**) fifth instar

## 6.2.6 Natural Enemies

*C. punctiferalis* has many natural enemies such as larval predators (adult ant lions, Myrmeleontidae, and birds), moth predators (mantis and spiders), and larval parasitic nematodes (*Steinernema glaseri*). In castor fields, several natural enemies of *C. punctiferalis* were recorded such as wasps, *Trathala flavoorbitalis*, *Brachymeria atteviae*, *Chelonus blackburni, Anthocephalus decipiens, Epitranus erythrogaster*, and Phorid flies (Maruthi et al. 2009).

In Australia, flies *Argyrophylax proclinata* parasitized *C. punctiferalis* larvae, and the parasitization rate goes up to 40% (Pena et al. 2002). According to Evangelista (1995), natural enemies of *C. punctiferalis* include larval and pupal parasitic wasp *Suallonius* sp. and larval predator springtails *Euborellia annulata*. There may also be other parasitoids (Fig. 6.2).

## 6.3 Status of *Conogethes punctiferalis* (Guenée) on Crops in South Vietnam

### 6.3.1 Durian

Durian (*Durio zibethinus* Murray) is the fruit of tree species belonging to the genus *Durio*. There are 30 recognized *Durio* species, at least nine of which produce edible fruit. There are over 300 named varieties in Thailand. *Durio zibethinus* is the only species available in the international market: other species are sold only locally. There are hundreds of durian cultivars; many consumers express preferences for specific cultivars, which fetch higher price in the market.

In the world, durian is a specialty of Malaysia. In the Far East, durian is popular in this region. Particularly in Thailand, the durian is one of the fruits of the country's strengths, especially variety Monthong, which is of good quality. In Vietnam, durian can be grown from the plains to the 1000 m as in Bao Loc, Di Linh. Durian is one of the fruit trees grown quite popularly in provinces in the Mekong Delta and the Southeast of Vietnam.

Common insect pests attacking durian are the fruit borers (*Tonica lagaropis* and *C. punctiferalis*), shot-hole borer, and psyllids. Although it is only the fruit borer that has a significant contribution in directly reducing durian yield, the other insect pests can play an indirect role in increasing the incidence of *Phytophthora* infection either by creating entry ways for the fungal pathogen or by decreasing the resistance of the trees against the disease. The shot-hole borer (*Xyleborus* sp.) is a tiny black or brown beetle which bore holes in the bark and feed on the cambium layer. It is of particular significance to the subject of disease control in durian because it is associated with *Phytophthora*. The psyllids (*Allocarsidara incognita*) lay eggs on the unopened leaves. The nymphs have body coverings that appear cottony white. They suck the young leaves causing yellowish spots and, if not controlled, severe infestation may cause dying of the tree.

#### 6.3.1.1 Damage Symptoms

Eggs are laid scattered on the young fruit. After hatching, larvae move on the surface of fruit and then bore into the fruit. *C. punctiferalis* attacks durian fruit from young to mature fruits. They attack more on bunch of fruit than lonely fruit. Damaged young fruits will be deforming or falling. Damaged mature fruits affect commercial value of fruit. The boring of caterpillars inside the fruit causes secondary infections of fungi and bacteria. This results in fruit rot and fruits drop-off. Larvae pupate between the spines of the fruits by making cocoons from their silk and frass.

*C. punctiferalis* attack the durian flower clusters also. Moth lays eggs on the flower clusters. After hatching, larvae feed on the stalk of flowers and then bore into stalk and bud of flowers for feeding. Damaged flowers dry and then fall off. The damage of *C. punctiferalis* on the durian flower is easily recognized through the perforated holes and clusters of dark brown frass on the surface of flower stalk clusters. Pupation was recorded in the damaged flower cluster (Fig. 6.3).

**Fig. 6.3** Damage symptoms of *C. punctiferalis* on durian fruit and flowers

#### 6.3.1.2 Management

**Cultural Control**

- Monitoring the orchard regularly in flowering stage, fruiting to detect early damage of *C. punctiferalis* on flowers and fruits
- Collection and destruction of the damaged flowers and fruits
- Removing fruits which are deformed from bunches
- Using a piece of carton to wedge inside bunches of fruits to limit the damage by fruit borer

**Chemical Control**

When necessary, chemical insecticides can be used in heavily infected areas. The chemicals can be sprayed at 10–15 days intervals during the flowering and fruit setting to prevent attack of *C. punctiferalis* on flowers and fruits. Insecticides can be used to control *C. punctiferalis* such as *Bacillus thuringiensis* var. *kurstaki* (Biobit 32 B FC, Crymax 35 WP, WP 3.2 Dipel) or spinosad (Success 25 SC) or emamectin benzoate (Acplant 1.9EC) or abamectin (Abatin 5.4 EC) according to the recommended dosage and safety on the label.

Huynh Thanh Loc et al. (2006) showed that *Bacillus thuringiensis* var. *kurstaki* (Biobit 32 B FC), spinosad (Success 25 SC), and Lambda-cyhalothrin (Karate 2.5 EC) were effective against *C. punctiferalis* on durian. Fruits infested by *C. punctiferalis* at 10 days after second treatment were 0.87, 0.00, and 1.59%, respectively, compared to 27.24% in control (water treatment).

### 6.3.2 Guava

Common insect pests attacking guava are the fruit borers (*Conogethes punctiferalis*), red-banded thrips (*Selenothrips rubrocinctus*), chilli thrips or yellow tea thrips (*Scirtothrips dorsalis*), fruit flies (*Bactrocera dorsalis*, *Bactrocera correcta*, *Bactrocera carambolae*), mealybugs (*Planococcus lilacinus*, *Planococcus* sp.,

**Fig. 6.4** Damage symptoms of *C. punctiferalis* on guava fruit

*Pseudococcus* sp.), scales (*Coccus viridis*, *Aonidiella aurantii*, *Lepidosaphes becckii*), *Archips micaceana*, whitefly (*Aleurodicus dispersus*), aphid (*Aphis gossypii*), mosquito bug (*Helopeltis* sp.), and mites.

#### 6.3.2.1 Damage Symptoms

Eggs are laid scattered on the young fruits. After hatching, larvae bore into the fruit. They attack guava fruit from young to mature fruits. On the young fruit, larvae feed inside the fruit; fruit can turn black and dry and falls off, or the growth gets retarded. Fruit borer preferred to lay eggs and damage fruits that are held in bunches. The damage of fruit borer on the guava fruits can be detected through the perforated dark brown frass adhering on the damaged fruits with holes.

Adults are nocturnal. They hide undersides of leaves during the day. Pupa is brown, about 10–13 mm in length. Pupation takes place inside the damaged fruits or at bunches of fruit by making cocoons from their silk and frass (Fig. 6.4).

#### 6.3.2.2 Management

**Cultural Control**

- Sanitation of the orchard by collecting and destructing the infested fruits.
- Pruning the canopy after harvest.
- Using black light traps to trap moths.
- Fruit bagging is one of the best practices to prevent fruit borer because this method is safe and effective.

**Chemical Control**

Monitoring the orchard regularly at stage of fruiting to detect early damage of the fruit borer. When the infested fruit rate up to 5%, chemical insecticides could be used to control *C. punctiferalis* such as *Bacillus thuringiensis* var. *kurstaki* (Biobit 32 B FC, Crymax 35 WP, WP 3.2 Dipel) or spinosad (Success 25 SC) or emamectin benzoate (Acplant 1.9EC) or abamectin (Abatin 5.4 EC) according to the recommended dosage and safety on the label.

**Fig. 6.5** Damage symptoms of *C. punctiferalis* on longan fruits

### 6.3.3 Longan

Longan (*Dimocarpus longan* Lour.) is a tropical tree that produces edible fruits. It is one of the better-known tropical members of the soapberry family (Sapindaceae), to which the litchialso belongs. It is native to Southern Asia. In Vietnam, longan is grown in Vung Tau, Dong Nai, Tien Giang, Vinh Long, and Ben Tre province.

The fruit is sweet, juicy, succulent, and superior in quality. Apart from being eaten fresh, it is also often used in Asian soups, snacks, desserts, and sweet-and-sour foods, either fresh or dried, sometimes canned with syrup. The seed and the shell are not consumed. Dried longan is often used in Chinese cuisine and Chinese sweet dessert soups, food therapy, and herbal medicines.

The most important pest is the Longan stink bug, *Tessaratoma javanica*. Other pests include erinose mite, scales, fruit flies, aphids, stem borers, fruit borer, leaf-eating caterpillars, flower-eating caterpillars, mealybug, fruit-spotting bug, elephant beetles, and fruit-piercing moths.

#### 6.3.3.1 Damage Characteristics

Females lay eggs on the fruits, especially at the junction between the two fruits. After hatching, larvae bore into the fruits and feed on the inside of the fruits. *C. punctiferalis* attack the fruits from young to mature ones, especially when the fruit starting flesh setting. Damaged young fruit will be distorted, dry, and fall down. Damaged mature fruits will affect commercial value of the fruits (Fig. 6.5).

#### 6.3.3.2 Management

**Cultural Control**

- Pruning the canopy after harvest to clear the orchard
- Sanitation of the orchard by collecting and destroying the infested fruits

**Chemical Control**

Monitoring the orchard regularly on different stages of fruiting to detect early damage of the fruit borer. When the infested fruit rate goes up to 20%, insecticides could be used to control *C. punctiferalis* such as *Bacillus thuringiensis* var. *kurstaki* (Biobit 32 B FC, Crymax 35 WP, WP 3.2 Dipel, etc.) or spinosad (Success 25 SC) or emamectin benzoate (Acplant 1.9EC) or abamectin (Abatin 5.4 EC) according to the recommended dosage and safety on the label.

Nguyen Thi Kim Thoa et al. (2009) showed that *Bacillus thuringiensis* var. *kurstaki* (Biobit 32 B FC), spinosad (Success 25 SC), and abamectin + *Bacillus thuringiensis* var. *kurstaki* (Kuraba WP) were highly effective on *C. punctiferalis* on longan. Fruit infested by *C. punctiferalis* at 14 days after second treatment were 1.74, 1.61, and 1.77%, respectively, compared to 2.98% in control (water treatment).

### 6.3.4 Rambutan

Rambutan (*Nephelium lappaceum* L.) is a tropical tree species in Southeast Asia, family Sapindaceae. Rambutan is native to Malay-Indonesian region and other regions of tropical Southeast Asia. It is closely related to several other edible tropical fruits including the litchi, longan, and mamoncillo. Rambutan fruit contains diverse nutrients but in modest amounts, with only manganese having moderate content at 16% of the daily value per 100 g consumed (right table; note data are for canned fruit in syrup, not as raw which may have different nutrient contents).

As an unpigmented fruit flesh, rambutan does not contain significant polyphenol content, but its colorful rind displays diverse phenolic acids, such as syringic, coumaric, gallic, caffeic, and ellagic acids having antioxidant activity in vitro. Rambutan seeds contain equal proportions of saturated and unsaturated fatty acids, where arachidic (34%) and oleic (42%) acids, respectively, are the highest in fat content. The pleasant fragrance of rambutan fruit derives from numerous volatile organic compounds, including beta-damascenone, vanillin, phenylacetic acid and cinnamic acid. In Vietnam, rambutan is grown in Ben Tre, Dong Nai, and other provinces, with high economic value, can be eaten fresh or canned.

Few insect pests have been reported by rambutan growers: fruit borers (*Conogethes punctiferalis, Acrocercops* sp., *Deudorix epijarbas amatius*, *Tirathaba ruptilinea*) and leaf-eating insects – the mealybug, *Pseudococcus lilacinus,* and the giant bug, *Tessaratoma longicorne* – the oriental fruit fly attack very ripe fruits and may require control measures.

#### 6.3.4.1 Damage Symptoms

Females lay eggs on the fruits. After hatching, larvae feed on the skin of fruit, and then larvae bore and feed into the fruits. Fruit borer can attack the fruit from young to mature fruits, especially when the fruit starts flesh setting. Through these holes, the disease agents such as bacteria and fungi affect the fruit. Larvae often join some fruits together by their silk and then feed on inside. They feed on all seeds of the

**Fig. 6.6** Damage symptoms of *C. punctiferalis* on rambutan fruit

young fruit; fruit is deformed, is dry, and falls down. Damaged mature fruits will affect commercial value (Fig. 6.6).

#### 6.3.4.2 Management

**Cultural Control**

- Pruning unproductive pests and diseases affected old branches aids in reducing the borer population.
- Maintaining field sanitation in the orchards is also important.
- Affected fruits should be collected and destroyed.
- Bagging of fruit saves yields loss.

**Chemical Control**

Monitoring the orchard regularly for borers on fruits to detect early damage of the fruit borer is important. When the infested fruit rate goes up to 5%, chemical insecticides could be used to control *C. punctiferalis* such as *Bacillus thuringiensis* var. *kurstaki* (Biobit 32 B FC, Crymax 35 WP, WP 3.2 Dipel) or spinosad (Success 25 SC) or emamectin benzoate (Acplant 1.9EC) or abamectin (Abatin 5.4 EC) according to the recommended dosage and safety on the label. Insecticides should be sprayed three to four times/fruiting season at 10–15 days intervals.

### 6.3.5 Soursop

Soursop (*Annona muricata* L.) belongs to the family Annonaceae. It is generally known in most Spanish-speaking countries as *guanabana*; in Salvador, as *guanaba*; in Guatemala, as *huanaba*; in Mexico, often as *zopotedeviejas* or *cabezade negro*; in Venezuela, as *catoche* or *catuche*; in Argentina, as *anona de puntitas* or *anona de broquel*; in Bolivia, *sinini*; in Brazil, *araticum do grande, graviola* or *jaca do Para*; and in the Netherlands Antilles, *sorsaka* or *zunrzak*; the latter name is also used in Surinam and Java; in French-speaking areas of the West Indies, West Africa, and Southeast Asia, especially North Vietnam, it is known as *corossol*, *grand corossol,*

**Fig. 6.7** Damage symptoms of *C. punctiferalis* on soursop fruit

*corossol epineux*, or *cachiman epineux*. In Malaysia it is called *durian belanda, durian maki*, or *seri kaya belanda*; in Thailand, it is called *thu-rian-khack*.

In Vietnam, soursop is not only used fresh but also used as a beverage processing and for confectionery. At Tet lunar holiday, soursop is also valuable especially in fruit tray on family altar in Vietnam. Soursop is a delicious fruit with high nutritional value.

Few insect pests have been reported from soursop orchards by Huynh Thanh Loc et al. (2015) through monitoring in Tien Giang province. The pests are mealybugs, scales, aphids, green leafhopper, mosquito bug, fruit borer, trunk borers, fruit flies, and fruit skin-eating caterpillars.

#### 6.3.5.1 Damage Symptoms

Eggs are laid scattered on the young fruit; each female lays about 30 eggs on an average. After hatching, larvae bore into the fruit. Fruit borer attacks soursop fruit from young to mature fruits. The larvae prefer to attack fruits in bunch than single fruit. Usually, there are about two larvae/infested fruit, but the high density can reach to 6–12 larvae/infested fruit. Damaged young fruits will be deformed or fall off. Damaged mature fruits will affect commercial value of fruits. The boring of caterpillars inside the fruit can cause secondary infections of fungi and bacteria. This can result in fruit rot causing fruits to drop off. Pupation takes place in the damaged portion or on the skin of the fruit by making cocoons from their silk and frass (Fig. 6.7).

#### 6.3.5.2 Management

**Cultural Control**

- Monitoring the orchard regularly at fruiting stage to detect early damaged fruit
- Collection and destruction of the damaged fruits
- Removing fruits which are deformed ones in bunches of fruits
- Using a piece of carton to wedge inside bunches of fruits to limit the damage by the fruit borer
- Fruit bagging with suitable, locally available material

**Chemical Control**

Monitoring the orchard regularly to detect early damaged fruits. When the infested fruit rate goes up to 1%, chemical insecticides could be used to control *C. punctiferalis* such as *Bacillus thuringiensis* var. *kurstaki* (Biobit 32 B FC, Crymax 35 WP, WP 3.2 Dipel) or spinosad (Success 25 SC) or emamectin benzoate (Acplant 1.9EC) or abamectin (Abatin 5.4 EC.) as per the recommended dosage and safety on the label. One can spray three to four times per season of fruit at 10–15 days intervals, depending on the level of infestation.

### 6.3.6 Ginger

Ginger (*Zingiber officinale* Roscoe) is a flowering plant whose rhizome, ginger root or simply ginger, is widely used as a spice or a folk medicine. It is a herbaceous perennial which puts forth annual stems about a meter tall-bearing narrow green leaves and yellow flowers. Ginger plant originated in the tropical rainforest in Southern Asia. Although ginger no longer grows in the wild, it is thought to have originated in the Indian subcontinent. The ginger plants found in India show the largest amount of genetic variation. The larger the number of genetic variations, the longer the plant is thought to have grown in that region. Ginger was exported to Europe via India in the first century AD as a result of the lucrative spice trade and was used extensively by the Romans.

Ginger is infested by many species of insects, among which the shoot borer (*Conogethes* sp.) and rhizome scale (*Aspidiella hartii* Sign.) are major pests in the field and during storage of rhizomes, respectively. Other insects that have been reported to affect ginger belong to diverse families and can be classified into sap feeders, leaf feeders, and rhizome feeders. Dry ginger is also infested by many species of insects, most importantly the cigarette beetle (*Lasioderma serricorne* (Fab.), the drug store beetle (*Stegobium paniceum* L.), and the coffee bean weevil (*Araecerus fasciculatus* DeG.).

#### 6.3.6.1 Damage Symptoms

The larvae bore into pseudostems and feed on internal tissues resulting in yellowing and drying of leaves of infested pseudostems. The presence of a bored hole on the pseudostem through which frass is extruded and the withered and yellow central shoot is a characteristic symptom of this pest infestation (Fig. 6.8).

#### 6.3.6.2 Management

**Cultural Control**

- Use the attractant wild plants for natural biocontrol conservation in the cultivated ecosystems.
- Cut open the shoot, and pick out the caterpillar and destroy. Mulching with green *Vitex negundo* leaves at 2 t/acre at 40 and 90 days after planting is effective in deterring the pest.

**Fig. 6.8** Damage symptoms of *Conogethes sp.* on ginger pseudostem

### Chemical Control

- Fruit borer usually appears early in the raining season. Neem oil (0.5%) at fortnightly intervals if found necessary can be applied.
- Prevention and treatment: The use of systemic insecticides such as Diazinon (Vibasu 10GR), Fipronil (Regent 0.3 G) is useful. When larvae are seen at 1st–2nd instars in the ginger fields, apply insecticides immediately, if not, control of the pest becomes difficult.

**Acknowledgment** Authors are greatfully thankful to the authorities of the SOFRI, Vietnam, ICAR-IIHR, Bengaluru, and ICAR, New Delhi. All photographs in this chapter are by the senior author.

## References

CAB International (2011) *Conogethes punctiferalis* datasheet. In: Crop protection compendium. CAB International, Wallingford

CABI (2005) *Conogethes punctiferalis* datasheet. In: Crop protection compendium. CAB International, Wallingford

Evangelista CC (1995) Biology, ecology and control of durian fruit borers, *Tonica lagaropis* Meyeick (Oecophoridae: Lepidoptera) and *Conogethes punctiferalis* (Guen.) (Lepidoptera: Pyralidae). University of the Philippines Los Banos, Ph D. thesis, PCARRD. http://www.sel.barc.usda.gov/scalenet/scalanet.htm

Huynh Thanh Loc, Le Quoc Dien and Nguyen Van Hoa (2006) Intergrated controling fruit borer (Conogethes punctiferalis Guen.) on Durian in the Mekong Delta. Results of the technological and scientific research 2004-2005 of the Southern Horticultural Research Institute. Agricultural Publishers: 214–222

Huynh Thanh Loc, Nguyen Van Hoa, Vo Huu Thoai, Luong Ngoc Trung Lap, Dang Thuy Linh, Nguyen Thanh Tung, Nguyen Thi Kim Thoa, Nguyen Van Thanh, Pham Thi Thu Hong and Nguyen Van Hai (2015) Study on technical solutions to improve productivity, quality and build models follow VietGAP on soursop at Tan Phu Dong (2011–2015). Report of Project

Maruthi SJ, Chandrasekaran S, Preetha G (2009) *Conogethes punctiferalis* (Lepidoptera: Pyralidae) its biology and field parasitization. Indian J Agric Sci 79(11)

Nguyen Thi Kim Thoa, Nguyen Thanh Hieu, Nguyen Van Hoa and Nguyen Huu Hoang (2009) The survey results on fruit borer species and effect of some biochemicals in controlling fruit borer on Tieu da bo Longan. Results of the technological and scientific research 2008-2009 of the Southern Horticultural Research Institute. Agricultural Publishers: 94–98

Pena JE, Nadel H, Barbosa-Pereira M, Smith D (2002) Pollinators and pests of annona species. In: Pena J, Sharp J, Wysoki M (eds) Tropical fruit pests and pollinators: biology, economic importance, natural enemies and control. CABI Publication, UK, p 208

Shashank PR, Kammar V, Mally R, Chakravarthy AK (2018) A new Indian species of shoot and capsule borer of the genus *Conogethes* (Lepidoptera: Crambidae), feeding on cardamom. Zootaxa 4374(2):215–234

# Status of Shoot and Fruit Borer, *Conogethes punctiferalis*, in Sri Lanka

7

Amani Mannakkara, A. D. N. T. Kumara, N. I. Suwandharathne, I. Hettiarachchi, and H. T. R. Wijesekara

**Abstract**

The genus *Conogethes* is a complex insect taxon that is widely distributed in South and East Asia, and its range extends to Australia, Papua New Guinea, and Pacific regions. Fruit and shoot borer, *Conogethes punctiferalis* Guenée (Lepidoptera: Crambidae), is considered as one of the most widely spread and economically important lepidopteran pests inflicting direct injury to tender vegetative and reproductive plant parts. The *Conogethes* species are emerging as an important group because of speciation, ability to disperse, and wide host range comprising more than 120 species of diverse host plants. *C. punctiferalis* is reported as an economically important pest of several crops in Sri Lanka. But, scientific records of the pest status are limited, and there is no information on crop loss. Bioecology of the pest is highly variable and depends upon the host plant and habitat. Therefore, any formulation of general management strategy for the pest under different cultivated ecosystems is difficult.

**Keywords**

*Conogethes punctiferalis* · Durian · Pest status · Sri Lanka · Teak

---

A. Mannakkara
Department of Agricultural Biology, Faculty of Agriculture, University of Ruhuna, Matara, Sri Lanka

A. D. N. T. Kumara (✉) · N. I. Suwandharathne · H. T. R. Wijesekara
Crop Protection Division, Coconut Research Institute, Lunuwila, Sri Lanka

I. Hettiarachchi
Department of Agriculture, Fruit Crop Research and Development Institute, Horana, Sri Lanka



A. K. Chakravarthy (ed.), *The Black spotted, Yellow Borer, Conogethes punctiferalis Guenée and Allied Species*,
https://doi.org/10.1007/978-981-13-0390-6_7

## 7.1 Introduction

Increasing food production for the growing world population is one of the key challenges in agriculture. Agriculture plays a vital role in the economy of Sri Lanka. Being a tropical agricultural country, the possibility of growing a wide range of crops throughout the year in Sri Lanka is very much feasible. But, the yield losses due to pest attack are rampant, which reduces farmers' agricultural income.

The genus *Conogethes* comprised 24 species (Shashank et al. 2015), and *Conogethes punctiferalis* is one of the widely distributed pest of broad host range in South and East Asia, Australia, and Papua New Guinea (CAB International 2011). Conogethes commonly known as yellow peach moth is named by many other names by local peoples generally indicative of host plant and affected plant parts. The larva of *C. punctiferalis* is polyphagous and has been reported on several host plants with economic significance in Sri Lanka (USDA 1957; Mannakkara 2006). *C. punctiferalis* is prevalent in Sri Lanka, and little information of pest status is available (USDA 1957; Mannakkara 2006; Dharmadasa et al. 2008).

## 7.2 Life History

Published information on life cycle, biology and ecology are scanty in Sri Lanka. However, it has been shown that the size, growth, and development of life cycle are greatly influenced by different host plants (Bilapate and Talati 1978; Jacob 1981; Twine 1971). Moths lay eggs singly or in masses, and larvae feed on flower buds, developing fruits, or growing shoots of the host pants.

Eggs are oval, flat, and pinkish. Newly hatched larvae are white with a brown head. Mature larvae are reddish brown with dark brown spots. Before pupation larvae became pink fully grown larva is 1.5–2.0 cm long. The larvae pupate in a silken cocoon among or inside the damaged fruits. Pupae are about 1.0 cm long and brown in color. Adult are medium-sized yellowish moths with black spots on wings, and wingspan is about 2 cm (Mannakkara 2004, 2006).

The larval period is about 15 days in laboratory conditions at $27 \pm 2$ °C and $75 \pm 5\%$ relative humidity on the natural host, and under same condition, pupation takes place about 8 days. However, it is reported that the biology of *C. punctiferalis* undergo host relation speciation (Stanley et al. 2009).

## 7.3 Host Plant and Nature of Damage

*C. punctiferalis* is highly polyphagus insect damaging many parts of the plant such as fruits, shoots, buds, and flowers. *Conogethes* is an economically important pest of South Asia with wide host range, but little information are available on pest status of different host plants in Sri Lanka. According to the literature, *C. punctiferalis* is reported on cocoa, *Theobroma cacao L.* (Devasahayam and Koya 2005); sunflower, *Helianthus annuus* L. (Devasahayam and Koya 2005); turmeric, *Curcuma longa* L (Devasahayam and Koya 2005); cardamom, *Elettaria cardamomum* Maton

**Fig. 7.1** Larva feeding on teak fruit and poor fruit setting after feeding. (Source: Mannakkara 2006)

**Fig. 7.2** Hollowed fruits by larval feeding. (Source: Mannakkara 2006)

(Dharmadasa et al. 2008; Devasahayam and Koya 2005); ginger, *Zingiber officinale* Rosc. (Balasuriya and Kelaniyangoda 2010; Devasahayam and Koya 2005); castor bean, *Ricinus communis* (USDA 1957); and teak, *Tectona grandis* (Mannakkara 2006).

## 7.4 *Conogethes* on Teak

Teak (*Tectona grandis*) is an introduced timber crop to Sri Lanka, and it is extensively grown mainly in dry and transitional zones of the country. Among the several insect pests that attack teak, *C. punctiferalis* was reported in mid-2000s. *C. punctiferalis* larvae attack teak fruits and flowers in major teak growing areas of the country. The newly emerged larvae bore into developing fruits and feed on the inner parts (Figs. 7.1, 7.2 and 7.3). Often, a larva completely hollows the fruits by feeding (Figs.7.2 and 7.3). Fecal matter pellets pushed out of fruits are an indication of the larvae feeding inside the fruit. The damage results in immature fruit fall (Fig. 7.1). The larval boring results poor fruit setting, and it became an important insect pest of teak seed orchards causing considerable losses in fruit production (Mannakkara 2006).

**Fig. 7.3** Empty fruits by larval boring. (Source: Mannakkara 2006)

*C. punctiferalis* being a serious pest of teak is essential to control the pest to ensure high-quality seeds and to fulfill the seed requirements. Being tall nature and the plantation forest crop, the use of resistant or tolerant varieties is the key management strategy, but the use of biocontrol agents as parasitoids, predators, and pathogens is the potential management strategies for this pest (Mannakkara 2006).

## 7.5 *Conogethes* on Zingiberaceae

Turmeric, cardamom, and ginger are perennial herbaceous spice of Zingiberaceae, which have been grown commercially in Sri Lanka. Cardamom is an introduced perennial herbaceous plant with a pseudostem and thick irregular-shaped rhizomes. Many historical texts mention cardamom as a spice (flavoring) and medicinal crop. Dried fruit or cardamom capsule is the commodity of trade. In Sri Lanka, it can be grown in central hill country where elevation is 600 m above mean sea level. Cardamom has been grown mainly under the under forest cover as a plantation crop in the country. Three types of cardamom are found in Sri Lanka and are common in India too, i.e., Malabar, Mysore, and Vazhukka.

The shoot and capsule borer *C. punctiferalis* is (now identified as C. sahyadriensis (Shashank et al. 2018) on zingibers) a serious pest (Dharmadasa et al. 2008). Larvae emerged from eggs laid on leaf sheaths eat the internal parts of the pseudostem which leads to drying up of central spindle. Immature capsules can also be attacked causing empty capsules. Fecal matter of larva can be seen coming out through holes in the stem.

Use of integrated pest management approaches including mechanical and cultural practices like cut and destroy the infested plant parts to kill living larvae is practiced to manage the pest. Use of systemic insecticides also contributes to pest management programs, but applications of synthetic chemicals are not encouraged as a reliable method due to adverse environmental consequences.

Ginger (*Zingiber officinale*) is an important herbaceous spice crop grown mainly in Central, Western, and North-Western provinces of Sri Lanka. Three major cultivars named Local, Chinese, and Rangoon are mainly grown in Sri Lanka (Kaluarachchi et al. 2007). Several pest species are reported from the ginger cultivation in Sri Lanka. Among them shoot borer (*C. punctiferalis*) and ginger maggots (*Eumerus figurans*) are the important insect pests (Balasuriya and Kelaniyangoda 2010). *C. punctiferalis* is reported as an important pest of ginger cultivation in the North-Western Province of Sri Lanka (Balasuriya and Kelaniyangoda 2010). Characteristic symptoms of the *Conogethes* on ginger are presence of bored holes on the pseudostems having frass extruded and yellow, central shoots. Larva lives inside the pseudostem and feed on the growing shoots of ginger and cause dead heart. Turmeric is one of the major spicy crops grown in the Sri Lankan home gardens. Stem borer is the major pest attacking to turmeric cultivation. Adult lays eggs in leaf sheaths and caterpillar enters into the pseudostem and damages the internal tissues of the plant. Initially, plants become yellow then turn into brown and die. Dead heart symptom is the clear evidence to identify the presence of stem borer.

## 7.6 *Conogethes* on Fruit Crops

In Sri Lanka, *C. punctiferalis* is reported as a pest of fruits (USDA 1957). But no or very little information are available on the *C. punctiferalis* damage in local conditions. It is observed that the pest damage several fruit crops including durian, jackfruits, breadfruits, rambutan, avocado, and guava in other countries. However, there are no proper studies or information on those fruits in Sri Lanka. *C. punctiferalis* is a minor pest of durian cultivating areas in wet zone of Sri Lanka. It has been identified from Pasyala and Urapola areas of Gampaha district which is one of the major durian cultivating area. It has been recorded from places, and the damage has been observed from durian fruits of imported durian varieties cultivated in home gardens. It seems that the adult prefers the varieties with thin outer coverings. They prefer to lay eggs in bunches of fruits rather than individual fruits. In fruit bunch with two, three, or more fruits, adult insects lay eggs in surfaces touching each other. These places are good for them to protect from enemies and to safeguard from environmental influences. The damage in early stage of fruit development caused fallen of the fruits when they get matured. Those fruits will not ripe properly, and internal fruit pulp will deteriorated. The damaged areas of the fruit can be seen from outside as brown or black colored areas with their feces deposited among spines (Fig. 7.4). No chemical pesticides recommendation given to manage this pest. The collection and destruction of fallen infested fruits is the recommended practice to manage under local conditions (Hettiarachchi, personnel comm 2016).

## 7.7 Management Practices

*C. punctiferalis* attacks a wide range of taxonomically unrelated plant species causing economic losses in the country. Therefore, practical methods to overcome the problem are urgently required.

**Fig. 7.4** *C. punctiferalis* infested fruit. (Source: Hettiarachchi I)

Adult moths are attracted to fire, and farmers burn the plant debris in the evening near the fields to destroy the attracting adults to the fire. The larvae and pupae can be destroyed by burning and burying fallen and rotting fruits due to borer attack beneath the trees. To reduce the borer attack in fruits, green fruits can be covered by fine mesh cloth or polythene bag to prevent eggs laid on the fruits.

Synthetic chemicals could be used to reduce the pest damage, but it is difficult to recommend single product because of the nature of the broad host rang of the pest. Insecticides could be used as the last resort when infestation is severe, but it is not encouraged as a regular method due to adverse environmental effects. Another problem with insecticide usage is the adverse effect on natural enemies and nontargeted insects like honey bees. Neem-based formulation can be used as alternative to synthetic insecticides.

There is a great potential to use natural enemies that include parasitoids, predators, and pathogens to minimize the pest attack. *Dolichurus* sp. (Sphingidae), *Xanthopimpla* sp. (Ichneumonidae), and *Phanerotoma hendecasisella* Cam. (Braconidae) were recorded as parasitoids of shoot borer from Sri Lanka (Rodrigo 1941). But there are no recent studies on natural enemies of *C. punctiferalis* in Sri Lanka. Cultural methods, such as pruning of freshly infested shoots at the initial stage, are practiced by local farmers to reduce the pest damage especially in Zingiberaceae crops. Mating disruption and trapping of moths using pheromones is one of the other good options to manage this pest. However, there is no experience under local condition to utilize pheromone to manage *C. punctiferalis*. Use of integrated approach has added advantages than use of single method for the control of this pest. The research for new innovative pest management measures to reduce the pest should be the focus in future.

## References

Balasuriya LBAKM, Kelaniyangoda DB (2010) Identification and management of pest and diseases of ginger. Proceedings of 10th agricultural research symposium, pp 74–78

Bilapate GG, Talati GM (1978) Some studies on bionomics of castor short and fruit borer *Dichocrocis punctiferalis* from Gujarat. J Maharashtra Agric Univ 3(1):75–76

CAB International (2011) *Conogethes punctiferalis* datasheet. In: Crop protection compendium. CAB International, Wallingford, UK.USDA. 1957. Insects not known to occur in the United States. Yellow peach moth (*Dichocrocis punctiferalis* (Guenee)), pp 37–38

Devasahayam S, Koya KMA (2005) Insect pests of ginger. In: Ravindran PN, Babu KN (eds) Ginger: the genus. Zingiber. CRC Press, Boca Raton, pp 367–390

Dharmadasa M, Nagalingam T, Seneviratne PHM (2008) Identification and screening of new generation insecticide against cardamom thrips (*Sciothrips cardamom*) in cardamom cultivations in Sri Lanka. Ceylon J Sci (Bio Sci) 37(2):137–142

Jacob SA (1981) Biology of *Conogethes punctiferalis* on turmeric. J Plant Crop 9(2):119–123

Kaluarachchi KAJC, Kelaniyangoda DB, Ranaweera B (2007) Production of bacterial – wilt free ginger planting material grown under protected culture. In: Proceedings of 7th agricultural research symposium. Wayamba University of Sri Lanka, pp 62 – 64

Mannakkara A (2004) Incidence of fruit borer damage in teak. Forest research Newsletter. Forest Research Center, Sri Lanka

Mannakkara A (2006) The teak fruit borer *Conogethes punctiferalis* (Lepidoptera: Pyralidae). The Sri Lanka Forester 29:47–59

Rodrigo, E (1941) Administration report 1940, director of agriculture, Colombo, Ceylon, pp 1–18

Shashank PR, Doddabasappa B, Kammar V, Chakravarthy AK, Honda H (2015) Molecular characterization and management of shoot and fruit borer *Conogethes punctiferalis* Guenee (Crambidae: Lepidoptera) populations infesting cardamom, castor and other hosts. In: Chakravarthy AK (ed) New horizons in insect science: towards sustainable pest management. Springer, New Delhi. https://doi.org/10.1007/978-81-322-2089-3_20

Shashank PR, Kammar V, Mally R, Chakravarthy AK (2018) A new Indian species of shoot and capsule borer of the genus *Conogethes* (Lepidoptera: Crambidae), feeding on cardamom. Zootaxa 4374(2):215–234

Stanley J, Chandrasekaran S, Gnanadhas P (2009) *Conogethes punctiferalis* (Lepidoptera: Pyralidae) its biology and field parasitization. Indian J. Agric Sci 79(11):906–610

Twine PH (1971) Insect pests of maize. Queens Land Agric J 97(10):523–531

USDA (1957) Insects not known to occur in the United States. Yellow peach moth (*Dichocrocis punctiferalis* (Guenee), pp 37–38

# Status and Management of *Conogethes* spp. in Malaysia

# 8

A. Sivapragasam, A. Badrulhadza,
and M. N. Mohamad Roff

**Abstract**

In Malaysia, four species of *Conogethes* have been recorded. *C. punctiferalis* is the most important with wide host range and distribution. It is a major pest on durian, carambola and other tropical fruits. Fruit thinning and fruit bagging are important practices as single fruits are disfavoured for oviposition and feeding. Bagging obviates the need for insecticidal applications. Intercropping and orchard/planting sanitation are another important measure to suppress the borer populations. The use of biopesticides like entomopathogens and neem formulations and encouragement of natural enemies can further bring down the *Conogethes* populations and infestation. Research on sex pheromone is in progress.

**Keywords**

Biointensive; *Conogethes*; Durian; Integrated management

A. Sivapragasam (✉)
CABI South East Asia, Building A19, Malaysian Agricultural Research and Development Institute (MARDI), Serdang, Selangor, Malaysia
e-mail: a.siva@cabi.org

A. Badrulhadza
Crops and Soil Science Research Center, Malaysian Agricultural Research and Development Institute (MARDI), Serdang, Selangor, Malaysia
e-mail: bhadza@mardi.gov.my

M. N. Mohamad Roff
Malaysian Agricultural Research and Development Institute (MARDI), Serdang, Selangor, Malaysia
e-mail: roff@mardi.gov.my





A. K. Chakravarthy (ed.), *The Black spotted, Yellow Borer, Conogethes punctiferalis Guenée and Allied Species*,
https://doi.org/10.1007/978-981-13-0390-6_8

## 8.1 Introduction

There are a number of insect pests damaging major tropical fruits in Malaysia. Several authors such as Yunus and Ho (1980), Yusof and Khoo (1989) and, more recently, Ithnin et al. (2008) listed the major fruit crops and their key insect pests. Durian, *Durio zibethinus*, is one of the main fruits of increasing economic importance as an export commodity. In durian, the fruit borers are important pests which comprise of the seed borer *Mudaria* spp. and the husk borer, *Conogethes punctiferalis*, *Tonica* sp. and *Terrasella* (Ithnin et al. 2000).

According to Mohamed et al. (1999), *C. punctiferalis* is among the key pests of the durian. However, despite its key status, there is not much published information on *Conogethes* in Malaysia or for that matter in the South East Asian (SEA) region. In the backdrop of this scenario, published and unpublished information available in Malaysia has been collated on *Conogethes* and presented. These are supported with pertinent references to literature available elsewhere outside the SEA region.

## 8.2 Key Species

In Malaysia, in addition to *C. punctiferalis*, three other species of *Conogethes* (they mentioned it under the genus *Dichocrocis)* had been recorded by Yunus and Ho (1980). These were *C. surusalis*, *C. clioalis* and *C. evakalis*. *C. surusalis* is found on *Hibiscus* sp., while *C. clioalis* is found on *Dillenia indica*. As for *C. evakalis*, this species is found on *Curcuma domestica* and *Wormia suffruticosa*.

However, as in other countries, the most significant species in economic literature is *C. punctiferalis*. This is essentially due to its expanding host range, geographical occupancy and complexity involved in species identifications. Further, since the number of species within the complex is unknown and their biology cannot be distinguished, the species *C. punctiferalis* is generally considered the putative species of the genus. Globally, 24 named species of genus *Conogethes* have been known through molecular identification (Shahshank et al. 2015). Robinson et al. (1994) had mentioned that at least seven of which may occur in SEA. Thus, whether the species recorded in Malaysia by Yunus and Ho (1980) exists as separate species or otherwise needs to be confirmed. For now, it is practical to consider them under the *Conogethes* spp. complex.

Significantly due to its wide host range (see below) and to some extent its feeding behaviour, this moth species is known by several names in literature depending on its host. In Malaysia, it is commonly known as the durian fruit or husk borer (Ithnin et al. 2000). In other countries, it is known as the shoot and fruit borer or shoot and capsule borer or yellow peach moth or cardamom stem borer or cardamom shoot and capsule borer or castor shoot and capsule borer or maize moth or durian shoot and fruit borer or black spotted yellow shoot and fruit borer.

## 8.3 Biogeography

According to Pena et al. (2002), the genus is an important pest in tropical as well as subtropical countries and distributed in Asia and Australia. It is also now reported to be a new and introduced pest in Britain and Europe (Stanley et al. 2009). Besides Malaysia, *C. punctiferalis* is found in Asia (from India eastwards) and Australasia. As stated by USDA (1957), CABI (2011) and EPPO (2013), the full list of countries the *C. punctiferalis* species complex has been recorded from is as follows: in Asia (Bangladesh, Brunei Darussalam, Cambodia, China, India, Indonesia, Iraq, Japan, Laos, Malaysia, Myanmar, North Korea, Philippines, South Korea, Sri Lanka, Taiwan, Thailand and Vietnam), in Oceania (Australia and Papua New Guinea) and in North America (Hawaii) (Nishida 2002).

## 8.4 Host Range

*C. punctiferalis* is a highly polyphagous pest, and it will be interesting to investigate the reason for polyphagy. In Malaysia, it had been recorded infesting at least 20 host plants that include durian, carambola, cocoa, jackfruit, rambutan and pulasan (Ithnin et al. 2000; Yunus and Ho 1980). Outside Malaysia, it has been recorded infesting 65 host plants from 30 different families (Molet 2015), and many of these plant species are economically important (Devasahayam and Abdulla Koya 2005). As stated and reviewed by Shashank et al. (2015), these are the major and minor host plants of *C. punctiferalis*:

Major hosts: On *Zingiber* plants, a new species *C. sahyadriensis* has been reported from India (Shashank et al. 2018). *Alpinia* spp., *Ammomum* spp. *Curcuma longa* (turmeric), *Dimocarpus longan* (longan), *Dimocarpus longan* subsp. *Malesianus* (mata kucing), *D. zibethinus* (durian), *Elettaria cardamomum* (cardamom), *Gossypium* spp. (cotton), *Hedycium* spp., *Nephelium lappaceum* (rambutan), *Prunus persica* (peach), *Ricinus communis* (castor bean), *Sorghum bicolor* (sorghum), *Sorghum bicolor* subsp. *bicolor* (great millet), *Theobroma cacao* (cocoa) and *Zingiber officinale* (ginger).

Minor hosts: *Artocarpus heterophyllus* (jackfruit), *Averrhoa carambola* (carambola), *Brachychiton acerifolius* (flame tree), *Caesalpinia bonduc* (yellow nicker), *Carica papaya* (papaya), *Castanea mollissima* (Chinese chestnut), *Ceiba pentandra* (kapok tree), *Citrus limon* (lemon), *Citrus sinensis* (orange), *D. indica* (elephant apple), *Diospyros virginiana* (American persimmon), *Eriobotrya japonica* (loquat), *Etlingera elatior* (kantan), *Ficus carica* (fig), *Flemingia* spp., *Glycine max* (soybean), *Helianthus* spp. (sunflower), *Livistona humilis* (livistona palm), *Macadamia integrifolia* (macadamia), *Macadamia ternifolia* (macadamia), *Malus* spp. (apple), *Mangifera indica* (mango), *Morus* spp. (mulberry), *Morus alba* (white mulberry), *Persea americana* (avocado), *Psidium guajava* (guava), *Planchonia careya* (billy goat plum), *Prunus domestica* (plum), *Punica granatum* (pomegranate), *Quercus virginiana* (live oak), *Sapindus* spp., *Solanum melongena* (eggplant), *Tectona grandis* (Bankok teak), *Vitis vinifera* (grape) and *Zea mays* (corn).

On most hosts, *C. punctiferalis* damages the fruit. However, in some cases (e.g. durian), the seed and flowers are also damaged. In *Alpinia purpurata*, Mohamed et al. (1999) reported that the moth damages the stem and rhizome.

## 8.5 Economic Importance

In Malaysia and in SEA, this pest is serious on durian (Fig. 8.1). The larva bores into fruit, flower buds, seeds and stems of the plants. Infestation starts in young fruits which are about 2–3 weeks old, and larval frass is spread over a large area of the fruit surface. In Malaysia, Mohamed (1998) recorded an infestation of about 4% in durian. The young larvae feed on the skin of young and mature durian fruit. When they get to latter instar stages, the larvae bore into the fruit where they feed. Infestation of fruits can often be recognized by the faeces of the larvae that can be seen on the outside of the durian fruit. Chong et al. (1991) mentioned that boring usually occurs under a mass of frass, webbed together in between thorns of the durian fruit. Injury due to the boring activities of the larvae can cause secondary infections of fungi or other pathogens. This can result in fruit rot and can cause the fruit to drop off. In crops such as cocoa, the larva bores into the husk of the fruits which often starts in regions of pods that meet with other pods or the bark. This

**Fig. 8.1** Developmental stages of *Conogethes punctiferalis* and damage on durian fruit; (**a**) adult, (**b**) nature of damage on durian fruit, (**c**) larva, (**d**) pupa

situation is similar to durian. The pest is also often associated with severe plant bug damage (Chong et al. 1991). According to Molet (2015), *C. punctiferalis* has been described as a destructive pest of peaches in China. In Australia, this species is a pest on cotton and also has occasionally caused almost complete loss of grain sorghum in coastal areas of the country (USDA 1957). More recent reports refer to it as minor and irregular pest of sorghum, corn and cotton (Korycinska 2012). Astridge (2001) considers this species to be a major pest of rambutan and durian in North Queensland. It is also the most serious insect pest of papaya in Australia as stated by Chay-Prove et al. (2000).

The insect has not been encountered on papaya in Malaysia. Devasahayam and Abdulla Koya (2005) stated that this species is the most serious pest of ginger, especially in India. Gahukar (2011) also mentioned that *C. punctiferalis* is the most destructive pest of ginger and turmeric. The larvae bore into pseudostems and feed on central growing shoot causing dead hearts. This species has caused yield losses of 25% when 23–24% of ginger pseudostems were infested and may go up to 40%. Crop yields can be significantly affected when more than 45% of shoots in a clump are damaged (Devasahayam et al. 2010). Varadasan (2001) mentioned that *C. punctiferalis* infests shoot, stem and panicle and capsule of cardamom that can cause yield losses up to 80%. In the nursery, the young larva bores into and feeds on the internal contents of unopened leaf buds, panicles and immature capsules, while later instars bore into pseudostems and cause dead hearts due to drying of growing shoot (Gahukar 2011). In India and Sri Lanka, it is a serious pest of castor bean and fruit (USDA 1957). The damage due to this pest on castor was 16–72% (Bilapate and Talati 1977). It attacks castor from flowering stage onwards and continues till maturity of the crop. Based on the review by Shashank et al. (2015), in India, capsule yield losses in cardamom were recorded between 6.8% and 9.2%, while castor capsule damage was recorded between 11% and 27%.

## 8.6 Bioecology

Ithnin et al. (2008) reported that on durian, eggs are laid on the young fruit, and the eggs hatch in about 4–5 days. The larval period ranges from 14 to 21 days and is spent mostly on the surface of the fruit below the frass (larval excreta). The pupal stage lasts about 6–10 days and pupation takes place on the fruit in pupal cocoons. The total life cycle takes about 24–36 days from the egg to the pupal stage. The conditions of the study on durian are not however known. On carambola, the biology of *C. punctiferalis* that bred on mature starfruit at a constant temperature of 25 ± 1 °C and 80 ± 5% relative humidity in the laboratory revealed that the mean egg, larval, prepupal, pupal and total developmental period (egg to adult emergence) was about 4.0, 23.0, 2.0, 11.9 and 41 days, respectively (Sivapragasam 2004).

In Malaysia, studies were done to compare biological parameters, such as longevity (male and female) and eggs laid by *C. punctiferalis* adult (Fig. 8.1) that emerged from infested fruits of carambola and durian as part of the study towards developing pheromones for the species (Sivapragasam 2004). The male longevity was 8.4 and 9.5 days for laboratory and field reared carambola population, respectively, whereas for the durian population, it was 10.7 and 9.0 days, respectively. The female longevity was 8.7 and 5.7 days for laboratory and field reared carambola population, respectively, whereas for the durian population, it was 11.8 and 6.3 days, respectively. In oviposition studies, and considering that not all females were fecund ones, the number of eggs laid per female during its lifespan was 2.08 and 0.5 for carambola and durian populations, respectively. Data suggested that the host (fruit) influenced the number of eggs laid. For example, the maximum number of eggs laid per female was higher for the carambola population (n = 25) compared to the durian population ($n$ = 11).

Details of the biological and life history parameters are abundant in literature and largely corroborated or fall within a range with those reported in Malaysia (Sivapragasam 2004). Based on published reports, after mating, females lay the small, oval eggs on or near fruit or seeds of hosts (USDA 1957). Females lay from 20 to 30 eggs (MAF Biosecurity New Zealand 2009). The egg stage lasts between 4 and 6 days. Once they hatch, the larva (Fig. 8.1) feed on or in seeds, seed capsules and young shoots (USDA 1957). The larva then undergoes five instar stages (Devasahayam and Abdulla Koya 2005) in approximately in 16–21 days. Pupation usually occurs in the larval tunnels within a silk cocoon, surrounded by shelters of webbing and frass (Chong et al. 1991; MAF Biosecurity New Zealand 2009). The adult emerges about a week later (Chong et al. 1991). Adults live approximately between 8 and 12 days. Adults are active mainly at night and during daytime; they spend their time hiding in foliage. The complete life cycle of this species is between 35 and 46 days.

In Malaysia, *C. punctiferalis* has been successfully reared on carambola until the 8th generation. The relative viability of eggs to adults under laboratory conditions was about 13%. Rearing on the durian fruit and *Zingiber* stems produced relatively very low populations of adults in the laboratory (Sivapragasam 2004). The suitability of *A. carambola* as a food source for *C. punctiferalis* was evaluated by studying the biology of this pest on fresh *A. carambola* fruits under laboratory conditions. Later studies by Nur Atiqah (2014) also supported the fact that carambola is a suitable host for mass rearing. Table 8.1 summarizes the results of the life cycle

**Table 8.1** Duration of life stages of *C. punctiferalis* reared on carambola under laboratory conditions

| Insect stages | Range | Mean ± SD ($n$ = 20) |
|---|---|---|
| Incubation period | 2.29–3.12 | 2.71 ± 0.15 |
| 1st instar | 1.90–2.65 | 2.28 ± 0.79 |
| 2nd instar | 2.10–2.95 | 2.53 ± 0.50 |
| 3rd instar | 1.22–3.40 | 2.31 ± 0.80 |
| 4th instar | 2.48–2.87 | 2.68 ± 0.87 |
| 5th instar | 2.10–3.40 | 2.75 ± 0.48 |
| Total larval period | 11.00–13.10 | 12.05 ± 1.80 |
| Pre pupal period | 2.18–2.90 | 2.54 ± 0.25 |
| Pupal period | 9.45–11.90 | 10.68 ± 0.75 |
| Total life cycle | 25.00–28.10 | 26.55 ± 0.50 |

Source: Nur Atiqah (2014)

parameters. The complete life cycle of *C. punctiferalis* ranged between 25 and 28 days which was much shorter than the one reported by Sivapragasam (2004). There were five larval instars based on measurements of the head capsules. The mean pupal weight was 0.023 g ± 0.02, and the mean number of eggs laid by a female was 102.5 ± 11.41 with 85–120 eggs range. The average egg viability under laboratory conditions was 79.80%. It was evident that the life cycle durations reported by Nur Atiqah (2014) differed with those observed by Sivapragasam (2004). Differences in life cycle duration may be due to environmental conditions.

## 8.7 Seasonality

There is no specific seasonality of occurrence noted in Malaysia as the presence very much depends on the availability of fruits. Fig. 8.1 shows the temporal changes in infestation of *C. punctiferalis* with time within the same durian crop. It is suggested that the population, based on incidence of infested fruits, tended to increase from cycles 1 to 3 during the fruit season (Fig. 8.2).

## 8.8 Management

Generally, the control of *C. punctiferalis* is considered collectively as for all the fruit borers in durian. There is no single method that can completely control fruit borers, and thus a holistic approach is generally recommended.

Insecticides are the preferred approach for management. A timely application, when the first infestations are seen (as there are no specific threshold levels for initiating sprays), is recommended. In addition, the alternate use of insecticidal sprays, e.g. of fenthion, dimethoate and deltamethrin, has also been recommended (Nik

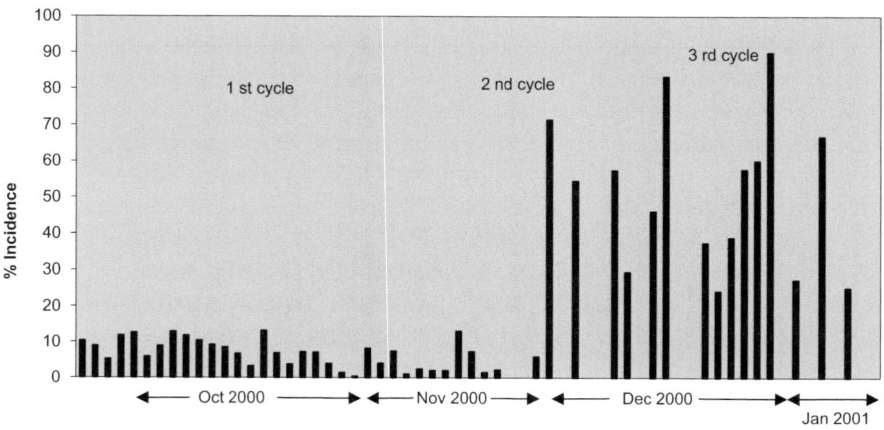


**Fig. 8.2** Incidence of *Conogethes punctiferalis* during one durian crop season

**Fig. 8.3** Bagging of durian to prevent borer infestation in a commercial farm

Masdek et al. 1991; Mohd Shamsudin 1994; Mohamed et al. 1996; Ithnin 1997). One key challenge in durian is the height of the trees which makes it difficult for insecticidal sprays to reach the target fruits. In such a case, a motorized spray equipped with an extended lance that can reach 25–30 m is used. However, prudent use is necessary due to the presence of a number of natural enemies for this pest.

Physical control measures advocated include the use of light traps to capitalize on the nocturnal behaviour of the moths (Ithnin et al. 2000). The purpose of light trapping is to catch and kill both male and female moths thus reducing their numbers and fruit infestation. The traps should be strategically located in the field to increase efficiency. However, there are limitations to this approach as all kinds of nocturnal insects, including natural enemies, are caught indiscriminately in the trap. Besides that, the moths (females) probably had already laid their eggs on the fruits before getting trapped.

The bagging of the fruit (Fig. 8.3) is also a physical measure that fully protects the fruit from insect damage. Fruit bagging provides almost 100% protection against *C. punctiferalis*. Once the fruits are bagged, no spray of insecticides is required, and this can reduce cost as well as the residues which are safe to the environment and health. However, fruit bagging is only feasible for low-lying fruits.

Cultural control measure, e.g. fruit thinning, is done when fruits touch each other on the branches. The touching of fruits is conducive for egg-laying moths. Single fruits are less favoured than compact fruits. This technique gives the chance for the farmers to allow only well-shaped and healthy fruits to develop on maturity, while the infested ones are removed, and this results in larger fruit production. Another cultural measure, i.e. orchard sanitation, involves the collection and destruction of infested fruits, by burning or burying them. When infested durian husks and seeds are left on the ground, mature larvae crawl out, pupate in the soil and then emerge as moths to damage healthy fruits on the trees. According to Sharma et al. (1992), the pest can be managed through cultural practices such as by increasing the planting distance. Other cultural practices recommended are using reduced amount of nitrogen fertilizer (Chakravarthy and Thyagaraj 1999) and practise clean cultivation and good field sanitary practices (Sharma et al. 1992). Intercropping also can be one of the tools in managing *Conogethes*. In semiarid tropics in India, introducing cluster bean, cowpea, blackgram and/or groundnut as intercrops in castor was able to reduce *C. punctiferalis* infestation and manage to build up natural enemies' populations (Rao et al. 2012).

Biopesticides include several approaches like using neem formulations (Rajabaskar and Regupathy 2013); *Bacillus thuringiensis* (Devasahayam 2000), entomopathogenic nematodes (Choo et al. 2001) and sex pheromones (Rajabaskar and Regupathy 2012) were reported for management of *C. punctiferalis*. Sex pheromone method is a combination of pheromone lure and trap. The trap used is the large plastic delta trap that is available in the market along with the pheromone lure product called '*C. punctiferalis* lure' (Molet 2015). The lure is effective for around 1 month. Rajabaskar and Regupathy (2013) developed IPM modules to control this pest on cardamom and found out that sequential application of neem, profenofos, diafenthiuron, neem and profenofos was the most effective.

Yunus and Ho (1980) and Mohamed et al. (1999) reported a few species of natural enemies on *C. punctiferalis* such as *Trathala* sp., *Xanthopimpla stemmator* Thumberg, *Xanthopimpla* sp., *Apanteles* sp. and *Chelonus* sp. However, there were no parasitization rates provided. Based on a study by Stanley et al. (2009) in India, parasitoids like *Trathala flavoorbitalis*, *Brachymeria atteviae*, *Chelonus blackburni*, *Anthrocephalus decipiens*, *Epitranus erythrogaster*, *Bassus* sp., *Bracon* sp. and phorids were found to parasitize *C. punctiferalis* larvae in castor crop. Among the different parasitoids, 47.3% of parasitization was done by *T. flavoorbitalis*, followed by *B. atteviae* (14.8%). Parasitization by phorids was found to be 11.5%, and all the others mentioned above are of less than 5%. *Bassus* sp. was rarely found to parasitize *C. punctiferalis*. Pena et al. (2002) reported that *Argyrophylax proclinata* is able to parasitize this pest up to 40% in Australia.

The use of natural enemies, especially parasitoids, in particular to egg, larval-pupal and pupal stages is needed for efficient management of *Conogethes* populations in the field (Chakravarthy et al. 2015). Parasitoid species *Trichogramma chilonis*, *T. pretiosum* and *Chelonus blackburni* are known to be potential agents against *Conogethes* spp. At present, the importance of other biocontrol agents from fungus species such as *Beauveria* sp., nematodes, bacteria and viruses (nuclear polyhedrosis virus) for managing *Conogethes* spp. has been tested out.

The use of semiochemicals, e.g. pheromones, is also being studied. In Malaysia, GC analysis of the pheromone extract on the polar DB23 column confirmed the presence of four compounds with retention times similar to that of hexadecenal isomers (Sivapragasam 2004). The GC and GC-MS analyses of the sex pheromone extracts were consistent with earlier work reported by Konno et al. (1982) who identified the pheromone of *C. punctiferalis* as a 9:1 blend of E10-16:Ald and Z10-16:Ald. There was no evidence of other pheromone components, but in the absence of live insects to conduct linked analyses with electroantennography, this cannot be ruled out, particularly in view of the large number of compounds present in the extracts. However, trapping studies were not successful as expected. Nur Atiqah et al. (2015) noted that the sex pheromone profile of *C. punctiferalis* was a blend of 9-hexadecenal and 11-hexadecenal.

**Acknowledgement** All the figures and photographs in this chapter are of the authors only.

## References

Astridge D (2001) Insect fauna surveys on rambutan, durian and mangosteen in North Queensland. Technical bulletin Department of Primary Industry and Fisheries, Northern Territory of Australia, pp 75–79

Bilapate GG, Talati GM (1977) Studies on the life tables of *Dichocrocis punctiferalis* Guenee on castor. J Maharashtra Agric Univ 2:246–248

CAB International (2011) *Conogethes punctiferalis* datasheet. In: Crop protection compendium. CAB International, Wallingford

Chakravarthy AK, Thyagaraj NE (1999) Effect of different NPK fertilizer levels on the incidence of cardamom pests. Insect Environ 4(4):139–140

Chakravarthy AK, Kamar V, Lokeshwari D, Rani AT, Nagaraj T, Varun Rajan V (2015) The black spotted yellow shoot and fruit borer, *Conogethes* spp. (Crambidae: Lepidoptera) – a global perspective. Curr Sci 109(6):1017–1018

Chay-Prove P, Ross P, O'Hare P, Macleod N, Kernot I, Evans D, Grice K, Vawdrey L, Richards N, Blair A, Astridge D (2000) Agrilink series: your growing guide to better farming. Queensland Horticulture Institute and Department of Primary Industries, Nambour

Chong KK, Ooi PAC, Tuck HC (1991) Crop pests and their management in Malaysia. Tropical Press Sdn. Bhd, Kuala Lumpur, p 242

Choo HY, Kim HH, Lee DW, Lee SM, Park SH, Choo YM, Kim JK (2001) Practical utilization of entomopathogenic nematodes, *Steinernema carpocapsae* Pocheon strain and *Heterorhabditis bacteriophora* Hamyang strain for control of chesnut insect pests. Korean J Appl Entomol 40:69–76

Devasahayam S (2000) Evaluation of biopesticides for the management of shoot borer (*Conogethes punctiferalis* Guen.) on ginger (Zingiber officinale Rosc). In: Proceedings of the centennial conference on spices and aromatic plants, Indian Society for Spices, Calicut, India, pp 276–277

Devasahayam S, Abdulla Koya KM (2005) Insect pests of ginger. In: Ravindran PN, Babu KN (eds) Ginger: the genus Zingiber. CRC Press, Boca Raton, pp 367–390

Devasahayam S, Jacob TK, Abdulla Koya KM, Sasikumar B (2010) Screening of ginger (Zingiber officinale) germplasm for resistance to shoot borer (*Conogethes punctiferalis*). J Med Aromat Plant Sci 32(2):137–138

EPPO (2013) PQR–EPPO database on quarantine pests. Available at: http://www.eppo.int

Gahukar RT (2011) Use of indigenous plant products for management of pests and diseases of spices and condiments: Indian perspective. J Spices Aromat Crop 20(1):1–8

Ithnin B (1997) Makhluk perosak tanaman durian. Paper presented at simposium Durian Negeri Perak, 21 July 1997. Batu Gajah, Perak. (in Bahasa Malaysia)

Ithnin B, Mohd Shamsudin O, Mohamed MS (2000) Management of fruit and stem borers of durian. In: Zainal AM et al (eds) Proceedings of national durian seminar 2000, 1–3 August, 2000 MARDI

Ithnin B, Mohd Shamsudin O, Mohamed MS (2008) Serangga perosak buah-buahan tropika Malaysia. Published by: Institut Penyelidikan dan Kemajuan Pertanian Malaysia (MARDI). 1st edn. (in Bahasa Malaysia), pp 260

Konno Y, Arai K, Sekiguchi K, Matsumoto Y (1982) (E)-10-hexadecenal, a sex pheromone component of the yellow peach moth, *Dichocrocis punctiferalis* Guenée (Lepidoptera: Pyralidae). Appl Entomol Zool 17(2):207–217

Korycinska A (2012) Rapid assessment of the need for a detailed pest risk analysis for *Conogethes punctiferalis* (Guenée). The Food and Environment Research Agency, North Yorkshire, United Kingdom, p 7

MAF Biosecurity New Zealand (2009) Import risk analysis: table grapes (Vitis vinifera) from China. MAF Biosecurity New Zealand, Wellington, p 322

Mohamed MS (1998) Integrated management of durian fruit borer. In: technical report of horticulture research center 1998. MARDI 173:122–123
Mohamed MS, Mohd Hanifah Y, Norsiah MJ (1996) Biology and control of durian seed borer, Mudaria magniplaga Walker in Malaysia. Paper presented at the 12th meeting of the joint Thailand/Malaysia plant quarantine working group, Nongkhai Thailand, 5–7 March 1996
Mohamed MS, Mohd Hanifah Y, Norsiah MJ (1999) Alternate hosts and natural enemies of durian fruit pests. In: Proceedings of the national horticultural conference, 16–17 November, 1999. MARDI-Tropical Fruits Network, Kuala Lumpur, pp 500–606
Mohd Shamsudin O (1994) Pengurusan serangga perosak durian. In: Mohamad O et al (eds) Recent developments in durian cultivation: proceedings of the durian seminar, 25 June 1992. MARDI-MAPPS, Serdang (in Bahasa Malaysia), pp 89–94
Molet T (2015) CPHST pest datasheet for *Conogethes punctiferalis*. USDA-APHIS-PPQ-CPHST
Nik Masdek NH, Ithnin B, Mohd Shamsudin O (1991) Penyakit dan makhluk perosak. In: Zainal AM et al (eds) Penanaman durian. MARDI Serdang, (in Bahasa Malaysia), pp 36–45
Nishida GM (2002) Hawaiian terrestrial arthropod checklist, 4th edn. Bishop Museum Press, Honolulu, p 313
Nur Atiqah MK (2014) Use of Averrhoa carambola (star fruit) as a food source for rearing Conogethes punctifelaris (Lepidoptera: Pyralidae), durian fruit husk borer. In: Proceedings of the 8th international conference on plant protection in the tropics, Kuala Lumpur, Malaysia, pp 144–147
Nur Atiqah MK, Ihnin B, Razali M, Suhana Y, Mohd Shahhizal AW, Jamaliah J, Khalissani K, Farah Huda SS, Mohamad Roff MN, Ullia I, Nooraris (2015) A sex pheromone profile for management of durian fruit husk borer, Conogethes punctiferalis. Poster presented in MARDI science and technology exhibition 2015, 25–27 August 2015, MAEPS, Serdang, Selangor, Malaysia
Pena JE, Nadel H, Pereira MB, Smith D (2002) Pollinators and pests of Annona Sp. In: Pena J, Sharp J, Wysoki M (eds) Tropical fruit pests and pollinators: biology, economic importance, natural enemies and control. CABI Publication, Wallingford, p 208
Rajabaskar D, Regupathy A (2012) Calling behaviour and attractive response of cardamom shoot and capsule borer *Conogethes punctiferalis* Guenee (Lepidoptera: Pyralidae) to the female crude extract and synthetic blend. J Entomol 9:296–301
Rajabaskar D, Regupathy A (2013) New based IPM modules for control of Sciothrips cardamom Ramk and *Conogethes punctiferalis* Gunee in small cardamom. Asian J Biol Sci 6(3):142–152
Rao MS, Rao CAR, Srinivas K, Pratibha G, Sekhar SMV, Vani GS, Venkateswarlu B (2012) Intercropping for management of insect pests of castor in the semi-arid tropics of India. J Insect Sci USA 12:14
Robinson GS, Tuck KR, Shaffer M (1994) A field guide to the smaller moths of South-East Asia. Malaysian nature society and the Natural History Museum. ArtPrinting Works Sdn Bhd, Kuala Lumpur
Sharma YR, Devasahayam S, Anandraj M (1992) Black pepper and cardamom: problems and prospects. In: Proceedings of national seminar on black pepper and cardamom. May, 17–18 1992, Calicut, Kerala, India
Shashank PR, Doddabasappa B, Kammar V, Chakravarthy AK, Honda H (2015) Molecular characterization and management of shoot and fruit borer *Conogethes punctiferalis* Guenee (Crambidae: Lepidoptera) populations infesting cardamom, castor and other hosts. In: Chakravarthy AK (ed) New horizons in insect science: towards sustainable pest management. Springer, New Delhi, pp 207–227
Shashank PR, Kammar V, Mally R, Chakravarthy AK (2018) A new Indian species of shoot and capsule borer of the genus *Conogethes* (Lepidoptera: Crambidae), feeding on cardamom. Zootaxa 4374(2):215–234
Sivapragasam A (2004) Development of pheromones for durian fruit borers, *Mudaria* spp. and *Conogethes punctiferalis,* End of project report presented to MARD. (mimeo.), pp 8
Stanley J, Chandrasekaran S, Preetha G (2009) *Conogethes punctiferalis* (Lepidoptera: Pyralidae) its biology and field parasitization. Indian J Agric Sci 79(11):906–909

USDA (1957) Insects not known to occur in the United States. Yellow peach moth (Dichocrocis punctiferalis (Guenee)), pp 37–38

Varadasan S (2001) Insect pest management in cardamom – key to reduce cost of production. Spice India 14:19–22

Yunus A, Ho TH (1980) Lists of economic pests, host plants, parasites and predators in West Malaysia (1920–1978). Published by Ministry of Agriculture, Malaysia Bull 153: 583

Yusof I, Khoo KC (1989) Serangga perosak utama tanaman di Malaysia. Dewan Bahasa dan Pustaka, Kementerian Pendidikan Malaysia, Kuala Lumput. (in Bahasa Malaysia), pp 157

# Predicting Potential Global Distribution of The Black Spotted Yellow Borer, *Conogethes punctiferalis* Guenée (Crambidae: Lepidoptera) by CLIMEX Modelling

9

V. Sridhar, Darren Kriticos, K. S. Nitin, A. K. Chakravarthy, and P. Swathi

**Abstract**

*Conogethes punctiferalis* Guenée is a serious invasive pest of several horticultural, agricultural and forestry plants and presently distributed in Asia and Australia. CLIMEX simulation was applied for *C. punctiferalis* to predict its potential geographic distribution in the world under present climate scenario. Ecoclimatic index (EI), which describes the differential climate suitability for the establishment of the pest, was assessed for different locations in the globe. The map comparisons show good agreements between simulated and present distribution of this pest, indicating that the CLIMEX model has promising potential for prediction of future distributions of this species globally. As the present pest distribution is restricted to Southeast Asia and Australia, potential areas where the pest can establish if inadvertently or via transit introduced are stated in this chapter.

**Keywords**

CLIMEX · Climate change · Future distribution · Potential regions · Ecoclimatic Index

V. Sridhar (✉) · K. S. Nitin · A. K. Chakravarthy · P. Swathi
Division of Entomology and Nematology, ICAR-Indian Institute of Horticultural Research, Bengaluru, Karnataka, India
e-mail: vsridhar@iihr.res.in

D. Kriticos
Health and Biosecurity CSIRO, Canberra, ACT, Australia
e-mail: Darren.Kriticos@csiro.au



A. K. Chakravarthy (ed.), *The Black spotted, Yellow Borer, Conogethes punctiferalis Guenée and Allied Species*,
https://doi.org/10.1007/978-981-13-0390-6_9

## 9.1 Introduction

The Black spotted yellow borer, *Conogethes punctiferalis* (Crambidae, Lepidoptera), commonly also known as castor capsule borer or durian borer is an invasive pest and currently restricted to Asia and Australia. There may be more than 20 species found in the genus, of which 7 at least may occur in Southeast Asian countries (Robinson et al. 1994). As the species number within the genus is unknown, their bioecology cannot be distinguished easily. The prediction on possible distribution of *Conogethes* has been made based on available information from all recognized species within *C. punctiferalis* species complex.

*C. punctiferalis* is a polyphagous pest on various crops infesting more than 30 plant species belonging to 23 different families in India (Devasahayam and Abdulla Koya 2005). Presently, it is restricted to Asia, Australia and Papua New Guinea (Fig. 9.1). In Asia, it is found in China (AQSIQ 2007), India, Indonesia, Japan, Korea, Malaysia, Taiwan, Thailand and Vietnam (Gour and Sriramulu 1992; Hang et al. 2000; Kang et al. 2002; CABI 2011). As *C. punctiferalis* can feed on broad range of host plants, its fitness differs on various host plant species (Honda et al. 1979; Li et al. 2015).

In Australia, it is a major pest on cotton and sorghum (USDA 1957). It is a major pest in North Queensland on *Nephelium lappaceum* (rambutan) and *Durio zibethinus* (durian) fruits (Astridge 2001). In Asian countries, like India and Sri Lanka, this is a serious pest of castor, tropical and sub-tropical fruits (USDA 1957), ginger and cardamom followed by *Hedycium* spp., *Alpinia* spp. and *Ammomum* spp. (Devasahayam and Abdulla Koya 2005; Shashank et al. 2015) in India. The borer *C. punctiferalis* is reported as major pest on peaches in China (USDA 1957). Yellow peach is preferred host in Japan (Konno and Shishido 1980; Konno et al. 1981; Kadoi and Kaneda 1990; Abe and Sanari 1992; Kimura and Honda 1999).

There is no evidence or published data regarding the existence of the borer in countries like Africa and the USA, but it was intercepted through international

**Fig. 9.1** *Conogethes punctiferalis* world distribution map (CABI 2011)

trades (AQAS 2014). Countries like England, Wales and Netherlands have intercepted this pest while importing agricultural and horticultural commodities from the countries where *C. punctiferalis* was already established. Thus, there is a need for the identification of potential areas for the establishment of this pest as a part of pest risk analysis.

In order to predict the potential favourable regions for the establishment of *C. punctiferalis* in the world, CLIMEX *v.4.02*, a bio-climatic modelling tool and its two components, viz. 'compare locations' and 'climate matching', were used.

## 9.2 Compare Locations

In the present study, 'compare location' function of CLIMEX was used to generate potential distribution maps of *C. punctiferalis* for current climatic conditions with special reference to India. CLIMEX uses two constraints to estimate the potential growth and survival of a population at a given location, i.e. growth (mainly temperature and soil moisture) and stress indices (cold, heat, wet and dry stress). The values of these two indices are clubbed to generate the ecoclimatic index (EI), generally scaled between 0 and 100 (Kriticos et al. 2015). An EI close to 0 indicates location not favourable for long-term survival of a species and EI nearer to 100 as highly suitable. As EI of 100 is not possible under natural systems, in the present study, an EI of >20 was considered as highly suitable for survival and establishment of the pest, based on the current distribution and occurrence of the pest

The 'compare location' function uses meteorological database consisting of monthly long-term average climatic variables, viz. maximum and minimum temperatures, rainfall, rainfall patterns, relative humidity (RH) and soil moisture for any number of locations worldwide. An iterative process comparing the known and predicted distributions for the same region was adopted to arrive at the parameter fitting. After adjusting the parameter values, the data were used to run the model for predicting the potential distribution of the species.

$$\mathbf{Ecoclimatic\ Index}(\mathbf{EI}) = \mathrm{TGI\ A} \times \mathrm{SI} \times \mathrm{SX}$$

where

GI A, the annual growth index, = $100 \sum_{52}^{i=1} TGI_{Wi} / 52$

**SI**, the annual stress index, = (1−CS/100)(1−DS/100)(1−HS/100)(1−WS/100)

**SX**, the stress interaction index, = (1−CDX/100)(1−CWX/100)(1−HDX/100)1−HWX/100).

CS, DS, HS and WS are the annual cold, dry, heat and wet stress indices, respectively.

CDX, CWX, HDX and HWX are the annual cold-dry, cold-wet, hot-dry and hot-wet

stress interaction indices (Kriticos et al. 2015).

**Climate Matching** Climate matching consists of selecting a location and then looking for similar locations elsewhere with similar climatographic conditions. It simply enables the user to compare the meteorological data from different locations, with no reference to the preferences of a given species. It answers questions such as the following: Are the extreme minimum and maximum temperatures similar? Do the two locations have similar amounts of rainfall and similar seasonal rainfall patterns? Do the two locations share similar climatography?

Match climate function is used for rough assessment of the risk of a pest establishing in a new location (country, continent), with implications for quarantine and pest risk analysis (Sutherst et al. 2004). The match climate analysis can be conducted with no knowledge of the species, except that it does occur in certain locations. Thus, if there are no biological data or distribution map of the species, the match climate analysis nonetheless enables the user to qualitatively assess the risk.

Weather conditions of different locations have implications on species. For example, if user knows about a beneficial species that occurs in a given location in any country, and looking to import that species to any other country of interest, as a biological control agent, a match climate analysis can be used for selecting suitable locations. Climate match index (CMI) obtained from match climate function shows whether the climate of the destination is similar to that of the area of collection of the species and the extent of similarity in the importing country. If the analysis indicates that there is only a small area in importing country, that has a similar climate to the home location for the natural enemy to be exported, there are high risks of failure of establishment by the species in the importing country. Conversely, it can be used to identify which region of a pest species occurrence has to be targeted for prospecting activities related to biological control agents (Dhileepan et al. 2006; Robertson et al. 2008).

The composite match index (CMI) is sensitive to the number of factors included in the analysis. It uses default setting of four factors, i.e. minimum temperature, maximum temperature, total rainfall and rainfall pattern; experience has shown that a CMI value of 0.7 is a rough threshold that delimits biological reasonability. CMI values below this threshold indicate little climatic similarity (Sutherst et al. 2007).

Note that the climatic match between a 'home location' and a set of 'away location' does not accord with climatic suitability for a species. If the reference location is at the edge of a species range, a poorer match may be obtained for locations that are beyond the suitable range of the species, as well as those that are closer to the core of the species distribution.

## 9.3 Materials and Methods

### 9.3.1 Climatic Preference of the Pest, Fitting CLIMEX Parameters

CLIMEX, which includes the weather data (monthly long-term average maximum and minimum temperatures, rainfall and relative humidity) from several

**Table 9.1** CLIMEX parameter values for *C. punctiferalis*

| Abiotic factor | Characteristic | Value |
|---|---|---|
| Moisture[a] | Lower limit of soil moisture necessary for growth (SM0) | 0.25 |
| | Lower limit of optimal soil moisture for growth (SM1) | 0.40 |
| | Upper limit of optimal soil moisture for growth (SM2) | 0.90 |
| | Upper limit of soil moisture necessary for growth (SM3) | 1.30 |
| Temperature (°C) | Lower temperature threshold for growth (DV0) | 8.00 |
| | Lower limit of optimal temperature for growth (DV1) | 24.00 |
| | Upper limit of optimal temperature for growth (DV2) | 26.00 |
| | Upper temperature threshold for growth (DV3) | 36.00 |
| | Degree-day threshold above DV0 (8 °C) to complete one generation (PDD) | 509.00 |
| Stress indices | Cold stress temperature threshold (TTCS) (°C) | 0.00 |
| | Weekly rate of cold stress accumulation (THCS) | −0.01[b] |
| | Heat stress temperature threshold (TTHS) (°C) | 38 |
| | Weekly rate of heat stress accumulation (THHS) | 0.0015 |
| | Dry stress threshold (SMDS) | 0.05 |
| | Weekly rate of dry stress accumulation(HDS) | −0.005[b] |
| | Wet stress threshold (SMWS) | 1.50 |
| | Weekly rate of accumulation of wet stress (HWS) | 0.002 |
| TDD | Thermal degree-days for the development | 509 |

[a]Moisture parameters were in units measuring the proportion of soil moisture-holding capacity
[b]Indicates the stress factor

meteorological stations located worldwide from 1931 to 1960, was used for the study. In order to overcome the spatial limitation of the limited meteorological database, a 10° grid of long-term average climate surface variables for the world terrain was used (New et al. 1999).

Iterative parameter fitting was used to develop the CLIMEX model for *C. punctiferalis*. The native range, relative abundance and seasonal phenology in India, other Southeastern Asia and Australia were used to infer climatic requirements. The adjusted parameter values were then visually validated against the reported distribution of *C. punctiferalis* around the world. In this process, the parameter values were compared favourably with the observed native distribution as summarized in Table 9.1.

**Growth Indices** The growth indices (on a scale from 0 to 100) indicate how favourable each location is for population growth; they were calculated weekly and annually after Sutherst and Maywald (1985).

**Temperature** The temperature parameter values for *C. punctiferalis* were initially inferred from available data (Bilapate and Talati 1978).

**Threshold Heat Sum (PDD)** The minimum thermal accumulation necessary to complete a generation (the sum of degree-days above DV0) was set at 509 degree-days for *C. punctiferalis* (Du et al. 2012).

Moisture and stress indices (cold, heat, dry, wet) were also iteratively adjusted so that the modelled distribution of *C. punctiferalis* exactly matches the present distribution of the pest. Soil moisture parameters in CLIMEX are calculated proportionally from the plant-available water capacity within the effective rooting zone. The irrigation function in CLIMEX was not used in the present context.

**Stress Indices** The stress indices indicate the limited population growth during unfavourable seasonal conditions. The annual stress values, on a scale from 0 (no stress) to 100 (lethal conditions), were calculated by the weekly value multiplied by the number of weeks subsequent to the stress first exceeding zero (Sutherst et al. 2007).

## 9.4 CLIMEX Model Matching

**Climate Matching** In addition to compare locations, climate match function of CLIMEX was also used by taking Hyderabad (17.37°N 78.48°E, 505 m MSL) located in Telangana state of South India, as 'home' location. This is one of the important regions with severe incidence of *C. punctiferalis*. In this region, it is a major pest on castor, pomegranate, grapes and other fruit and field crops. For predicting the favourable climatic conditions for the establishment of *C. punctiferalis*, climatic conditions of "home" location were compared with other regions of the world ('away locations'), and the climate match model was run using base climate data for the years 1961–1990. The data was obtained from CliMond (Kriticos et al. 2012). The resulting modelling output was mapped as climate match index (CMI) with different per cent match. In general, CMI of 85% and above is considered as a good match with the 'home location', favouring the establishment of the pest species.

## 9.5 Results and Discussion

**Climatic Preference of the Pest** Ecoclimatic index (EI) obtained through compare location function indicated the favourable regions for *C. punctiferalis* in all the continents except Antarctica and isolated patches in different continents as shown in the Fig. 9.2. The map denotes the favourable EI for the pest, wherein regions depicted in shades of red are more suitable for the pest compared to the regions with blue shades. The potential areas for the establishment of *C. punctiferalis* are presented below.

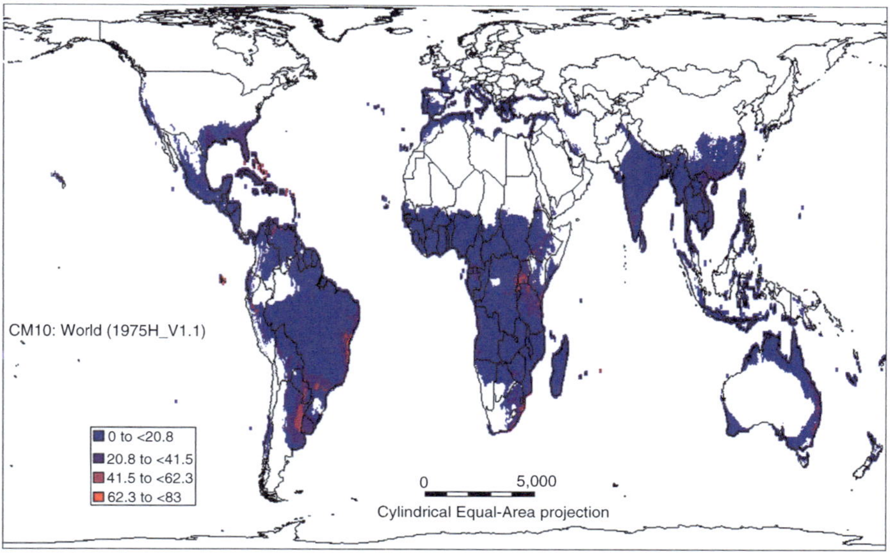


**Fig. 9.2** Potential favourable regions for *C. punctiferalis* establishment based on ecoclimatic index values

**North America** Southeastern states of the USA like North Carolina, South Carolina, Florida, Georgia, Alabama, Luciana, Mississippi and Texas and western parts of the USA like California are favourable for the borer pest. Other countries like Mexico, Guatemala, Nicaragua, Cuba and Puerto Rico are also favourable for the pest occupancy and damage to cultivated crops. However, northern and eastern USA are not suitable for this pest establishment.

**South America** Major portions of South America are favourable for the pest establishment except southern and south western parts like Argentina, Chile and parts of Peru. Entire Brazil is favourable for the pest, *C. punctiferalis*, except state of Parana and Sao Paulo. Similarly the zones in southern Colombia, northern and north eastern parts of Peru and north-western parts of Brazil, i.e. State of Amazonas, are not favourable for the establishment of the pest.

**Europe** Western parts of Spain, Palma, Portugal, western and southern parts of France, coastal region of Italy, Crose, Sardegna, Sicilica, Malta and Greece are favourable for the pest. Parts of southern Ukraine, Moldova and parts of Romania are also favourable.

**Middle East** Major portions of the Middle East are not favourable for the pest, *C. punctiferalis*. However, southern parts of Yemen and south western parts of Iran are favourable for pest establishment and population development under favourable weather conditions and presence of preferred food plant.

**Asia** South East Asia is highly favourable for establishment of the pest embracing India, southern parts of China, southern Nepal, Bhutan, Sri Lanka, Vietnam, Thailand, Myanmar, Malaysia and regions like Indonesia, the Philippines, northern Pakistan and eastern parts of Afghanistan. The above countries share almost the same climatography and cropping situations.

Entire India is suitable for the establishment of pest, *C. punctiferalis*, except north-western parts of Gujarat, Rajasthan, Srinagar and sub-Himalayan region. However, north eastern states are also favourable for the pest infestation. All the southern states like Kerala, Tamil Nadu, Karnataka, Andhra Pradesh, Goa and eastern states like Orissa, West Bengal and Bihar provide favourable conditions for establishment of the pest and its proliferation.

**Australia** An analysis of pest and population growth factors for *C. punctiferalis* revealed that coastal regions of northern, eastern, southern, south western parts of Australia were also found favourable for pest establishment and population development. Obviously, these areas also have host plants like cotton, maize, sorghum and temperate fruits like peach, pear, apricots and plum and in some parts forest trees that the pest prefers to feed and continue generations. The growth index reveals wide fluctuations in growth ranging from 0.2 to almost 1, the peak. Monthly rainfall pattern shows fluctuations from no rains to 50 mm.

## 9.6 Climate Matching

When validation of climate matching model was tested within India for the borer, Maharashtra, Gujarat, parts of Uttar Pradesh and in general west part of India showed a good match. Crops like castor and arid tropical fruits are extensively grown in these regions. The area also enjoys four seasons with varying climatographic conditions. The borer, *C. punctiferalis*, completes 6–7 generations provided preferred host plants and favourable weather conditions prevail.

Outside India, for the borer, Myanmar, Malaysia, Thailand, Vietnam, North Australia and South East Asian countries showed a good match. These are the countries where durian, rambutan, sour sop, litchi, teak, etc. are the major host plants. The borer completes 6–7 generations and incurs heavy yield losses in fruit crops. From these regions fruit crops are also exported to other countries. Therefore, monitoring *Conogethes* populations in these countries is crucial.

Based on the CMI, highest similarity between Hyderabad (home) was observed in the following region of the world (Fig. 9.3). Within India, parts of states, viz.

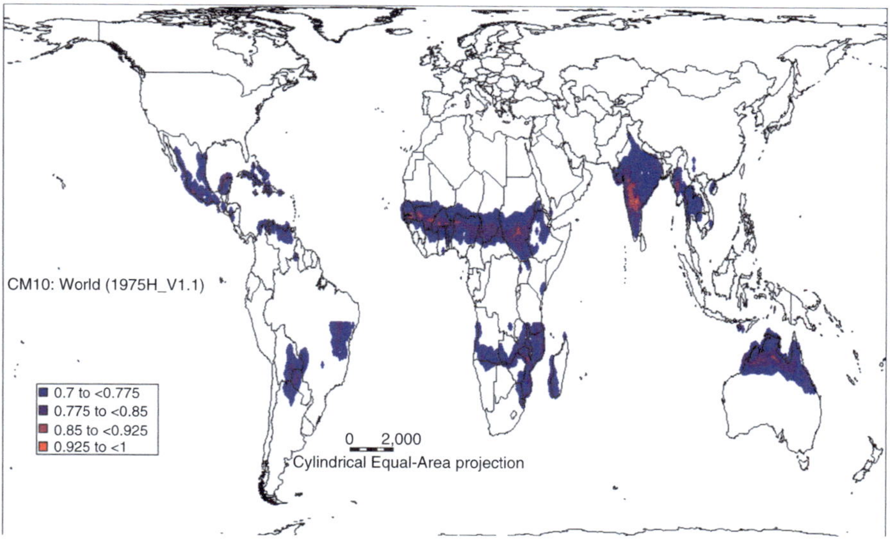


**Fig. 9.3** Potential favourable regions for *C. punctiferalis* establishment based on Climate Match Index

Maharashtra, Karnataka and Gujarat, showed highest similarity (>85%) with home location. Outside India, Burma, Malaysia, Thailand, Vietnam in Southeastern Asia, Northern territories of Australia, parts of Central America, South America (parts of Brazil and Argentina) and central and southern African countries are also found similar to the home location (Hyderabad).

## 9.7 Validation of the Models

In contrast to the growth promoting factors, the stress factors for population growth of *C. punctiferalis* were also examined. The three major stress factors considered are dry and wet conditions, cold and hot stress factors and rainfall. The depicted patterns of dry and wet stress factors and hot and cold stress factors show wide fluctuations and extremities that are unfavourable for the borer, *C. punctiferalis*. Under these wide fluctuation conditions, the pest cannot establish, grow and develop populations. Areas experiencing such conditions are not suitable for the borer growth and population build-up. Similarly, prolonged dry conditions or no rainfall period is unsuitable for borer population to grow. Such unsuitable areas include northern and eastern USA, desert areas and equatorial parts of Africa and a large portion of the Middle East, where neither the weather conditions nor suitable host plants for the borer occur. In such areas, pockets or patches may provide favourable conditions for the borer, *C. punctiferalis*. In such patches, the shoot and fruit borer undergoes only 2–3 generations a year. For example, in some parts of China, only 2–3 generations of *C. punctiferalis* are recorded (FAO document 2007).

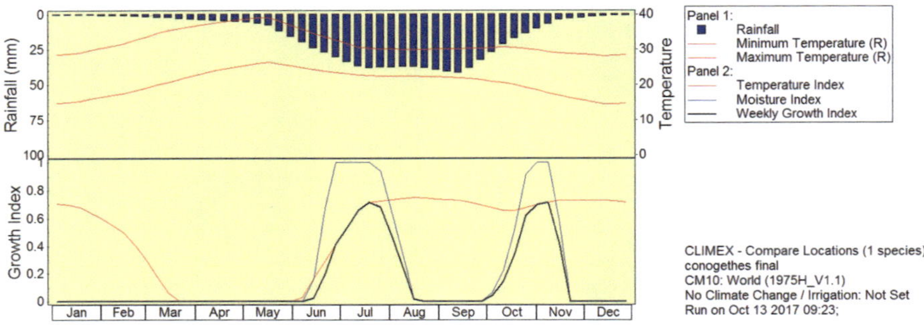


**Fig. 9.4** Weekly growth index of *C. punctiferalis* at Hyderabad, South India

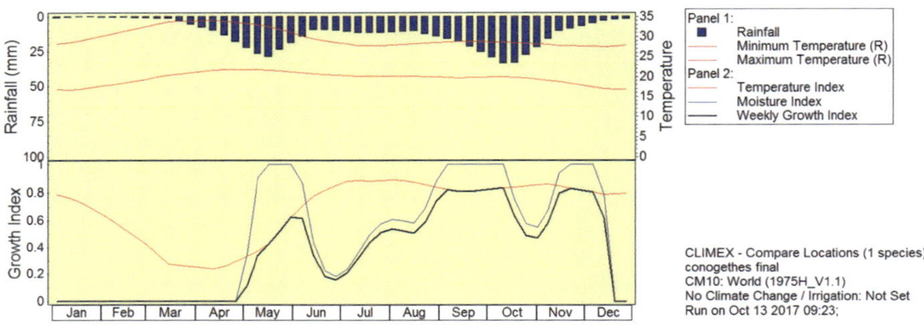


**Fig. 9.5** Weekly growth index of *C. punctiferalis* at Mysore, Karnataka, South India

The annual fluctuation in the abiotic factors in Hyderabad (India) is depicted in Fig. 9.4. In Hyderabad, annual temperature fluctuates from a minimum of 15–27° C to a maximum of 30–39° C. Rainfall pattern shows fluctuation between no rainfalls to 50 mm. Generally Hyderabad has warm, dry arid conditions, and the borer is active from June to August and October to November. Custard apple, ber, pomegranate, grapes, guava and other arid fruits form major host plants for the borer, *C. punctiferalis*, in this region. Areas in other parts of India with closely related climatic conditions may favour the pest establishment provided hosts are available.

CLIMEX output revealed, in Mysore region of Karnataka (12.2958° N, 76.6394° E), the weekly growth index was observed highly favourable from May to December and is coinciding with the higher temperature index and moisture index. This has been validated through the number of generations of *C. punctiferalis* from the adjoining locations like Bengaluru (12.9716° N, 77.5946° E). The positive weekly growth index from May to mid-December indicates the favourable weather conditions for the pest supporting at least five generations/year (Krishnamurthy et al. 1989). This figure indicates the favourable abiotic factors promoting the development of the pest, i.e. moisture, temperature and weekly growth index (Fig. 9.5).

On the contrary, In Rampur region of Madhya Pradesh (24.5095° N, 81.0577° E), the stress factors played an important role in suppression of the establishment of

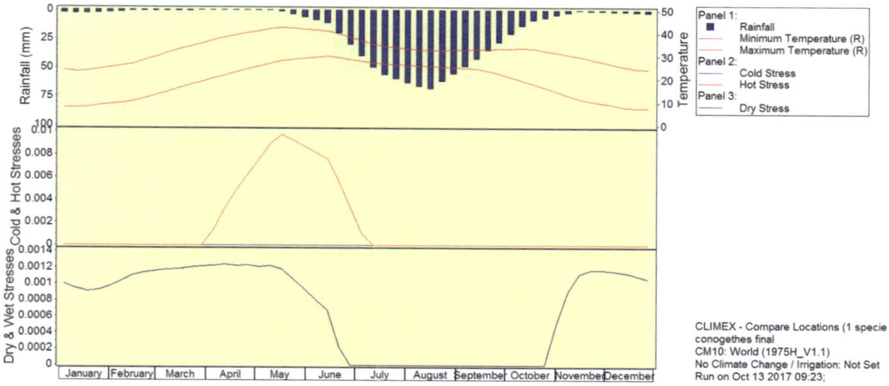


**Fig. 9.6** Stress index of *C. punctiferalis* at Rampur, India

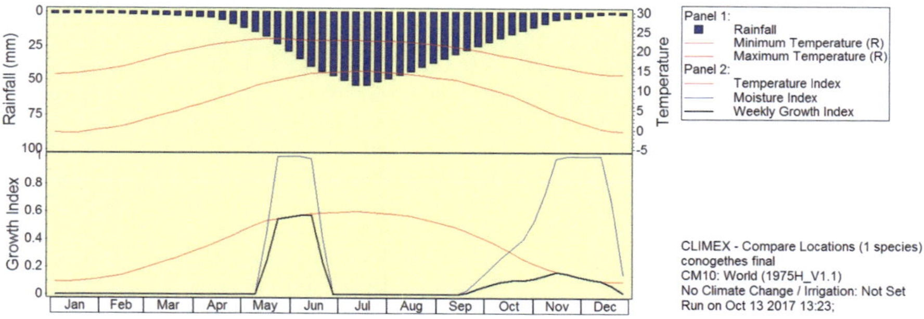


**Fig. 9.7** Weekly growth index of *C. punctiferalis* in China

the *C. punctiferalis* indicated by lower EI. The stress factors (Fig. 9.6), viz. hot stress during April to July and dry stress during the months of January to June and mid-October to December, made the climate suitable only during August to October resulting to lower EI. These adverse climatic conditions in this region may not support the establishment of the pest even when the hosts are available.

Current analysis showed that in China *C. punctiferalis* can complete 2–5 generations during May to June and again during September to December (Fig. 9.7), which is validated with the observations made by Yang and Shaw (1962) who also reported 2–5 generations/year from China.

## 9.8 Match Climate Output

The CMI with reference to Hyderabad (home) for the rest of the world is depicted in Fig. 9.8. Amdapur located in the Nizamabad district of Telangana state in India (18.6290° N, 77.9325° E) is considered as ‘away’ location, which showed similar

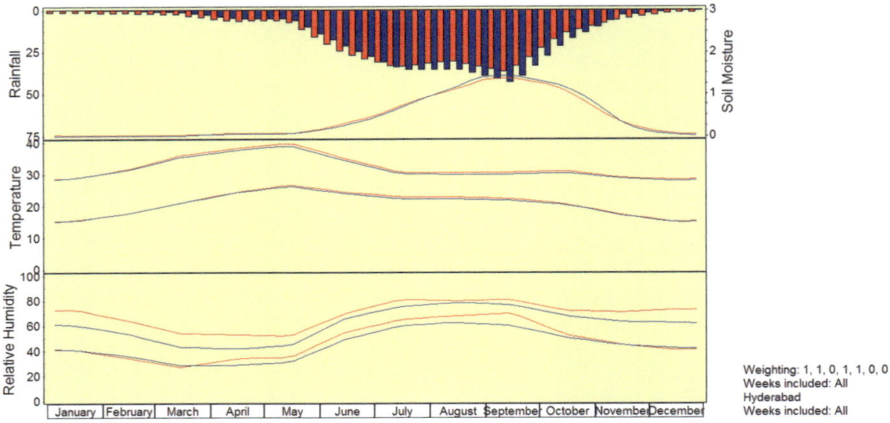


**Fig. 9.8** Abiotic factors in Amdapur similar to Hyderabad which showed CMI of 0.93–1

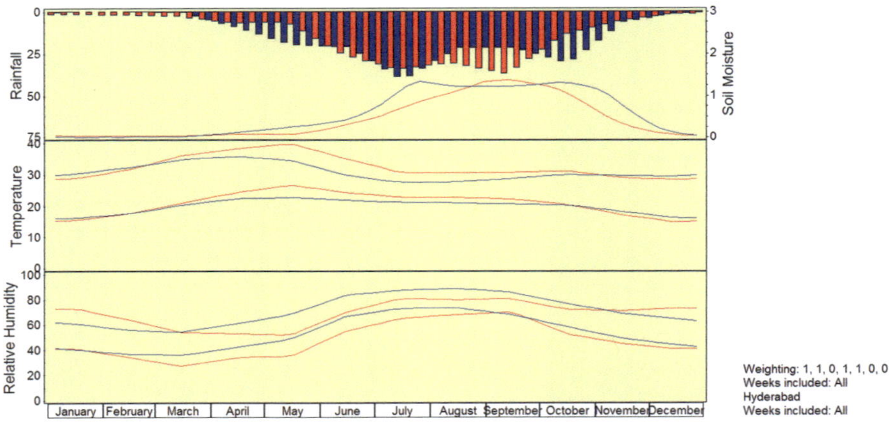


**Fig. 9.9** Abiotic factors in Haveri, Karnataka, South India

CMI as 'home' location, i.e. Hyderabad with a CMI of 0.93–1 (Fig. 9.8), where favourable climatic conditions are suitable for the establishment of *C. punctiferalis*, provided suitable hosts are available. Thus, locations spread over in different parts of the globe are more suitable for *C. punctiferalis* attack.

However, Haveri (away), located in Karnataka state (14.6610° N, 75.4345° E), is compared with the Hyderabad (home) location, and it showed lower CMI of 0.7–0.77 with home location (%) and thus may not be suitable for the establishment of the pest (Fig. 9.9).

Keeping in view the importance of pest risk analysis as a part of plant quarantine measures, CLIMEX can be a prominent tool for guiding identification of potential areas for the establishment of the borer pest, and importing countries can monitor whether the exporting materials are from the pest free areas. As there are several potential areas for the establishment of *C. punctiferalis*, this can help partly PRA for importing countries of agri-horticultural commodities.

**Acknowledgment** Authors also thank the ICAR-NICRA project for providing the facilities to carry out CLIMEX simulations during our studies.

## References

Abe Y, Sanari G (1992) Notes on the peach moth *Conogethes punctiferalis* attacking cynipid and aphid galls. Japanese J Entomol 60(1):108

AQAS (2014) *Conogethes* sp. interceptions. Agricultural Quarantine Activity Systems. https://aqas.aphis.usda.gov/aqas/HomePageInit.do#defaultAnchor

AQSIQ (2007) Chinese table grape export technical Information. Animal and plant quarantine Service of the people's Republic of China, pp 99

Astridge D (2001) Insect fauna surveys on rambutan, durian and mangosteen in North Queensland. In: Proceedings of the sixth workshop for tropical agricultural entomologists, Darwin. Northern territory Department of Primary Industry and Fisheries, technical bulletin no. 288, pp 75–79

Bilapate GG, Talati GM (1978) Some studies on bionomics of castor shoot and fruit borer *Dichocrosis punctiferalis* from Gujarat. J Maharashtra Agric Univ 3(1):75–76

CAB International (2011) *Conogethes punctiferalis* datasheet. In: Crop protection compendium. CAB International, Wallingford

Devasahayam S, Abdulla Koya KM (2005) Insect pests of ginger. In: Ravindran PN, Babu KN (eds) Ginger: the genus. Zingiber. CRC Press, Boca Raton, pp 367–390

Dhileepan K, Senaratne KADW, Raghu S (2006) A systematic approach to biological control agent exploration and prioritisation for prickly acacia (*Acacia nilotica* ssp. *indica*). Aust J Entomol 45:303–307

Du Y-L, Guo H-M, Sun S-L, Zhang M-Z, Zhang A-H, Wang J-B, Qin L (2012) Effects of temperature on the development and reproduction of the yellow peach moth, Conogethes punctiferalis (Lepidoptera: Pyralidae). Acta Entomol Sin 55(5):561–569

FAO document (2007) Overview of forest pests- People's Republic of China. http://www.fao.org/forestry/12286-09d6c18bf6807d46832a5f6c6357dab3c.pdf

Gour TB, Sriramulu M (1992) Grapevine, *Vitis vinifera* Linn. A new host of castor shoot and capsule borer, *Conogethes punctiferalis* (Guenée). Trop Pest Manag 38(4):459

Hang HL, Yan KF, Sun XH, Ma JX (2000) Investigation on the kinds of fruit moth in the western part of Henan province and their control. China Fruits 2:44–45

Honda H, Kaneko J, Konno Y, Matsumoto YA (1979) Simple method for mass-rearing of the yellow peach moth, *Dichocrocis punctiferalis* Guenée (Lepidoptera: Pyralidae), on an artificial diet. Appl Entomol Zool 14:464–468

Kadoi M, Kaneda M (1990) Development of yellow peach moth Conogethes punctiferalis on apple fruit. Res Bull Plant Prot Serv Japan 26:61–63

Kang CH, Lee SM, Chung YJ, Choi KS, Park CG (2002) Overwintering ecology of the peach pyralid moth, *Dichocrosis punctiferalis* in southern regions of Korea. Korean J Appl Entomol 43(3):201–209

Kimura T, Honda H (1999) Identification and possible functions of the hairpencil scent of the yellow peach moth, *Conogethes punctiferalis* (Guenée) (Lepidoptera: Pyralidae). App Entomol Zool 34(1):147–153

Konno Y, Shishido T (1980) Glutathione – dependent o-alkyl and oral conjunctions for dicapthon and fertikothine in several insects. J Pestic Sci 21(4):430–433

Konno Y, Honda N, Matsumoto Y (1981) Mechanism of reproductive isolation between the fruit feeding and pin ace feeding type of yellow peach moth, *Dichocrosis punctiferalis*. Jpn J Appl Entomol Zool 25(4):253–258

Krishnamurthy K, Khan MM, Avadhani KK, Venkatesh J, Siddaramaiah L, Chakravarthy AK, Gurumurthy BR (1989) Three decades of cardamom research, regional research station (1958–1988). Station Technol Bull 2:44–68

Kriticos DJ, Webber BL, Leriche A, Ota N, Macadam I, Bathols J, Scott JK (2012) CliMond: global high resolution historical and future scenario climate surfaces for bioclimatic modelling. Methods Ecol Evol 3:53–64. https://doi.org/10.1111/j.2041-210X.2011.00134.x

Kriticos DJ, Maywald GF, Yonow T, Zurcher EJ, Herrmann NI, Sutherst RW (2015) CLIMEX version 4: exploring the effects of climate on plants, animals and diseases. CSIRO, Canberra, p 184

Li DY, Ai PP, Du YL, Sun SL, Zhang MZ (2015) Effects of different host plants on the development and reproduction of yellow peach moth, *Conogethes punctiferalis* (Lepidoptera: Crambidae). Austral Entomol 54:149–153

New M, Hulme M, Jones PD (1999) Representing twentieth century space-time climate variability. Part 1: development of a 1961–90 mean monthly terrestrial climatology. J Clim 12:829–856

Robertson MP, Kriticos DJ, Zachariades C (2008) Climate matching techniques to narrow the search for biological control agents. Biol Control 46:442–452

Robinson GS, Tuck KR, Shaffer M (1994) A field guide to the smaller moths of South-East Asia. Malaysian nature society and the Natural History Museum. Art Printing Works Sdn Dhd, Kuala Lumpur, Malaysia

Shashank PR, Doddabasappa B, Kammar V, Chakravarthy AK, Honda H (2015) Molecular characterization and management of shoot and fruit borer *Conogethes punctiferalis* Guenee (Crambidae: Lepidoptera) populations infesting cardamom, castor and other hosts. In: Chakravarthy AK (ed) New horizons in insect science: towards sustainable pest management. Springer, New Delhi, pp 207–227

Sutherst RW, Bottomley W, Yonow T, Maywald GF (2004) Use of CLIMEX in pest risk analysis for quarantine. http://www.cabi.org/isc/FullTextPDF/2009/20093238355

Sutherst RW, Maywald GF (1985) A computerised system for matching climates in ecology. Agric Ecosyst Environ 13:281–299

Sutherst RW, Maywald GF, Kriticos DJ (2007) CLIMEX Version 3: user's guide. Hearne Scientific Software Pty Ltd [WWW document]. URL www.hearne.com.au

USDA (1957) Insects not known to occur in the United States. Yellow peach moth (*Dichocrocis punctiferalis* (Guenee)). pp 37–38

Yang HY, Shaw KY (1962) Preliminary studies on the peach borer, *Dichocrosis punctiferalis* Guenee. Acta Ocean Ent Sin 1:119–134

# Bioecological Studies on *Conogethes sahyadriensis* in South India

# 10

Kiran S. Kasareddy, L. Hanumantharaya, and K. P. Kumar

**Abstract**

Studies on the biology of *Conogethes* sp. on cardamom, turmeric and ginger in Mudigere, Karnataka, South India, revealed that eggs were oval and translucent, with an incubation period of 4.00 ± 0.20, 4.00 ± 0.36 and 4.00 ± 0.05 days in cardamom, turmeric and ginger, respectively. Average duration of fifth instar larva reared on cardamom, turmeric and ginger was 7.50 ± 0.82, 6.90 ± 0.69 and 8.75 ± 0.81 days, respectively. Average pre-oviposition period of adult female reared on cardamom, turmeric and ginger was 3.12 ± 0.20, 2.20 ± 0.20 and 1.41 ± 0.14 days, respectively, and that of oviposition was 2.25 ± 0.30, 2.57 ± 0.36 and 2.76 ± 0.30 days, respectively. Post-oviposition period was 2.50 ± 0.10, 1.89 ± 0.18 and 2.15 ± 0.18 days, respectively. Total development period of male and female moths reared on turmeric was maximum (42.45 ± 5.38 and 44.05 ± 5.24 days, respectively) compared to those reared on cardamom (37.80 ± 1.91 and 39.25 ± 1.75 days, respectively) and ginger (40.10 ± 1.37 and 42.50 ± 3.50 days, respectively).

**Keywords**

Biology · Cardamom · *Conogethes sahyadriensis* · Ginger · Turmeric

K. S. Kasareddy (✉) · L. Hanumantharaya
College of Horticulture, University of Agricultural and Horticultural Sciences -Shimoga, Mudigere, Karnataka, India

K. P. Kumar
Department of Entomology, University of Agricultural Sciences, Gandhi Krishi Vignan Kendra (GKVK), Bengaluru, Karnataka, India


A. K. Chakravarthy (ed.), *The Black spotted, Yellow Borer, Conogethes punctiferalis Guenée and Allied Species*,
https://doi.org/10.1007/978-981-13-0390-6_10

## 10.1 Introduction

The Malabar coast of India with its abundance of the world's best quality cardamom, ginger, turmeric, cinnamon and pepper is a vast emporium of spices, which once attracted foreign trade. Among the spices grown in India, ginger (*Zingiber officinale* Rosc.), turmeric (*Curcuma longa* L.) and cardamom (*Elettaria cardamomum* Maton) have great economic value and are distributed throughout tropical Africa, Asia and America. Small cardamom is grown only in India and Sri Lanka (Chomchalow 2001).

Among the factors affecting the yield and quality of cardamom, ginger and turmeric, the damage caused by insect pests is considered as a major contributor and poses considerable threat to spice production in Karnataka. Among the insects, shoot and capsule borer is considered a major pest of spice crops. Shoot and capsule borer infests shoot, stem, panicles, flowers and capsule of cardamom and causes yield loss up to 70–80%;(Varadarasan 2001), whereas in ginger and turmeric, the shoot borer causes yield loss up to 50% (Senthil et al. 2015). Therefore, there is a need to study in detail the basic aspects of shoot borer, *Conogethes* sp., for its effective management. Farmers apply the conventional insecticides available in the market indiscriminately, which results in higher costs of cultivation and lower-quality produce. This also causes adverse effects on beneficial arthropods and deteriorates environmental quality. Besides, as cardamom is cultivated in tropical evergreen forests, conservation of biodiversity is crucial.

## 10.2 Material and Methods

### 10.2.1 Biology

Larvae of *C. sahyadriensis* were collected from cardamom plantations in Mudigere. Larvae were brought to the laboratory with infested cardamom suckers and reared to pupae in reckangular plastic boxes. Minute holes were provided on the lids of the boxes. Larvae were pupated inside the infested shoot. fully formed pupae were and maintained in small plastaler cup (4 cm dia.) moist cotton wad was provided inside the cup. Emerging moths were collected with the help of plastic tube and transfered to oviposition cages. 10 % honey in cotton wad aws provided as a food for moth. Young three months old cardamom seedlings were provided in conical flask oviposition and red bulb (20 watt) for oviposition.

**Egg** Ten eggs were observed under a stereozoom microscope for morphological characteristics. Eggs were removed carefully from the substrate using fine camel hair brush, and their size was measured using stage and ocular micrometre. The number of eggs freshly laid by each female was recorded everyday until the moths lived, and the incubation period was observed from the date of egg laying up to the date of hatching at 24-hour intervals.

**Larva** Freshly hatched (neonate) larvae were transferred on tender cardamom, ginger and turmeric pseudostem in the petri plates separately. Observations on larval growth and development of each instar were recorded separately on the crops until pupation. In order to determine the larval instars, hatched larvae of uniform age were released individually on cardamom, ginger and turmeric shoots. Cardamom shoot pieces of 10–12 cm length were kept in petri dishes (0.4 cm diameter) for larval feeding, and fresh food was placed for each observation. The observations on larval length, width and head capsule size were recorded at each instar ($n = 10$).

**Prepupa** The time period between the larval stage and the stage at which larva stopped feeding and became inactive to pupate was considered as the prepupal period. The length and breadth of prepupae was measured using a millimetre scale. Length and breadth of ten prepupae of *Conogethes* sp. were measured.

**Pupa** After pupation, the male and female pupae were separated based on the distance between the anal and genital slit. Pupal period was recorded from the date of formation of pupa to the date of emergence of the adult. Further, pupal period of male and female pupae was recorded and maintained separately to determine the sex ratio. The length and breadth of pupa were measured using a millimetre scale, and totally ten pupae were considered to measure length and breadth and to determine the sex ratio.

**Adult Morphology** The freshly emerged male and female moths were separately killed using chloroform and were properly mounted to record wing spread, type of antenna, colour and markings on the body.

**Sex Ratio** Twenty (10 each male and female) pupae were randomly selected and observed under microscope for the determination of sex and sex ratio. The sex was determined by observing the distance between genital and anal slit which was less in male pupa than that of the female.

**Pre-Oviposition, Oviposition, Post-Oviposition and Fecundity** In an oviposition cage, a pair of freshly emerged male and female moths was released on cardamom pseudostem with tender leaves. Five such replications were maintained to study the pre-oviposition, oviposition, post-oviposition and fecundity of shoot and capsule borer. The period between the emergence of the female and deposition of eggs was considered as pre-oviposition period. Period between deposition of eggs until cessation of egg laying was considered as oviposition period, while period between the cessation of egg laying and death of the female was considered as post-oviposition period. Further, the number of eggs laid by each female during oviposition period was recorded as total fecundity.

**Adult Longevity** In order to determine adult longevity, freshly emerged adult moths (five male and female, five each) were released separately into rearing cages and were provided with 10% sugar solution in a cotton wad, and their activates were observed every day. The food (10% sugar solution) was replaced once in 2 days, and the number of days the moths survived was recorded.

**Total Life Cycle** The period from freshly laid egg up to the death of adult was recorded and considered as total life cycle.

## 10.3 Results and Discussion

The borer on zingibers is now identified as *C. sahyadriensis* (Shashank et al. 2018).

## 10.4 Biology on Cardamom, Ginger and Turmeric

The biology of *Conogethes* sp. was studied under laboratory conditions at the College of horticulture, Mudigere. The growth parameters and developmental periods of egg, larvae, prepupae, pupae and adults are presented in Tables 10.1, 10.2, 10.3, 10.4, 10.5 and 10.6.

**Egg** The female moths laid their eggs either singly or in small groups in between the leaf axils or on unopened leaves. The freshly shoot borer eggs laid on cardamom, ginger and turmeric were milky white in colour initially and turned into dark yellow colour, and on the second day, a pink eye spot appeared at the anterior end of the egg. The eggs were flat and hemispherical in shape, and the tip of the egg turned light brown when about to hatch. The eggs were laid singly sticking to the sides of veins and midrib of unopened leaves. The moth laid eggs both on upper and lower surfaces of tender leaves under laboratory conditions. Almost all eggs were found to hatch during night time under laboratory conditions. The developing larva could be seen inside the egg through the chorion under a stereo binocular zoom microscope. Unfertilized eggs were milky white and transparent and did not show any change in colour but gradually shrivelled.

The average incubation period of *C. sahyadriensis* under laboratory conditions was maximum (4.00 ± 0.36 days) when eggs were laid on turmeric compared to those laid on cardamom and ginger (4.00 ± 0.20 and 4.00 ± 0.05 days, respectively). The average lengths of eggs laid on cardamom, turmeric and ginger were 0.62 ± 0.05 mm, 0.62 ± 0.50 mm and 0.62 ± 0.10 mm, respectively, and the average breadths were 0.39 ± 0.01 mm, 0.41 ± 0.01 mm and 0.46 ± 0.02 mm, respectively (Tables 10.1, 10.3 and 10.5).

**Table 10.1** Developmental biology of *Conogethes sahyadriensis* on cardamom in laboratory

| Stage of insect | | Duration[a] (days) | Length[a] (mm) | Breadth[a] (mm) | Head capsule width (mm)[a] |
|---|---|---|---|---|---|
| Egg | | 4.00 ± 0.20 | 0.62 ± 0.05 | 0.39 ± 0.01 | – |
| Larva[a] | First instar | 2.50 ± 0.55 | 2.76 ± 0.06 | 0.48 ± 0.02 | – |
| | Second instar | 3.25 ± 0.60 | 4.45 ± 0.12 | 0.62 ± 0.05 | 0.22 ± 0.05 |
| | Third instar | 4.90 ± 0.62 | 9.68 ± 0.82 | 1.48 ± 0.02 | 0.32 ± 0.07 |
| | Fourth instar | 6.00 ± 0.72 | 16.80 ± 0.30 | 2.50 ± 0.08 | 0.56 ± 0.16 |
| | Fifth instar | 7.50 ± 0.82 | 24.65 ± 0.68 | 2.98 ± 0.03 | 0.86 ± 0.15 |
| Total larval period | | 26.70 ± 2.30 | – | – | – |
| Prepupa | | 4.25 ± 0.71 | 15.45 ± 1.15 | 2.66 ± 0.64 | – |
| Pupa | Male | 7.90 ± 0.72 | 15.45 ± 1.15 | 3.06 ± 0.15 | – |
| | Female | 8.10 ± 0.30 | 17.80 ± 1.30 | 3.60 ± 0.36 | – |
| Adult | Male | 6.90 ± 0.68 | 12.40 ± 0.26 | 25.54 ± 1.50[b] | – |
| | Female | 7.95 ± 0.77 | 14.50 ± 1.26 | 27.60 ± 1.62[b] | – |
| Total development period | Male | 37.80 ± 1.91 | – | – | – |
| | Female | 39.25 ± 1.75 | – | – | – |

*Note*: Figures in each cell are mean ± standard deviation values
[a]Number observed (N) = 10 per observation
[b]Wing expansion.

**Table 10.2** Effect of cardamom on reproductive traits and longevity of *Conogethes sahyadriensis*

| Reproductive stages | Duration (days) | | |
|---|---|---|---|
| | Minimum | Maximum | Mean ± SD |
| Pre-mating | 1.25 | 2.00 | 1.30 ± 0.40 |
| Mating | 0.20 | 0.30 | 0.13 ± 0.03 |
| Pre-oviposition | 2.50 | 3.75 | 3.12 ± 0.2 |
| Oviposition | 2.00 | 2.50 | 2.25 ± 0.30 |
| Post-oviposition | 2.25 | 2.75 | 2.50 ± 0.10 |
| Fecundity | 25.00 | 42.00 | 33.50 ± 5.00 |

Bilapate and Talati (1978) studied the characteristics of *C. punctiferalis* eggs laid on castor and reported that the average length and width of eggs were 0.52 mm and 0.34 mm, respectively, and the incubation period was 5.15 ± 0.36 days. These observations are in agreement with the findings of Stanley et al. (2009) and Jacob (1981).

**Larvae** Eggs hatched during the night, and the larva moulted four times during its development period and had five larval instars. The average total larval period was

**Table 10.3** Developmental biology of *Conogethes sahyadriensis* on turmeric in laboratory

| Stage of insect | | Duration[a] (days) | Length[a] (mm) | Breadth[a] (mm) | Head capsule width (mm)[a] |
|---|---|---|---|---|---|
| Egg | | 4.00 ± 0.36 | 0.62 ± 0.50 | 0.41 ± 0.01 | – |
| Larva[a] | First instar | 3.50 ± 0.30 | 2.57 ± 0.08 | 0.40 ± 0.01 | – |
| | Second instar | 3.95 ± 0.73 | 4.70 ± 0.42 | 0.75 ± 0.25 | 0.36 ± 0.06 |
| | Third instar | 4.25 ± 0.65 | 9.65 ± 0.42 | 1.39 ± 0.11 | 0.51 ± 0.03 |
| | Fourth instar | 4.90 ± 0.59 | 15.80 ± 0.78 | 2.45 ± 0.33 | 0.88 ± 0.06 |
| | Fifth instar | 6.90 ± 0.69 | 24.10 ± 1.19 | 2.90 ± 0.30 | 1.40 ± 0.04 |
| Total larval period | | 28.50 ± 3.32 | – | – | – |
| Prepupa | | 4.20 ± 0.72 | 15.60 ± 0.84 | 2.70 ± 0.50 | – |
| Pupa | Male | 9.50 ± 0.51 | 15.60 ± 0.61 | 3.10 ± 0.30 | – |
| | Female | 9.90 ± 0.55 | 17.60 ± 1.07 | 3.39 ± 0.29 | – |
| Adult | Male | 7.40 ± 0.70 | 12.00 ± 0.79 | 25.30 ± 1.20[b] | – |
| | Female | 8.10 ± 0.56 | 14.80 ± 1.48 | 27.40 ± 1.55[b] | – |
| Total development period | Male | 42.45 ± 5.38 | – | – | – |
| | Female | 44.05 ± 5.24 | – | – | – |

**Note:** Figures in each cell are mean ± standard deviation values
[a]Number observed (N) = 10 per observation
[b]Wing expansion

**Table 10.4** Effect of turmeric on reproductive traits and longevity of *Conogethes sahyadriensis*

| Reproductive stages | Duration (days) | | |
|---|---|---|---|
| | Minimum | Maximum | Mean ± SD |
| Pre-mating | 1.00 | 1.50 | 1.15 ± 0.24 |
| Mating | 0.10 | 0.15 | 0.11 ± 0.01 |
| Pre-oviposition | 2.00 | 2.50 | 2.20 ± 0.2 |
| Oviposition | 2.00 | 3.00 | 2.57 ± 0.36 |
| Post-oviposition | 1.50 | 2.00 | 1.89 ± 0.18 |
| Fecundity | 15.00 | 28.00 | 20.80 ± 5.63 |

**Table 10.5** Developmental biology of *Conogethes sahyadriensis* on ginger in laboratory

| Stage of insect | | Duration[a](days) | Length[a](mm) | Breadth[a](mm) | Head capsule width (mm)[a] |
|---|---|---|---|---|---|
| Egg | | 4.00 ± 0.05 | 0.62 ± 0.10 | 0.46 ± 0.02 | – |
| Larva[a] | First instar | 3.50 ± 0.51 | 2.48 ± 0.08 | 0.27 ± 0.01 | – |
| | Second instar | 3.95 ± 0.55 | 3.38 ± 0.15 | 0.37 ± 0.01 | 0.32 ± 0.01 |
| | Third instar | 4.20 ± 0.52 | 7.55 ± 0.24 | 0.89 ± 0.15 | 0.46 ± 0.08 |
| | Fourth instar | 6.25 ± 0.67 | 14.45 ± 0.28 | 1.59 ± 0.06 | 0.82 ± 0.06 |
| | Fifth instar | 8.75 ± 0.81 | 18.49 ± 1.31 | 2.46 ± 0.09 | 1.26 ± 0.06 |
| Total larval period | | 27.50 ± 2.30 | – | – | – |
| Prepupa | | 3.15 ± 0.80 | 13.49 ± 0.84 | 2.59 ± 0.50 | – |
| Pupa | Male | 6.90 ± 0.65 | 11.60 ± 0.30 | 2.74 ± 0.13 | – |
| | Female | 7.20 ± 0.40 | 13.60 ± 1.07 | 2.98 ± 0.25 | – |
| Adult | Male | 7.80 ± 0.77 | 12.10 ± 0.79 | 23.30 ± 1.80[b] | – |
| | Female | 8.60 ± 0.68 | 13.15 ± 0.50 | 25.40 ± 1.45[b] | – |
| Total development period | Male | 40.10 ± 1.37 | – | – | – |
| | Female | 42.50 ± 3.50 | – | – | – |

*Note*: Figures in each cell are mean ± standard deviation values
[a]Number observed (N) = 10 per observation
[b]Wing expansion

maximum (28.50 ± 3.32) in case of larvae reared on turmeric compared to those reared on cardamom and ginger with total larval periods of 26.70 ± 2.30 and 27.50 ± 2.30 days, respectively. The head width increased by about 1,4 times in a regular geometrical progression in the successive instars (Tables 10.1, 10.3 and 10.5).

**First Instar Larvae** The newly hatched larvae were sluggish in nature and became active after a few hours. The first instar larvae were translucent pale yellowish when reared on cardamom and translucent light pinkish when on turmeric and ginger with a number of short scattered hairs arising from dark coloured tubercles. The head

**Table 10.6** Effect of ginger on reproductive traits and longevity of *Conogethes sahyadriensis*

| | Duration (days) | | |
|---|---|---|---|
| Reproductive stages | Minimum | Maximum | Mean ± SD |
| Pre-mating | 2.40 | 2.95 | 2.65 ± 0.19 |
| Mating | 0.15 | 0.25 | 0.20 ± 0.01 |
| Pre-oviposition | 1.22 | 1.65 | 1.41 ± 0.14 |
| Oviposition | 2.30 | 3.20 | 2.76 ± 0.30 |
| Post-oviposition | 2.00 | 2.50 | 2.15 ± 0.18 |
| Fecundity | 20.00 | 32.00 | 24.80 ± 5.00 |

was blackish brown and prominently bigger in size. They were smooth-skinned with brown prothoracic shield, and the sclerites on the body were dark brown in colour. As the caterpillar completed its first moult, the abdomen became more or less cylindrical in shape.

The durations of the first instar larvae reared on cardamom, turmeric and ginger were 2.50 ± 0.55, 3.50 ± 0.30 and 3.50 ± 0.51 days, respectively. The average lengths of first instar larvae reared on cardamom, turmeric and ginger were 2.76 ± 0.06 mm, 2.57 ± 0.08 mm and 2.48 ± 0.08 mm, respectively, whereas the average breadths were 0.48 ± 0.02 mm, 0.40 ± 0.01 mm and 0.27 ± 0.01 mm, respectively (Tables 10.1, 10.3 and 10.5).

**Second Instar Larvae** As the larvae advanced from first to second instar, their body grew faster, and as a result, the body turned wider than the head. The second instar larvae were light yellowish with eye spots and dark mandibles. The larvae were translucent with a number of short scattered hairs arising from dark coloured tubercles, which were prominent and the head was black. They were smooth-skinned with a scattered brown colour tinge on the body which was not prominent and a brown coloured prothoracic shield. The spiracles were brown in colour and were more prominent compared to the first instar.

The average lengths of the second instar larvae reared on cardamom, turmeric and ginger were 4.45 ± 0.12 mm, 4.70 ± 0.42 mm and 3.38 ± 0.15 mm, respectively, whereas the average breadths were 0.62 ± 0.05 mm, 0.75 ± 0.25 mm and 0.37 ± 0.01 mm, respectively. The average durations of the second instar larvae reared on cardamom, turmeric and ginger were 3.25 ± 0.60, 3.95 ± 0.73 and 3.95 ± 0.55 days, respectively (Tables 10.1, 10.3 and 10.5).

**Third Instar Larvae** The third instar larvae were morphologically similar to the second instar with short scattered hairs, black head and brownish prothoracic shield. The larvae of third instar were light yellowish with dark brown head when reared on cardamom, whereas when reared on turmeric and ginger, they were light pinkish with dark brown head. The body was elongated and longer than the second instar larva. The spiracles were blackish in colour and were nine in number. Three pairs of prothoracic legs were distinct, and their tips were black. Prolegs were present on the abdomen, particularly on the sixth, seventh, eighth and ninth segments.

The average lengths of the third instar larva reared on cardamom, turmeric and ginger were 9.68 ± 0.82 mm, 9.65 ± 0.42 mm and 7.55 ± 0.24 mm, respectively, and the average breadths were 1.48 ± 0.02 mm, 1.39 ± 0.11 mm and 0.89 ± 0.15 mm, respectively. The average durations of the third instar larva reared on cardamom, turmeric and ginger were 4.90 ± 0.62, 4.25 ± 0.65 and 4.20 ± 0.52 days, respectively (Tables 10.1, 10.3 and 10.5).

**Fourth Instar Larvae** Fourth instar larva was comparatively stout and long with nonprominent scattered hairs arising from dark coloured tubercles. The larva was light yellowish when reared on cardamom, whereas when reared on turmeric and ginger, it was light pinkish with a scattered prominent brown tinge. Head and prothoracic shield were black, and the crochets on prolegs were distinct.

Fourth instar duration of larvae reared on cardamom was 6.00 ± 0.72 days, whereas for those reared on turmeric and ginger, it was 4.90 ± 0.59 days and 6.25 ± 0.67 days, respectively. The average lengths of fourth instar larvae reared on cardamom, turmeric and ginger were 16.80 ± 0.30 mm, 15.80 ± 0.78 mm and 14.45 ± 0.28 mm, respectively, and the breadths were 2.50 ± 0.08 mm, 2.45 ± 0.33 mm and 1.59 ± 0.06 mm, respectively (Tables 10.1, 10.3 and 10.5).

**Fifth Instar Larvae** Fifth instar larva was almost similar to the fourth instar, except for its size. Colour of the fifth instar larva was light brown when reared on cardamom and light pinkish on turmeric and ginger with a prominent scattered brown tinge; later, fully grown larva turned translucent light brown on cardamom and pinkish on turmeric and ginger with an absence of brown tinge and the head was reddish brown. The tubercles and thoracic legs were black, and the spiracles were prominent (Fig. 10.1).

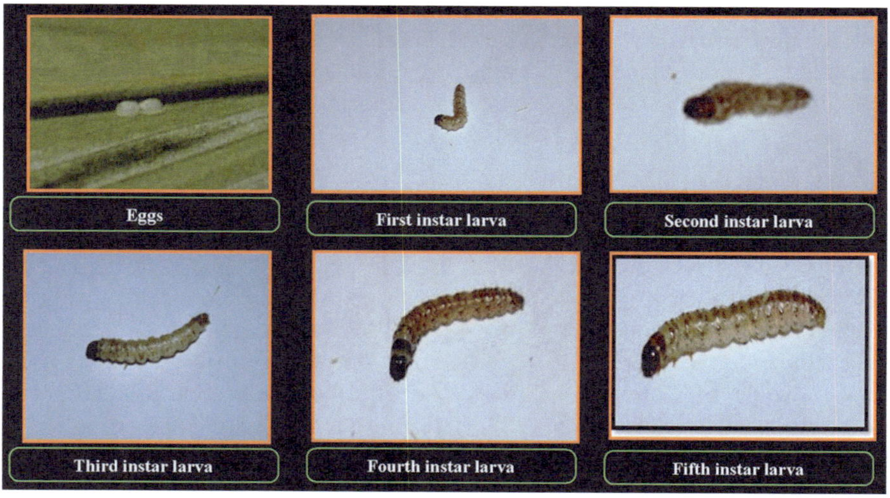


**Fig. 10.1** Different larval instars of shoot and capsule borer, *C. sahyadriensis*

**Fig. 10.2** Rearing techniques and biology studies of shoot and capsule borer, *C. sahyadriensis*

The average durations of fifth instar larva reared on cardamom, turmeric and ginger were 7.50 ± 0.82, 6.90 ± 0.69 and 8.75 ± 0.81 days, respectively. The average length and breadth of fifth instar larva were maximum (24.65 ± 0.68 mm and 2.98 ± 0.03 mm, respectively) on cardamom compared to that of the moths reared on turmeric and ginger, i.e. 24.10 ± 1.19 mm and 2.90 ± 0.30 mm and 18.49 ± 1.31 mm and 2.46 ± 0.09 mm, respectively (Tables 10.1, 10.3 and 10.5). The rearing method and materials used are indicated in Fig. 10.2.

Results of the present investigation are in accordance with the findings of Stanley et al. (2009) who reported that *C. sahyadriensis* larvae reared on cardamom took 25–40 days to complete larval duration and those raised on ginger to 24 days in Karnataka. Similarly larval duration of 26 days on turmeric was reported by Senthil et al. (2015). Further, Kadoi and Kaneda (1990) also observed that fully grown larvae took 32 days to complete the stage when reared on apples and 16 days on fresh maize. Besides, workers have also studied the biology of *C. punctiferalis* attacking several crops in different parts of the world (Twine 1971; Bilapate and Talati 1978; Jacob 1981; Kadoi and Kaneda 1990; Xi-ke et al. 1996; Wang and Cai 1997).

**Width of Larval Head Capsule after Moulting** Average head capsule widths of second, third, fourth and fifth instar larvae reared on cardamom were 0.22 ± 0.05 mm, 0.32 ± 0.07 mm, 0.56 ± 0.16 mm and 0.86 ± 0.15 mm, respectively. Average head capsule widths of second, third, fourth and fifth instar reared on turmeric were 0.36 ± 0.06 mm, 0.51 ± 0.03 mm, 0.88 ± 0.06 mm and 1.40 ± 0.04 mm, respectively. Average head capsule width of second instar larvae reared on ginger was

0.32 ± 0.01 mm, and that of third, fourth and fifth instars were 0.46 ± 0.08, 0.82 ± 0.06 and 1.26 ± 0.06 mm, respectively (Tables 10.1, 10.3 and 10.5).

**Behaviour of the Larva** Immediately after hatching, the larvae were sluggish and later became active and bored into the unopened leaves. At the time of moulting, they became inactive and stopped feeding. The head width increased by about 1.4 times in a regular geometrical progression in successive instars. During the larval period, the caterpillar moulted four times, and thus there were five instars.

**Prepupa** At the end of the fifth instar, the larvae became inactive and entered into the prepupa stage. Prepupal stage was characterized by the shortened larval length, suspended feeding and movement towards periphery. The colour of the larvae became milky white with a prominent tinge.

The average lengths of the prepupae reared on cardamom, turmeric and ginger were 15.45 ± 1.15 mm, 15.60 ± 0.84 mm and 13.49 ± 0.84 mm, respectively, and the average breadths were 2.66 ± 0.64 mm, 2.70 ± 0.50 mm and 2.59 ± 0.50 mm, respectively. The average durations of prepupae reared on cardamom, turmeric and ginger were 4.25 ± 0.71, 4.20 ± 0.72 and 3.15 ± 0.80 days, respectively (Tables 10.1, 10.3 and 10.5).

**Pupa** Pupation occurred in the tunnelled shoots. It was elongate and oval in shape with prominent eyes and antennal case. It was light brown when freshly formed but changed to dark brown within a few hours and became much darker prior to the emergence of the moth. The abdomen was distinctly marked into ten segments with a sharp dark brown spine on the terminal segment. The covering of the wing was similarly prominent and was lighter than the rest of the body. Six of these spiracles were visible on either side. Further, the female pupa was longer than male pupa. Pupa was obtect, with the anterior end broad, round and tapering posteriorly into a pointed tip (Fig. 10.3).

The sexes of *Conogethes* sp. could be easily differentiated at the pupal stage based on the position of the genital opening. In female, the genital opening was found on the eighth abdominal segment which was like a slit, and it was away from the anal slit. Whereas in males the genital opening with two raised pads was found on the ninth abdominal segment which was smaller and closer to the anal slit.

The average lengths of male pupa when reared on cardamom, turmeric and ginger were 15.45 ± 1.15 mm, 15.60 ± 0.61 mm and 11.60 ± 0.30 mm, respectively, and the average breadths were 3.06 ± 0.15 mm, 3.10 ± 0.30 mm and 2.74 ± 0.13 mm, respectively. The average lengths of female pupa reared on cardamom, turmeric and ginger were 17.80 ± 1.30 mm, 17.60 ± 1.07 mm and 13.60 ± 1.07 mm, respectively, and the average breadths were 3.60 ± 0.36 mm, 3.39 ± 0.29 mm and 2.98 ± 0.25 mm, respectively. The average durations of male pupa reared on cardamom, turmeric and ginger were 7.90 ± 0.72, 9.50 ± 0.51 and

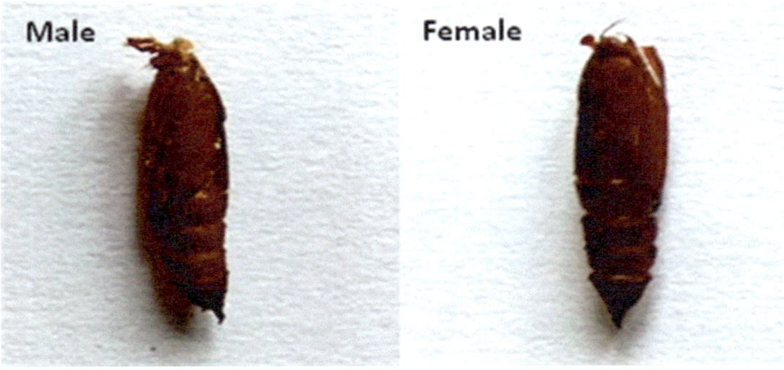


**Fig. 10.3** Male and female pupa of shoot and capsule borer, *C. sahyadriensis*

6.90 ± 0.65 days, respectively. The average durations of female pupa reared on cardamom, turmeric and ginger were 8.10 ± 0.30, 9.90 ± 0.55 and 7.20 ± 0.40 days, respectively (Tables 10.1, 10.3 and 10.5).

**Adult** Moths emerged during the night. They were medium sized and brownish yellow when reared on cardamom and light yellowish when reared on turmeric and ginger with a number of dark spots on its wings. The head, thorax and abdomen were distinct, and the antennae and legs were brownish yellow. Two long, segmented, setaceous antennae were located dorsally on the head close to the compound eyes. The lower edges of the wings were surrounded by hair-like structures, and it had golden margins with pale veins. The adults are generally active during night hours (Plate 4).

**Male** Male was narrower than the female with a brownish yellow body. The forewings were brownish yellow with on an average ($n$ = 50) 30 black dots, and the hindwings were also brownish yellow with on an average 15 ($n$ = 20) black dots. The abdomen was sharply tapered compared to the female.

The average durations of male moth reared on cardamom, turmeric and ginger were 6.90 ± 0.68, 7.40 ± 0.70 and 7.80 ± 0.77 days, respectively. The average lengths of male moth reared on cardamom, turmeric and ginger were 12.40 ± 0.26 mm, 12.00 ± 0.79 mm and 12.10 ± 0.79 mm, respectively, and breadths were 25.54 ± 1.50 mm, 25.30 ± 1.20 mm and 23.30 ± 1.80 mm, respectively (Tables 10.1, 10.3 and 10.5).

**Female** The female was similar to male in all aspects except body size, i.e. it was bigger with a stout abdomen which was relatively pointed in shape. The average durations of female moth reared on cardamom, turmeric and ginger were 7.95 ± 0.77, 8.10 ± 0.56 and 8.60 ± 0.68 days, respectively. The average lengths of female moth reared on cardamom, turmeric and ginger were 14.50 ± 1.26 mm, 14.80 ± 1.48 mm and 13.15 ± 0.50 mm, respectively, and breadths were 27.60 ± 1.62 mm, 27.40 ± 1.55 mm and 25.40 ± 1.45 mm, respectively (Tables 10.1, 10.3 and 10.5).

**Total Life Span** The total development period of male and female moth was maximum (42.45 ± 5.38 days and 44.05 ± 5.24 days, respectively) when reared on turmeric compared to those reared on cardamom and ginger (37.80 ± 1.91, 39.25 ± 1.75 days and 40.10 ± 1.37, 42.50 ± 3.50 days, respectively).

The trend of incubation and larval and pupal periods was consistent across three crops (cardamom, turmeric and ginger). The observation recorded showed that the shoot borer completed its life cycle earlier on cardamom than on turmeric and ginger. These findings are in close accordance with study of Bilapate and Talati (1978) and Stanley et al. (2009).

**Mating** Moths started copulating nearly 1–3 days after emergence (1.25–2.00; 1.00–1.50 and 2.40–2.95 days in moths reared on cardamom, turmeric and ginger, respectively). Mating occurred during night hours, and the duration lasted from 0.10–0.30 days in moths reared across the crops (Tables 10.2, 10.4 and 10.6). The male was very active during mating, and it mounted several times on the female. The male gradually moved by walking close to the female and touched it using the antennae. It quickly mounted the female, and soon there was a downward movement of the antennae. The male uncoiled its proboscis during mating which returned to its original form as soon as mating was completed.

The average durations of pre-mating of moths reared on cardamom, turmeric and ginger were 1.30 ± 0.40, 1.15 ± 0.24 and 2.65 ± 0.19 days, respectively, while that of mating were 0.13 ± 0.03, 0.11 ± 0.01 and 0.20 ± 0.01 days, respectively (Tables 10.2, 10.4 and 10.6).

**Oviposition** The oviposition site was first located by the female, and after identifying, the moth cleaned the leaf surface using the tip of the abdomen which was immediately followed by oviposition.

The average pre-oviposition periods of adult females reared on cardamom, turmeric and ginger were 3.12 ± 0.20, 2.20 ± 0.20 and 1.41 ± 0.14 days, respectively, and the oviposition periods were 2.25 ± 0.30, 2.57 ± 0.36 and 2.76 ± 0.30 days, while the post-oviposition periods were 2.50 ± 0.10, 1.89 ± 0.18 and 2.15 ± 0.18 days, respectively (Tables 10.2, 10.4 and 10.6).

The trend of incubation and larval and pupal periods was consistent across three crops (cardamom, turmeric and ginger). The observation recorded showed that the shoot borer completed its life cycle earlier on cardamom than on turmeric and ginger (Fig. 10.4).

**Fig. 10.4** Male and female adult of *C. sahyadriensis* reared on select hosts. (**a**) Male and female adult of *C. sahyadriensis* reared on cardamom. (**b**) Male and female adult of *C. sahyadriensis* reared on turmeric. (**c**) Male and female adult of *C. sahyadriensis* reared on ginger

**Acknowledgement** Authors thank the authorities of the University Agricultural and Horticultural Sciences (UAHS) and Shimoga for the encouragement and facilities. All photographs in this chapter are of the senior author.

## References

Bilapate GG, Talati GM (1978) Some studies on bionomics of castor shoot and capsule borer Dichocrocis *punctiferalis* (Guenee). J Maharashtra Agric Univ 3:47–49

Chomchalow (2001) Small cardamom grown only in India and Sri Lanka this exotic spice contains many plants derived chemical compounds that are known to have been anti-oxidant, disease preventing and health promoting properties. J Plant Crop 47:234–262

Jacob SA (1981) Biology of *Dichocrocis punctiferalis* (Guenee.) on turmeric. J Plant Crop 9:119–123

Kadoi M, Kaneda M (1990) Development of yellow peach moth, *Conogethes punctiferalis* (Guenee) on apple fruit. Res Bull Plant Prot Serv Japan 26:61–63

Senthil Y, Rajput SG, Pawar K (2015) Incidence of pests of turmeric in western Maharashtra. Indian Cocoa Arecanut and Spices J 12:17–19

Shashank PR, Kammar V, Mally R, Chakravarthy AK (2018) A new Indian species of shoot and capsule 0062orer of the genus *Conogethes* (Lepidoptera: Crambidae), feeding on cardamom. Zootaxa 4374(2):215–234

Stanley J, Chandrasekaran S, Preetha G (2009) *Conogethes punctiferalis* Guenee (Lepidoptera: Pyralidae) its biology and field parasitisation. Indian J Agric Sci 79:906–909

Twine PH (1971) Insect pests of maize. Queensland Agric J 97:523–531

Varadarasan (2001) Studies on the shoot and capsule borer, *Dichocrocis punctiferalis* Guenee. Acta Ocean Entmol Sin 24:11–13

Wang YX, Cai RT (1997) The emergence and control of *Dichocrocis punctiferalis* Guenee for younai plum variety. South China Fruits 26:45

Xi-ke JX, Zhou ZF, Xu RS, Ni DN, Wang H (1996) Preliminary experiment report on pest control by fenitrothion of *Castanea mollissima* nut. J Zhejiang Sci Tech 22:28–30

# Status of Shoot and Fruit Borer, *Conogethes punctiferalis* Guenee (Lepidoptera: Crambidae), in Central India

11

K. R. Yatish, V. J. Tambe, J. C. Ankush, and P. N. Guru

**Abstract**

Central India is of strategic importance for the *Conogethes* moths as the tropical and subtropical climes the region provides facilitates dispersal and expansion of its host range, thereby enabling the pest to feed and proliferate. Biological studies of *C. punctiferalis* on castor have revealed that the average life cycle of the borer lasts for 22–25 days. Life table analysis revealed that apart from some unknown factors, parasitization and nonavailability of alternate hosts, especially during off-season, contribute to the mortality of larvae and pupae. Off-season survival on alternate hosts plays a key role in the population and continuation of borer infestation in the subsequent seasons.

**Keywords**

Bioecology · Central India · *Conogethes* · Life table

K. R. Yatish
Division of Entomology and Nematology, ICAR-Indian Institute of Horticultural Research, Bengaluru, Karnataka, India

V. J. Tambe
Department of Agricultural Entomology, Dr. Panjabrao Deshmukh Krishi Vidyapeeth (PDKV), Akola, Maharashtra, India

J. C. Ankush (✉) · P. N. Guru (✉)
Department of Entomology, Mahatma Phule Krishi Vidyapeeth (MPKV), Ahmednagar, Maharashtra, India





A. K. Chakravarthy (ed.), *The Black spotted, Yellow Borer, Conogethes punctiferalis Guenée and Allied Species*,
https://doi.org/10.1007/978-981-13-0390-6_11

## 11.1 Introduction

Central India is considered to comprise of three states, namely, Maharashtra (21° 08′ N latitude and 79° 04′ E longitude with an altitude of 310 m MSL), Madhya Pradesh (22.9734° N, 78.6569° E with an 567 m MSL) and Chhattisgarh (21.2787° N, 81.8661° E with an 728 m above MSL). But for the sake of this chapter parts of Maharashtra stretching towards western India as well are included. This chapter assumes significance as it involves castor, *Ricinus communis* L., which is one of the principal host plants of the shoot and fruit borer, *C. punctiferalis*, and originates from the tropical belt of India and Africa. According to a theory, the pest, *C. punctiferalis*, originated in Central and Northeast India; thus, it can survive and perpetuate on almost all classes of crops that are cultivated in this region, such as mango, cashew nut, jackfruit, banana, coconut, pineapple, clove, nutmeg, black pepper, niger, potato, and field crops like maize.

Maharashtra occupies parts of the Western and Central India with a long coastline of nearly 720 km along the Arabian Sea, and a major portion of the state is semiarid. Similarly in Madhya Pradesh, almost all types of crops are cultivated, and agriculture is the mainstay of the people. Further, Chhattisgarh, also known as the 'herbal state', has three agroclimatic zones wherein distinct crops like cereals, pulses, fruits and vegetables and plantation crops are cultivated.

Bilapate and Talati (1977) recorded that the borer caused about 16–72% damage to castor capsules in Maharashtra, and the peak of its activity and infestation was recorded during the period between September and November. A temperature of 21 °C to 33 °C, 80–90% relative humidity and 7–9 sunlight hours per day were found favourable for the population build-up of the borer species. On castor crop, the borer population was favoured by wet, humid and cloudy weather conditions, and sowing the seeds during the first fortnight of July was found to be ideal for reducing the infestation in Maharashtra (Akashe et al. 2015).

The bioecology of the pest has not been extensively examined with regard to central India, and its biology was studied under laboratory conditions from 1971 to 1973. Gour and Sriramulu (1992) who examined the biology of the pest under lab conditions at Parbhani, Maharashtra, found that on castor, the pest population increased at a daily rate of 1.09 per female. Further, the population was able to multiply by 20.90 times by the end of a generation, which was completed on an average of 34–46 days (Yathish 2012).

In the present study, the bioecology of *C. punctiferalis* was studied in detail in field and laboratory conditions. A summary of the study findings is presented below, with inputs from other parts of Maharashtra and Central India. Further, for details on the pest, the readers are directed to the other chapters in this book (Figs. 11.1, 11.2, 11.3, 11.4 and 11.5).

**Methods Adopted** The research work on the life table studies of *Conogethes punctiferalis* was conducted during 2011–2012 under field and laboratory conditions on castor crop at the insectary of the Entomology Section, College of Agriculture, Nagpur, at 21° 08′ N latitude and 79° 04′ E longitude with an altitude of 310 m above mean sea level (AMSL).

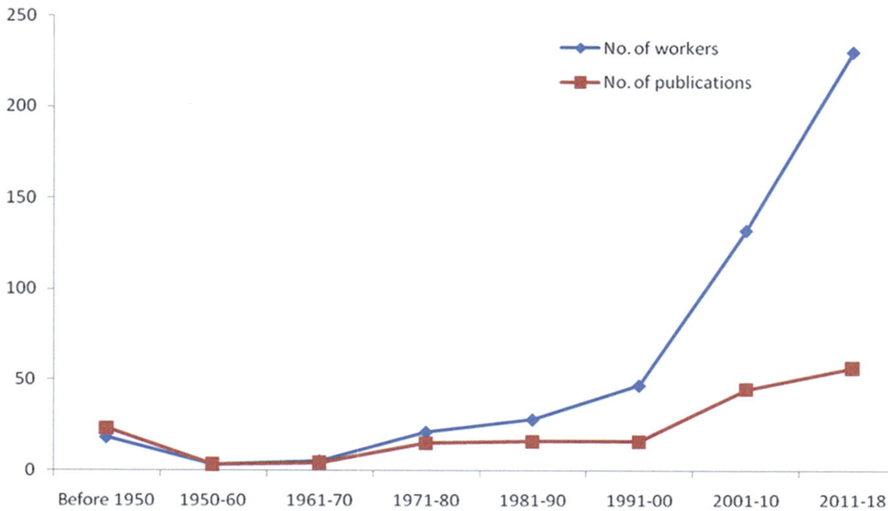


**Fig. 11.1** The number of researchers and publications on *Conogethes* is increasing. From the twenty-first century in view of the scientific advances, teams of workers are addressing issues on this important group of moths instead of individuals in the earlier decades. This figure has been projected from a random sample of 200 papers on *Conogethes* spp.

**Fig. 11.2** Local castor cultivated under coconut palms generally suffer less from insect pest attack in South India than sole castor crop

**Fig. 11.3** Local castor cultivated as a perennial crop exhibit 'recovery resistance' against semi-looper and borer damage in South India

**Fig. 11.4** Castor crop forms an important component of multiple cropping systems in South India and serves as a reservoir of beneficials

**Fig. 11.5** Local cultivar of castor is less preferred by the pests as it contains higher levels of phenols, flavonoids, anthocyanin pigments and other antimetabolites than certain cultivars and hybrids of castor in South India

The detailed studies on the biology of the pest were conducted in the laboratory during the post-monsoon and winter seasons of 2011 and 2012. Daily observations were made to determine the incubation period of egg. The data on the pest life history, such as the number of larval instars, based on moulting, duration of each larval instar, pupal period and the total developmental period, were collected. Further, the morphological characteristics, such as colour, shape and habits of larva were recorded. The sampling of *C. punctiferalis* was carried out from the untreated castor field from the beginning of pest incidence up to the seventh day, and the samples were collected in the morning hours.

**Eggs** Eggs were collected from untreated castor crops and were reared separately in glass tubes until they hatched.

**Larvae** The larvae belonging to younger age groups, i.e. I, II and III instars, and older age groups, i.e. IV and V instars, were collected separately from the castor crop and were reared in plastic trays until pupation.

**Pupae** The pupae were collected from the infected (webbed) castor capsules and were placed individually in plastic vials until the adults emerged and the sex ratio was determined based on the pupal morphology.

**Adults** The adults that emerged from the larvae and pupae were collected from the field. The known adult pairs were confined in oviposition chamber to record the female fecundity and adult longevity. Sponge pads soaked with 10% sugar solution were given as adult food. Castor inflorescence (panicles) with flowers and immature capsules were kept as ovipositional substrate with the cut end dipped in water soaked cotton in a small plastic container.

## 11.2 Laboratory Studies

### 11.2.1 Life Stages of *Conogethes*

1. Eggs were reared in laboratory until adult emergence and oviposition by providing larval food of castor.
2. Observations were recorded during each stage and each larval instar for mortality, survival and mortality factors by rearing few larvae separately in plastic trays.
3. Observations were recorded for pupal mortality, adult emergence, pupal deformities, sex ratio, etc.
4. A known adult pair was released in the mating chamber for mating and oviposition to record fecundity and adult longevity.
5. About 100 eggs of *C. punctiferalis* were taken for life table study.
6. Duration of different life stages was recorded for the eggs collected from the field.
7. Generation survival rate was worked out and life table was constructed for *C. punctiferalis*.

### 11.2.2 Construction of Life Table

Data regarding various stages and mortality at each stage was tabulated for the collected population under the following headings (Harcourt 1969; Atwal and Bains 1974).

| | |
|---|---|
| x | = Age interval |
| lx | = The number of individuals alive at the beginning of age interval, x |
| dx | = The number of individuals dying during age interval, x |
| d x f | = Mortality factors responsible for dx |
| 100 qx | = Percentage mortality during x |
| Sx | = Survival rate within x |

### 11.2.3 Criteria for Filling the Columns of Life Tables

The criteria proposed by Harcourt (1969) were utilized in the present case for filling data in the life table for each age interval (stage).

(a) Eggs: The lx for eggs was obtained by totalling the number of eggs recorded at various sampling dates during a season starting from the appearance of the insect on the corp.
(b) Larvae: The lx was obtained by direct sampling during pest infestation.
(c) Pupae: The lx was determined after subtracting the mortality due to various reasons from the population of (lx) the prepupal larvae.
(d) Adults: This lx represents the number of pupae emerging as adults. The monthly value for this was obtained by subtracting the mortality in the pupal stage due to various factors from lx of pupae.

**Trend Index (I)** This is simply 'lx' for eggs in the preceding months, expressed as:

$$\textbf{Trend index} = \textbf{N2 / N1}$$

Where,

N2 = No. of eggs during the next season.
N1 = No. of eggs during the same season.

**Generation Survival (SG)** This is an index of population trend without the effect of fecundity and adult mortality.

$$\textbf{Generation survival}(\textbf{SG}) = \textbf{N3 / N1}$$

Where,

N3 = Population of adults in a generation.
N1 = Population of eggs in the same generation.

## 11.3 Analysis and Identification of the Cause and Key Factors of Mortality

The most important step in explaining the population is determining that stage in the life of the pest, which contributes the most to the index of population trend (I) and generation survival (SG). Separate budget for each season was prepared to find out the key mortality factors that have influenced the population trend (I) in different months. The key factor analysis method developed by Varley and Gradwell (1960) was used to detect density relationship of mortality factors. Using this method, the killing power (K) of such mortality factors in each age group was estimated as the difference between the logarithm of population density before and after its action. As a series of mortality factor operating in succession on a population, the total killing power 'k' is equal to the sum of the killing power or 'k's.

| | | |
|---|---|---|
| If | k1 | =log lx of egg stage – lx of larval stage |
| | k2 | =log lx of larval stage (younger age group)–log lx of larval stage (older age group) |
| | k3 | =log lx of pupal stage – log lx of adult stage then K = k1 + k2 + k3 |

**Method of Observation** Observations were recorded for two seasons across two generations along with the approximate dates of egg laying, hatching, pupation and moth emergence for each generation. The duration of egg, larval and pupal periods was recorded, and total life cycle for each generation was determined. The extent of egg, larval and pupal mortality and parasitism at each stage was recorded. Meteorological data was obtained from meteorological observation unit, College of Agriculture, Nagpur, which was correlated with the natural mortality of *C. punctiferalis*.

A survey was undertaken at Junagadh, Gujarat, for detecting the *Conogethes*-infested fruit orchards for 4 days each during the period between May and June 2017. Although Gujarat is not a part of central India, the observations recorded are included in this chapter.

## 11.4 Results and Discussion

**Egg** The eggs that were singly deposited on capsules were very minute to be observed with naked eyes; however, they appeared pale yellowish white when viewed with the help of magnifying lens. On account of the difficulty in ascertaining their size, morphometric measurements were not made.

Bilapate and Talati (1978a) observed the morphometric measurements of the egg and recorded that they measured 0.59 mm and 0.39 mm in length and width, respectively. Incubation period varied from 2.66 ± 0.17 s/day which is in close agreement with the observations of Patel and Gangrade (1971), Stanley et al. (2009) and Kumar et al. (2016) who recorded an incubation period of 2–4 days.

**Larva** *C. punctiferalis* larvae moulted four times and thus there were five larval instars. Newly hatched first instar larvae were highly active and minute with dark head and prothorax and light pinkish brown body with pale black spots all over.

The length of the third instar larvae was between 5.3 mm and 5.7 mm with an average of 5.5.mm and width ranged from 0.7 mm to 0.9 mm with an average of 0.78 mm. The average duration of the third larval instar was 2.50 days (Table 11.1). The last two instars of the pest were similar to earlier instars in colour, and morphological characters expect the size. The larvae of these two instars were light brown with dark brown head and very dark spots on the body, and they hung on with a fine silken thread when disturbed.

**Table 11.1** Morphometric data on different life stages of *C. punctiferalis*

| | | Length (mm) | | Width(mm) | Duration(days) | | |
|---|---|---|---|---|---|---|---|
| Insect stages | | Range | Mean | Range | Mean | Range | Mean |
| Egg | | – | – | – | – | 3.9–59 | 4.7 |
| I instar | | – | – | – | – | 1.7–2.2 | 2.1 |
| II instar | | – | – | – | – | 2.6–3.0 | 2.8 |
| III instar | | 5.3–5.7 | 5.50 | 0.7–0.9 | 0.78 | 2.1–2.6 | 2.5 |
| IV instar | | 11.4–12.0 | 11.80 | 1.4–1.7 | 1.54 | 2.8–3.2 | 2.9 |
| V instar | | 15.6–15.9 | 15.70 | 2.3–2.8 | 2.52 | 2.3–28 | 2.5 |
| Prepupa | | 13.2–13.8 | 13.40 | 2.4–2.7 | 2.50 | 2.2–2.8 | 2 |
| Pupa | | 11.2–11.8 | 11.48 | 2.7–2.9 | 2.78 | 9.5–12.0 | 11.6 |
| Adult | Male | 10.2–10.8 | 10.54 | 21.5–22.7 | 22.04 | 8.5–10.0 | 9.4 |
| | Female | 11.0–11.8 | 11.10 | 20.1–20.9 | 2.56 | 9.0–11.0 | 10.5 |

Source: Yathish (2012)

The fourth instar larvae were11.4 mm to 12.0 mm long, with an average length of 11.8 mm and 1.4 mm to 1.7 mm wide, with an average width of 1.54 mm. The duration of fourth larval instar was 2.9 days (Table 11.1).

In the laboratory, the total larval period ranged from 12.50 to 13.40 days with an average of 12.78 days for both seasons. While total larval period on castor was reported as 12.73 days by Bilapate (1978), Gour and Sriramulu reported the same to be 17 days and Patel and Gangrade (1971) found it to last for 20–23 days. This variation could be due to the effect of host plant and the locality.

**Prepupa and Pupa** The colour of the prepupa was light greenish with distinct dark spots over. The prepupal length varied from 13.2 mm to 13.8 mm with an average of 13.4 mm, and width ranged from 2.4 mm to 2.7 mm with an average of 2.5 mm, and the prepupal period lasted for 2.25–2.88 days.

The freshly formed pupae were brownish yellow with dark compound eyes and they later turned light brown. Pupal length varied between 11.2 cm and 11.8 mm with an average of 11.48 mm, and width varied from 2.7 mm to 2.9 mm with an average of 2.78 mm, and the pupal period lasted for 9.50–12 days (Table 11.1). These findings are comparable with Bilapate and Talati (1978a) who reported the pupal duration to be 7–9 days and Gour and Sriramulu (1992) who reported a pupal period of 9–11 days. Thyagaraj (2003) reported differences in the size, shape and weight of the male and female pupa, and female pupae were bigger (17.81–6.29 mm) than male pupae (14.30–4.26 mm). Further, the biological studies of *C. punctiferalis* conducted by Stanley et al. (2009) on castor revealed that the number of days taken by a neonate larva to become adult was 27.76 d.

**Adult** The adults were medium sized, with a brownish yellow body and straw yellow wings with a number of dark spots. Generally, the female moths were bigger in size with a bulged abdomen, and male moths were smaller with no tufts of hairs on the tip of the abdomen.

**Life Table** Life tables were constructed during two different seasons in order to fill in the lacunae of harnessing the potentialities of natural enemies. So far, no work has been conducted on the life mortality tables of *C. punctiferalis* on castor from the Nagpur region. The present study throws light on the biology of the pest under laboratory and field conditions, the number of generations completed by the pest in each cropping season, the total duration of each generation and the population density of various stages of the pest in different seasons; it also undertakes construction and interpretation of life mortality tables through survivorship and death rate curves and finally determines the key mortality factors resulting in population fluctuation, and the same have been presented and discussed below.

**Egg Count, Survival and Mortality** The eggs and larval instars were collected from the beginning of pest incidence in the field. As the moth preferred to lay eggs on inflorescence (panicle) with flowers, such inflorescence were collected from the field and observed for the presence of eggs and the number of such eggs, but as the eggs were very minute, it was difficult to make them out on the inflorescence. However, the egg population was ascertained indirectly from the larval counts.

The data of the field sampling of eggs from two seasons have been presented in Table 11.2, and it was observed that the incidence of *C. punctiferalis* begins on castor from 42nd M) (15–21 October) and goes on up to 4th MW (22–28 January 2012). The peak egg count was observed in 52nd M and 1st MW during winter. Two peaks of egg laying were observed, i.e. from the 44th MW to 46th MW and from 51st MW for post-monsoon and winter seasons, respectively. The highest egg mortality rate of 50% was observed during the 48th MW which was due to unclear factors, such as sterility or infertility of the eggs (Table 11.3).

## 11.5 Larval Count, Survival and Mortality of Larvae Belonging to Older Age Group

The larvae belonging to the older age group (IV and V instars) counted the highest, i.e. 12 larvae per 10 plants during 2nd MW which was followed by 10 larvae per 10 plants during 52nd MW, and this was responsible for the peak incidence in the field. Another important finding was the very low to non-existent mortality in several MW ultimately increasing the survival rate of the larvae of this age group. However, the highest mortality rate of 66.67% was observed during 4th MW which might be due to an increase in temperature. Further, as the mortality in the larvae age group could not be traced, the factors responsible for mortality were considered unknown. Thus, the average percent mortality and survival rate observed during the study were 17.56% and 0.88, respectively (Table 11.4). Data on survival, mortality and sex ratio of *C. punctiferalis* are presented in Tables 11.5 and 11.6. Sambathkumar et al. (2017) reported that the sex ratio of *C. punctiferalis* was 1.0:1.92 on castor in Tamil Nadu, South India.

**Table 11.2** Egg count, survival and mortality

| Egg stage X | | | No. of eggs | No. dying during X | Factors responsible for dx | dx as % of lx | Survival rate at age X |
|---|---|---|---|---|---|---|---|
| MW | | | Lx | Dx | dxF | 100 qx | Sx |
| 08–14 | Oct 11 | 41 | – | – | – | – | – |
| 15–21 | | 42 | 3 | 1 | Unknown factor | 33.33 | 0.67 |
| 22–28 | | 43 | 6 | 1 | Unknown factor | 16.66 | 0.83 |
| 29–04 | Nov 11 | 44 | 10 | 2 | Unknown factor | 20.00 | 0.80 |
| 05–11 | | 45 | 10 | 4 | Unknown factor | 40.00 | 0.60 |
| 12–18 | | 46 | 11 | 2 | Unknown factor | 18.18 | 0.82 |
| 19–25 | | 47 | 9 | 1 | Unknown factor | 11.11 | 0.89 |
| 26–02 | Dec 11 | 48 | 6 | 3 | Unknown factor | 50.00 | 0.50 |
| 03–09 | | 49 | 9 | 3 | Unknown factor | 33.33 | 0.67 |
| 10–16 | | 50 | 11 | 2 | Unknown factor | 18.18 | 0.82 |
| 17–23 | | 51 | 13 | 1 | Unknown factor | 7.69 | 0.92 |
| 24–31 | | 52 | 20 | 3 | Unknown factor | 15.00 | 0.85 |
| 01–07 | Jan 12 | 1 | 20 | 2 | Unknown factor | 10.00 | 0.90 |
| 08–14 | | 2 | 10 | 1 | Unknown factor | 10.00 | 0.90 |
| 15–21 | | 3 | 7 | 3 | Unknown factor | 42.85 | 0.57 |
| 22–28 | | 4 | 4 | 2 | Unknown factor | 50.00 | 0.50 |
| 29–04 | Feb 12 | 5 | – | – | | | |
| Subtotal | | | 149 | 31 | | 22.14 | 0.66 |

Source: Yathish (2012)

**Table 11.3** Larval (I–III instar) count, survival and mortality

| Egg stage X | | | No. of eggs | No. dying during X | Factors responsible for dx | dx as % of lx | Survival rate at age X |
|---|---|---|---|---|---|---|---|
| MW | | | Lx | dx | dxF | 100 qx | Sx |
| 08–14 | Oct 11 | 41 | – | – | – | – | – |
| 15–21 | | 42 | 3 | 1 | Unknown factor | 33.33 | 0.67 |
| 22–28 | | 43 | 9 | 1 | Unknown factor | 11.11 | 0.89 |
| 29–04 | Nov 11 | 44 | 8 | 2 | Unknown factor | 25.00 | 0.75 |
| 05–11 | | 45 | 10 | 2 | Unknown factor | 20.00 | 0.80 |
| 12–18 | | 46 | 6 | 1 | Unknown factor | 16.67 | 0.83 |
| 19–25 | | 47 | 7 | 1 | Unknown factor | 14.29 | 0.86 |
| 26–02 | Dec 11 | 48 | 7 | 3 | Unknown factor | 42.86 | 0.57 |
| 03–09 | | 49 | 9 | 0 | Unknown factor | 0.00 | 1.00 |
| 10–16 | | 50 | 7 | 2 | Unknown factor | 28.57 | 0.71 |
| 17–23 | | 51 | 10 | 1 | Unknown factor | 10.00 | 0.90 |
| 24–31 | | 52 | 9 | 3 | Unknown factor | 33.33 | 0.67 |
| 01–07 | Jan 12 | 1 | 12 | 2 | Unknown factor | 16.67 | 0.83 |
| 08–14 | | 2 | 10 | 1 | Unknown factor | 10.00 | 0.90 |
| 15–21 | | 3 | 3 | 0 | – | 0.00 | 1.00 |
| 22–28 | | 4 | 1 | 0 | – | 0.00 | 1.00 |
| 29–04 | Feb 12 | 5 | – | – | – | – | – |
| Subtotal | | | 111 | 20 | | 15.40 | 0.74 |

Source: Yathish (2012)

**Table 11.4** Larval (IV–V instar) count, survival and mortality

| Egg stage X | | | No. of eggs | No. dying during X | Factors responsible for dx | dx as % of lx | Survival rate at age X |
|---|---|---|---|---|---|---|---|
| MW | | | Lx | dx | dxF | 100 qx | Sx |
| 08–14 | Oct 11 | 41 | – | – | – | – | – |
| 15–21 | | 42 | 1 | 0 | Unknown factor | 0.00 | 1.00 |
| 22–28 | | 43 | 3 | 1 | Unknown factor | 33.33 | 0.67 |
| 29–04 | Nov 11 | 44 | 2 | 0 | Unknown factor | 0.00 | 1.00 |
| 05–11 | | 45 | 6 | 2 | Unknown factor | 33.33 | 0.67 |
| 12–18 | | 46 | 3 | 1 | Unknown factor | 33.33 | 0.67 |
| 19–25 | | 47 | 3 | 0 | Unknown factor | 0.00 | 1.00 |
| 26–02 | Dec 11 | 48 | 8 | 2 | Unknown factor | 25.00 | 0.75 |
| 03–09 | | 49 | 9 | 1 | Unknown factor | 11.11 | 0.89 |
| 10–16 | | 50 | 6 | 1 | Unknown factor | 16.67 | 0.83 |
| 17–23 | | 51 | 7 | 0 | Unknown factor | 0.00 | 1.00 |
| 24–31 | | 52 | 10 | 3 | Unknown factor | 30.00 | 0.70 |
| 01–07 | Jan 12 | 1 | 8 | 2 | Unknown factor | 25.00 | 0.75 |
| 08–14 | | 2 | 12 | 2 | Unknown factor | 16.67 | 0.83 |
| 15–21 | | 3 | 4 | 1 | Unknown factor | 25.00 | 0.75 |
| 22–28 | | 4 | 3 | 2 | Unknown factor | 66.67 | 0.33 |
| 29–04 | Feb 12 | 5 | – | – | – | – | – |
| Subtotal | | | 85 | 18 | | 17.56 | 0.88 |

Source: Yathish (2012)

**Table 11.5** Matured larvae and pupal count, survival and mortality

| Egg stage X | | | No. of eggs | No. of dying during X | Factors responsible for dx | dx as % of lx | Survival rate at age X |
|---|---|---|---|---|---|---|---|
| MW | | | Lx | Dx | dxF | 100 qx | Sx |
| 08–14 | Oct 11 | 41 | – | – | – | – | – |
| 15–21 | | 42 | – | – | – | – | – |
| 22–28 | | 43 | – | – | – | – | – |
| 29–04 | Nov 11 | 44 | 2 | 0 | | 0.00 | 1.00 |
| 05–11 | | 45 | 2 | 2 | | 100.00 | 0.00 |
| 12–18 | | 46 | 3 | 1 | | 33.33 | 0.67 |
| 19–25 | | 47 | 3 | 0 | | 0.00 | 1.00 |
| 26–02 | Dec 11 | 48 | 4 | 2 | | 50.00 | 0.50 |
| 03–09 | | 49 | 8 | 1 | | 12.50 | 0.88 |
| 10–16 | | 50 | 10 | 1 | | 10.00 | 0.90 |
| 17–23 | | 51 | 9 | 0 | | 0.00 | 1.00 |
| 24–31 | | 52 | 11 | 3 | | 27.27 | 0.73 |
| 01–07 | Jan 12 | 1 | 9 | 0 | | 0.00 | 1.00 |
| 08–14 | | 2 | 8 | 2 | | 25.00 | 0.75 |
| 15–21 | | 3 | 3 | 1 | | 33.33 | 0.67 |
| 22–28 | | 4 | 1 | 0 | | 0.00 | 1.00 |
| 29–04 | Feb 12 | 5 | – | – | – | – | – |
| Subtotal | | | 80 | 16 | | 21.28 | 0.68 |

Source: Yathish (2012)

**Table 11.6** Male and female sex ratios

| MW | | | No. of moths emerged from pupa lx | No. of male: female | Simplified ratio | % Male | % Female |
|---|---|---|---|---|---|---|---|
| 29–04 | Nov 11 | 44 | – | – | – | – | |
| 05–11 | | 45 | – | – | – | – | |
| 12–18 | | 46 | 2 | 0:2 | 0:2 | 0.00 | 100.00 |
| 19–25 | | 47 | 4 | 2:2 | 1:1 | 50.00 | 50.00 |
| 26–02 | Dec 11 | 48 | 7 | 3:4 | 1:1.33 | 42.86 | 57.14 |
| 03–09 | | 49 | 9 | 3:6 | 1:2 | 33.34 | 66.66 |
| 10–16 | | 50 | 13 | 6:7 | 1:1.16 | 46.16 | 53.84 |
| 17–23 | | 51 | 11 | 5:6 | 1:1.2 | 45.46 | 54.54 |
| 24–31 | | 52 | 10 | 4:6 | 1:1.5 | 40.00 | 60.00 |
| 01–07 | Jan 12 | 1 | 12 | 5:7 | 1:1.4 | 41.67 | 58.33 |
| 08–14 | | 2 | 7 | 4:3 | 1:0.75 | 57.15 | 42.85 |
| 15–21 | | 3 | 4 | 2:2 | 1:1 | 50.00 | 50.00 |
| 22–28 | | 4 | – | – | – | – | – |
| Subtotal | | | | 34:44 | | 40.66 | 59.34 |

Source: Yathish (2012)

Seven varieties of castor were tested by Singhvi et al. (1972) at a field in Hissar, India, for susceptibility to attack by the *C. punctiferalis* larvae, and it was observed that on an average, the pest population ranged from 4 to 14.25 larvae per 5 spikes in all the 7 varieties. In study conducted by He-Rong Qiang (1997), it was observed that the shoots, flower clusters, young leaves and fruit of *Fortunella* trees are attacked by *C. punctiferalis* larvae throughout the year (except during winter and early spring) in Renshan, Sichuan, China. Finally, at Rahuri, Maharashtra, *Conogethes* larvae were found mainly infesting castor, pomegranate and guava, but the infestation was <5%.

## 11.6 Matured Larvae and Pupae Population Count, Survival and Mortality

The matured larval and pupal peaks were observed during the 52nd and 50th MW, owing to climatic factors with a peak count of 11 matured larvae and pupae per 10 plants on 52nd MW. After the 1st MW, the incidence of larval instars belonging to younger age group started declining. Survival rate of younger age group larvae was maximum (1.00) during 44th, 47th, 51st, 1st and 4th MW. Further, the data on adult longevity, life table analysis and budgeting are presented in Tables 11.7, 11.8, 11.9, 11.10, 11.11 and 11.12.

Even here the reason for most of the recorded mortality could not be ascertained, and the factors responsible for mortality were considered unknown. However, many of the pupae failed to emerge which was due to the natural enemies, i.e. the parasitoids parasitizing the larva-pupal instars as implied by the exit holes made by them;

**Table 11.7** Adult longevity and female fecundity

| | Adult longevity | | | | | |
|---|---|---|---|---|---|---|
| | Male | | Female | | Fecundity | |
| Season | Range | Mean | Range | Mean | Range | Mean |
| Post-monsoon | 8.5–9.5 | 9.2 | 9.0–10.5 | 10.3 | 74–96 | 81 |
| Winter | 9.0–10.0 | 9.7 | 9.5–11.0 | 10.8 | 88–116 | 102.3 |

Source: Yathish (2012)

**Table 11.8** Life table of *C. punctiferalis* in the post-monsoon season during 2011–2012 under laboratory conditions

| Age interval | No. alive at beginning of | No. dying during x | Factors responsible for dx | dx as % of ix | Survival rate at age x |
|---|---|---|---|---|---|
| x | Ix | dx | dxF | 100qx | Sx |
| Egg | 100 | 5.32 | Unknown factor and sterility | 5.32 | 0.95 |
| Larvae (small) | 94.68 | 9.63 | Unknown factor | 10.17 | 0.90 |
| Larvae (big) | 85.05 | 2.03 | Unknown factor | 2.39 | 0.98 |
| Pupae | 83.02 | 36.41 | Fail to emerge | 43.86 | 0.56 |
| Adult | 46.61 | 1.8 | Unknown factor | 3.86 | 0.96 |
| Femalex2 | 46.61 | | | | |
| ($N_3$) | | | | | |
| Generation | | | | | |
| Survival | 0.4661 | | | | |
| ($N_3/N_1$) | | | | | |

Source: Yathish (2012)

**Table 11.9** Life table of *C. punctiferalis* in the winter season during 2011–2012 under laboratory conditions

| Age interval | No. alive at beginning of X | No. dying during x | Factors responsible for dx | dx as % of ix | Survival rate at age x |
|---|---|---|---|---|---|
| x | Ix | dx | dxF | 100qx | Sx |
| Egg | 100 | 6.25 | Unknown factor and sterility | 6.25 | 0.94 |
| Larvae (I, II, III) | 93.75 | 13.01 | Unknown factor | 13.88 | 0.86 |
| Larvae (IV, V) | 80.74 | 1.31 | Unknown factor | 1.62 | 0.98 |
| Pupae | 79.43 | 20.86 | Emerge | 26.26 | 0.74 |
| Adult | 58.57 | 2.65 | Unknown factor | 4.52 | 0.95 |
| Femalex2 ($N_3$) | 58.57 | | | | |
| Generation | | | | | |
| Survival | 0.5857 | | | | |
| ($N_3/N_1$) | | | | | |

Source: Yathish (2012)

**Table 11.10** Life table of *C. punctiferalis* on castor during post-monsoon season

| Age interval | No. alive at beginning of X | No. dying during x | Factors responsible for dx | dx as % of ix | Survival rate at age x |
|---|---|---|---|---|---|
| x | Ix | dx | dxF | 100qx | sx |
| Egg($N_1$) | 12,286 | 2059 | Unknown factor | 16.76 | 0.83 |
| Larvae (I, II, III) | 10,227 | 1907 | Unknown factor | 18.65 | 0.81 |
| Larvae (IV, V) | 8320 | 1795 | Unknown factor | 21.57 | 0.78 |
| Pupae | 6525 | | Failed toemerge | | |
| | | 2257 | | 34.59 | 0.65 |
| Adult moth | 4268 | | | | |
| Female x2 ($N_3$) | 4268 | | | | |
| Generation total | | 8018 | | 91.57 | |
| No. of actual eggs ($N_2$) | 21,358 | | | | |
| Trend index ($N_2/N_1$) | 1.738 | | | | |
| Generation survival ($N_3/N_1$) | 0.347 | | | | |

Source: Yathish (2012)

**Table 11.11** Life budgeting of *C. punctiferalis* for post-monsoon season

| Age interval | No./Ac | Log No. Ac | K's |
|---|---|---|---|
| Egg | 12,286 | 4.0894 | |
| | | | 0.0797 |
| Larvae (I, II, III) | 10,227 | 4.0097 | |
| | | | 0.0896 |
| Larvae (IV, V) | 8320 | 3.9201 | |
| | | | 0.1055 |
| Pupae | 6525 | 3.8146 | |
| | | | 0.1844 |
| Adult | 4268 | 3.6302 | |
| | | | 0.3010 |
| Total | | | 0.4592 |

Source: Yathish (2012)

however, the role of the parasitoids could not be conformed because the pests had already emerged in the field, and it could not be identified using the pupal case and was designated as an unobserved parasitoid. The average percent mortality during the study was 21.28% with survival rate of 0.68. Besides, as per the findings of Bilapate and Talati (1977) regarding the incidence of *C. punctiferalis* on castor, the larvae and pupae reached peak numbers in November, and during the off-season they survived on perennial castor plants. Several workers have made similar observations in other parts of central India. Further, Suganthy (2011) screened castor germplasm against the borer and found RG2770, RG2776 and RG2778 to be free from borer damage.

**Table 11.12** Life tables of *C. punctiferalis* on castor during winter season

| Age interval | No. alive at beginning of X | No. dying during x | Factors responsible for dx | dx as % of ix | Survival rate at age x |
|---|---|---|---|---|---|
| x | Ix | dx | dxF | 100qx | Sx |
| Egg($N_1$) | 21,358 | 1899 | Unknown factor | 8.89 | 0.91 |
| Larvae (I, II, III) | 19,459 | 1962 | Unknown factor | 10.08 | 0.90 |
| Larvae (IV, V) | 17,497 | 1795 | Unknown factor | 10.26 | 0.90 |
| Pupae | 15,702 | 996 | Unobserved parasitoid | 6.34 | 0.94 |
| Adult moth | 14,706 | | | | |
| Female x2 ($N_3$) | 14,706 | | | | |
| Generation total | | 6652 | | 35.58 | |
| No. of actual eggs ($N_2$) | Nil | | | | |
| Trend index ($N_2/N_1$) | Nil | | | | |
| Generation survival ($N_3/N_1$) | 0.688 | | | | |

Source: Yathish (2012)

In the research study conducted by Kang et al. (2002) regarding the seasonal occurrence pattern of the peach pyralid moth (PPM), *C. punctiferalis* was studied by way of sex pheromone traps and mercury light traps in several Fuyu persimmon orchards under different control pressures in the southern region of Korea Republic in 2000 and 2001. Fruits damaged by the larvae were also checked during the harvest time from 1999 to 2001.

An intensive field survey was carried out in Gujarat, India, by Bilapate and Talati (1978b) between August 1971 and March 1973 on the parasites of the castor pest, *C. punctiferalis*, and the following were recorded as parasitizing the pest: a species of *Brachymeria* near *B. nephantidis* Gah. (pupal parasite), *Diadegma* (Nythobia) sp. (larval parasite) and an unidentified species of the family Eulophidae. The percentage parasitism of *C. punctiferalis* by the latter species was maximum (36.84%) during January 1972.

Surveys conducted in Junagadh, Gujarat, during 2017 revealed *Conogethes* infestation in jamun (6%), pomegranate (11%), guava (8%) and ber (2%). The percentage in the brackets indicates the mean fruit damage, where $n = 8$ trees. Extensive surveys are likely to reveal damage caused to more number of fruit trees. In the Kutch region of Gujarat, cultivation of grapes is being taken up on a large scale, and *C. punctiferalis* is likely to infest grapes in the future. Similarly in other regions of Gujarat, ginger is being cultivated on large scale, and infestation of *Conogethes* species has already been recorded, but it is not clear (Table 11.13).

**Table 11.13** Life budgeting of *C. punctiferalis* for winter season

| Age interval | No./Ac | Log No. Ac | K's |
|---|---|---|---|
| Egg | 21,358 | 4.3296 | |
| | | | 0.0404 |
| Larvae (I, II, III) | 19,459 | 4.2891 | |
| | | | 0.0462 |
| Larvae (IV, V) | 17,497 | 4.2430 | |
| | | | 0.0470 |
| Pupae | 15,702 | 4.1960 | |
| | | | 0.0285 |
| Adult | 14,706 | 4.1675 | |
| | | | 0.3010 |
| Total | | | 0.1621 |

Source: Yathish (2012)

**Acknowledgement** Authors are thankful to the authorities of the Dr. Panjabrao Deshmukh Krishi Vidyapeeth, Akola and Mahatma Phule Krishi Vidyapeeth, Rahuri for the facilities and encouragement.

## References

Akashe VB, Indi DV, Patil SR, Jadhav JD, Pawar PB (2015) Incidence of insect pest damage in castor in relation to meteorological parameters in the scarcity zone of Maharashtra. J Agrometeorol 17(1):139–141

Atwal AS, Bains SS (1974) Applied animal ecology. Kalyani Publishers, Ludhiana, p 245

Bilapate GG (1978) Life table for the caster capsule borer, *Dichocrocis punctiferalis* Guenee on different hosts. Proc Ind Acad Sci 7(9):217–220

Bilapate GG, Talati GM (1977) Field observations on the incidence of *Dichocrocis punctiferalis* Gueen on caster. Madras Agric J 64:407–408

Bilapate GG, Talati GM (1978a) Some studies on bionomics of castor shoot and capsule borer *Dichocrocis punctiferalis* Guenee. J Maharashtra Agric Univ 3(1):47–49

Bilapate GG, Talati GM (1978b) Record of parasites of *Conogethes punctiferalis* Guenee from Gujarat. J Maharashtra Agric Univ 3(1):75–76

Gour TB, Sriramulu M (1992) Grapevine, *Vitis vinifera* Linn. A new host of castor shoots and capsule borer, Conogethes punctiferalis (Guenee). Trop Pest Manag 38:459

Harcourt DG (1969) The development and use of life tables in the study of natural insect population. Annu Rev Entomol 25:197–208

Kang CH, Lee KC, Park CG, Lee DW (2002) Korean J Appl Entomol 41(2):107–112

Kumar KP, Nitin KS, Vasudev Kammar, Chakravarthy AK, Khader Khan H (2016) Bio-ecology of castor shoot and capsule borer, *Conogethes punctiferalis* Guenee (Lepidoptera: Crambidae). National conference on emerging challenges and opportunities in agriculture, social, plant, environment, co-operatives and technology ECOASPECT (10–11th September, 2016), GUARD, Telangana, India, p 202

Patel RK, Gangrade GA (1971) Note on the biology of castor capsule borer, *Dichocrocis punctiferalis* Gn. (Lepidoptera: Pyralidae). Ind J Agric Sci 41:443–444

Qiang H-R (1997) Study on the control of yellow peach moth. South-China-Fruits 26(4):33

Sambathkumar S, Durairaj C, Mohankumar S, Ganapathy N (2017) Studies on sex ratio of castor shoot and capsule borer, *Conogethes punctiferalis* Guenee (Lepidopteara: Crambidae) in Castor. J Entomol Zool Stud 5(1):206–208

Singhvi SM, Balan JS, Yadav TP (1972) Note on varietal susceptibility in castor to shoot and capsule borer, *Dichocrocis punctiferalis* Guenee (Lepidoptera: Pyralidae). Ind J Agric Sci 42:269–270

Stanley J, Chandrasekaran S, Preetha G (2009) Conogethes punctiferalis (Lepidoptera: Pyralidae) its biology and field parasitization. Ind J Agric Sci 79:906–909

Suganthy M (2011) Field screening of promising castor cultures against capsule borer, *Conogethes punctiferalis* and leafhopper, *Empoasca flavescens*. Madras Agric J 98(4–6):178–179

Thyagaraj NE (2003) Integrated management of some important cardamom pest of hill region of Karnataka, South India. Ph.D Thesis, Dr. BR Ambedkar University, Uttar Pradesh, India

Varley GC, Gradwell GR (1960) Key factors in population studies. J Anim Ecol 29:399–401

Yathish KR (2012) Life table studies of castor shoot and capsule borer, *Conogethes punctiferalis* Guenee. M Sc (agri.) Thesis, Dr Panjabrao Deshmukh Krishi Vidyapeeth, Krishinagar, Akola, Maharashtra India

# Shoot and Fruit Borer, *Conogethes* spp. (Crambidae: Lepidoptera) in Northeast India

12

P. Thiyagarajan, S. S. Bora, K. Dhanapal, B. A. Gudade, and A. B. Rema Shree

**Abstract**

Northeast India (NER) is characterized by rich biodiversity, and most of the farmers in the region practice organic farming. Ginger and turmeric are the principle hosts for *Conogethes* spp. in NER. *Conogethes* infestation starts in June and continues till October to November. Large cardamom, *Amomum subulatum*, remains uninfested. *Conogethes* sp. incurs yield loss of up to 35% on ginger and turmeric. Cultural practices and phytosanitary measures are adopted to contain the pest. Insecticides are seldom applied for management of the borer.

**Keywords**

Northeast India (NER) · Zingiberaceae · *Conogethes* spp. · Major pest

## 12.1 Introduction

The North Eastern Region (NER) comprises the states of Assam, Arunachal Pradesh, Manipur, Meghalaya, Nagaland, Mizoram, Tripura, and Sikkim. The NER is endowed with rich biodiversity of forest vegetation, crop species, and wildlife.

P. Thiyagarajan · A. B. Rema Shree
ICAR-Indian Cardamom Research Institute, Spices Board, Myladumpara, Idukki, Kerala, India

S. S. Bora · K. Dhanapal
Regional Research Station, ICAR-Indian Cardamom research Institute, Spices Board, Tadong, Sikkim, India

B. A. Gudade (✉)
Indian Cardamom Research Institute, Regional Research Station, Spices Board, (Ministry of Commerce & Industry, Govt. of India),
Tadong Gangtok, Sikkim, India





A. K. Chakravarthy (ed.), *The Black spotted, Yellow Borer, Conogethes punctiferalis Guenée and Allied Species*,
https://doi.org/10.1007/978-981-13-0390-6_12

About 40% of the total flora of the country is represented in this region. The heavy precipitation, soil conditions, and climate of the region are widely suited to produce and support a whole range of plantation crops, spices, fruits, vegetables, flowers, medicinal and aromatic plants/herbs, oil seeds, fodder, bamboo and canes, and dyes. NER by default is practicing organic cultivation and lesser amount of pesticides used for pest management compared to rest of the country. In view of this, the pest insects have exploited wild and cultivated plants as food plants. The information on bioecology and distribution of *Conogethes* spp. in NER are poorly documented. Issues on shoot and fruit borer, *Conogethes* spp., in NER are discussed.

## 12.2 Bioecology

The borer moths on turmeric and ginger which have now been identified as *Conogethes sahyadriensis* (Shashank et al. 2018) lay eggs on leaves and other soft parts of the plants. The eggs hatch in about a week. The larvae pass through five instars and are full-fledged in 2–3 weeks. The pupal stage lasts about 1 week. The life cycle is completed in 4–5 weeks, and three generations are completed in a year in NER. The full-grown caterpillar measures 25–30 mm in length and is reddish brown with black blotches all over the body and a pale stripe on the lateral side. The moths are orange yellow, with black markings on both wings (Darlong et al. 2006). The moth lays eggs on the growing bud, petiole, or leaf of the young plants of ginger (Anonymous 1954). The shoot and fruit borer, *Conogethes* spp., breeds throughout the year in India and parts of Australia (USDA 1957). In Japan, this species has two to three generations per year. In India, three generations in NER and five to six generations in other parts of India.

## 12.3 Damage

*C. sahyadriensis* is an oligophagous pest infesting *Zingiber* crops in India. The small cardamom is the major host of this species in India and also damages zinger, turmeric, etc.

Ginger and turmeric are the principal hosts for *C. sahyadriensis* in all the states of NER. The pest is most active from July to October in states of the North Eastern Region. The damage is caused by the caterpillar which bores into the main stem on the young plants causing their death (Darlong et al. 2006; Shashank et al. 2015) and causes severe production loss in ginger (Lalnuntlunga 2005). Infestations start in June and continue till October. The moths lay eggs on growing bud, petiole, or leaf of the young plants. Larvae bore through the central shoots of the plants and feed on the growing buds resulting in withered and dried shoot referred to as "dead heart." The presence of a bored hole on the pseudostem through which grass is extruded and withered and yellow central shoot is a characteristic symptom of this pest infestation.

## 12.4 Major and Minor Host Plants in NER

This species that has been recorded in Zingiberaceae family in all states of NER are *Curcuma longa* (turmeric) and *Zingiber officinale* (ginger) except large cardamom which belongs to the family Zingiberaceae (Figs. 12.1 and 12.2). So far, no minor hosts were recorded on this species in NER. A few studies on *Conogethes* spp. were conducted in NER, and further studies are required for finding alternate hosts. The information on organic cultivation of large cardamom is contained in Gudade et al. (2016).

## 12.5 Pathway

This species *C. sahyadriensis* can move through international trade. There are no records of this species being intercepted on unlike *C. punctiferalis* (MAF Biosecurity New Zealand 2009; Korycinska 2012). For *C. punctiferalis* interceptions in England and Wales occurred on sugar apple, mango, and guava (Korycinska 2012). The species *C. punctiferalis* has been intercepted and identified 104 times at US ports of entry (all interceptions were of larvae) (AQAS 2014; queried December 9, 2014). Within Asia, the species must have spread and found new hosts from India or Sri Lanka. The species must have spread through plants taken from one country to another. This may also be the case for *C. sahyadriensis.*

**Fig. 12.1** Large cardamom, *Amomum subulatum* Roxb. capsules at the base of stem, a *Zingiber* species alien to *Conogethes*? (Source: http://gernot-katzers-spice-pages.com/engl/Amom_sub.html)

**Fig. 12.2** A typical habitat (hilly terrain) of large cardamom in Sikkim, Northeast India. (Source: Gudade et al. 2016)

## 12.6 Crop Loss

The amount and type of damage caused by this species and most reports are limited to a specific or few crops (Korycinska 2012). *C. punctiferalis* has been documented as a serious pest of castor bean and fruits in tropical, subtropical, and temperate countries (USDA 1957). In India, Devasahayam and Abdulla Koya (2005) reported that *C. sahyadriensis* is the most serious pest of ginger, particularly in South India. Crop yield can be significantly affected when more than 45% of shoots in a clump are damaged (Devasahayam et al. 2010). In South India, small cardamom capsule yield loss was recorded between 6.8 and 9.2%, while castor capsule damage was between 11 and 27% (Shashank et al. 2015). In Guatemala and other countries, the borer damage on small cardamom varies widely (5–30%). In small cardamom, the yield loss was estimated to be more than 20% every year (Kapadia 1996; Thyagaraj 2003). However, the crop loss due to this pest was worked out in small cardamom, and economic threshold level was fixed at 10% (Anonymous 1954; Krishnamurthy et al. 1989; Ram et al. 1997). Suganthy (2011) estimated capsule yield loss in castor due to *C. punctiferalis* in Tamil Nadu which was 10.80–26.70% and estimated 42.30% crop loss in India (Kapadia 1996), and 50% yield reduction was in grapes (Ram et al. 1997).

In NER, *Conogethes* spp. is the most serious pest on ginger and turmeric, and this species causes damage up to 35% on ginger (Thakur et al. 2012) in Meghalaya; in Sikkim it causes 15–35% damage (Yadav et al. 2014). It causes severe production loss in ginger in Nagaland (Lalnuntlunga 2005) and is the most serious pest in Nagaland (Mhonchumo and Singh 2010) and other states of Northeast India. The economic threshold level (ETL) of this species worked out on ginger in NER. Management methods should be adopted at a stage when there is one egg mass per square meter (Darlong et al. 2006).

## 12.7 Pest Risk Assessment

Details concerning this aspect are presented in another chapter in this volume. The shoot and fruit borer, *C. sahyadriensis*, does not seem to occur in Europe, South Africa, Canada, and the USA, and it is not listed in the EC Plant Health Directive or in any of the EPPO lists. However, it may be a pest of quarantine importance, and there is a need to rapidly asses its importance. International Standards for Phytosanitary Measures (ISPM) No. 11 (Pest Risk Analysis for Quarantine Pests Including Analysis of Environmental Risks and Living Modified Organisms 2004) provides guidelines for the further analysis of organisms that are quarantine pests.

The pest risk assessment should be comprehensive embracing several factors or aspects of each pest including data on spread, geographical distribution, biology, and economic significance in localities where it occurs. Experts are then utilized to assess the probability that the pest will be introduced and its potential for establishment, spread, and economic importance in the PRA area. In characterizing the risk, the amount of data generated on each pest will vary with each pest and the locality. The quality of the assessment will also vary with the resources and resource persons. For instance, one NPPO may have comprehensive pest databases and geographical information systems; another may depend on books, printed soil maps, climate maps, and expert opinion. In some cases, no or scarce information may be available, or research data may be required. Countries where the pest is present may provide, upon request, required data for conducting the pest risk assessment in the nation needed. Since, to date, *C. sahyadriensis* was branded as *C. punctiferalis*, database on the new species has to be collected and analyzed.

The *Conogethes* moths are laying eggs on growing bud, petiole, or leaf of the young plants on ginger and turmeric, and this pest undergoes three generations per year in NER, and the full-grown larvae under favorable conditions undergo pupation. This suggests that eggs could be present at harvest time, and due to small size, they may not be detected during the harvest and packing processes. This suggests egg could be present at harvest time, and it may not be detected during the harvest and packing processes. Pupation takes place inside the pseudostem/rhizome or sometimes in the grass that collects after feeding. Larvae and pupae are more likely bigger than eggs and to be detected during harvest and intercepted in consignments. Hence, the likelihood of entry of *Conogethes* larvae is considered to be moderate and therefore non-negligible.

In ginger and turmeric, generally rhizomes are consuming, and the leaves and pseudostems are disposed of. The waste materials generated from ginger and turmeric are pseudostem and leaves, and it could allow some *Conogethes* larvae/pupae which are present inside the plant parts to disperse and find a suitable host. *Conogethes* spp. can cause significant damage to stems, pseudostems, rhizomes, fruits, and seeds of host plants (FAO 2007). In India, 50% yield loss is reported on grapes by capsule and shoot borer (Ram et al. 1997), and there is more than 50% of coca pods loss in coastal Karnataka, South India, during 2016. In NER, *C. sahyadriensis* is the most serious pest on ginger and turmeric, and this species cause yield loss up to 35% on ginger (Thakur et al. 2012), cause severe production loss in ginger (Lalnuntlunga 2005), and cause most serious damage in Nagaland Mhonchumo and Singh 2010) and other NER states.

## 12.8 Surveillance and Management

*Conogethes* spp. produces sex pheromones and this may facilitate targeted surveys of export crop areas to detect its presence. However, currently pheromone traps are not effective in detecting moths under field conditions where *Conogethes* population occurs. Infested plant material/parts from an infested area should not be permitted to enter the importing area or areas where the pest is absent.

**Acknowledgment** Authors are grateful to the Indian Cardamom Research Institute, Myladumpara and Sikkim, India, and to the Spices Board, Cochin, Kerala.

## References

Anonymous (1954) Pests of cardamom. Memories of the Department of Agriculture, Madras, pp 998–1000

AQAS (2014) *Conogethes* sp. interceptions. Agricultural quarantine activity systems. https://aqas.aphis.usda.gov/aqas/HomePageInit.do#defaultAnchor. Accessed Dec 9 2014

Darlong VT, Hmar M, Ng T, Meitei MI, Khanikar M, Chakraborty A (2006) Package of organic practices from Assam, Manipur, Meghalaya and Nagaland for arecanut, ginger, large cardamom, passion fruit and pineapple. North Eastern Region Community Resource Management project for Upland Areas (NERCORMP), Shillong, Meghalaya, pp 149

Devasahayam S, Abdulla Koya KM (2005) Insect pests of ginger. In: Ravindran PN, Babu KN (eds) Ginger: the genus. Zinger. CRC Press, Boca Raton, pp 367–390

Devasahayam S, Jacob TK, Abdulla Koya KM, Sasikumar B (2010) Screening of ginger (*Zingiber officinale*) germplasm for resistance to shoot borer (*Conogethes punctiferalis*). J Med Aromat Plant Sci 32(2):137–138

FAO (2007) Forest health and biosecurity working papers. Overview of forests pests. People's Republic of China. Food and Agriculture Organization of the United Nations – Forestry department. Rome, Italy

Gudade BA, Vijayan AK, Chhetri P, Bora SS, S.K. Bhat (2016) Importance of organic mulches in large cardamom (*Amomum subulatum* Roxburgh) in Sikkim. Ind Botanists: 3

ISPM No. 11 (2004) Pest risk analysis for quarantine pests including analysis of environmental risks and living modified organisms

Kapadia MN (1996) Estimation of losses due to pod borers in oilseed crops. J Oilseeds Res 13(1):139–140
Korycinska A (2012) Rapid assessment of the need for a detailed pest risk analysis for *Conogethes punctiferalis* (Guenée). The Food and Environment Research Agency, pp 7
Krishnamurthy K, Khan MM, Avadhani KK, Venkatesh J, Siddaramaiah L, Chakravarthy AK, Gurumurthy BR (1989) Three decades of cardamom research, regional Research Station (1958–1988). Station Technol Bull 2:44–68
Lalnuntlunga J (2005) Performance of certain chemicals and neem formulations against ginger shoot borer (Dichocrosis punctiferalis Guen.), M. Sc. (Agri.) thesis, Nagaland University, Medziphema
MAF Biosecurity New Zealand (2009) Import risk analysis: table grapes (*Vitis vinifera*) from China, pp 322
Mhonchumo, Neog P, Singh HK (2010) Eco-friendly management of shoot borer and rhizome fly of ginger in Nagaland. In: Dr. Sema A, Dr. Srinivasan V, Shitiri M (ed). National Symposium on spices and aromatic crops (SYMSAC - V) on harnessing the potential of north eastern state for spices production through technological inventions, 30–31, October 2009 at Central Institute of Horticulture, Nagaland, pp 117–120
Ram KT, Hugar PS, Somasekhar PBV (1997) Record of *Conogethes punctiferalis* (Guenee) on grapes. Pest Manag Hortic Ecosyst 3(1):37
Shashank PR, Doddabasappa B, Kammar V, Chakravarthy AK, Honda H (2015) Molecular characterization and management of shoot and fruit borer *Conogethes punctiferalis* Guenee (Crambidae: Lepidoptera) populations infesting cardamom, castor and other hosts. In: Chakravarthy AK (ed) New horizons in insect science: towards sustainable pest management. Springer, New Delhi, pp 207–227
Shashank PR, Kammar V, Mally R, Chakravarthy AK (2018) A new Indian species of shoot and capsule borer of the genus *Conogethes* (Lepidoptera: Crambidae), feeding on cardamom. Zootaxa 4374(2):215–234
Suganthy M (2011) Field screening of promising castor cultures against capsule borer, *Conogethes punctiferalis* and leafhopper, *Empoasca flavescens*. Madras Agric J 98(4/6):178–179
Thakur NSA, Firake DM, Behere GT, Firake PD, Saikia K (2012) Biodiversity of agriculturally important insects in north eastern Himalaya: an overview. Ind J Hill Farming 25(2):37–40
Thyagaraj NE (2003) Integrated management of some important cardamom pests of hill region of Karnataka, South India. Ph D Thesis, Dr BR Ambedkar University, Uttar Pradesh, India
USDA (1957) Insects not known to occur in the United States. Yellow peach moth (*Dichocrocis punctiferalis* (Guenee)), pp 37–38
Yadav A, Kumar A, Gopi B, Kalita H, Lepcha B (2014) Ginger (*Zingiber officinale* L.) hand book of organic crop production in Sikkim, pp 165–168

# 13 Bioecology and Management of Fruit and Shoot Borer, *Conogethes punctiferalis* Guenée (Crambidae: Lepidoptera), on Fruit Crops in Central India

S. Gundappa, H. Khader Khan, and A. K. Chakravarthy

**Abstract**

The shoot and fruit borer, *Conogethes punctiferalis* (Guenée) (Lepidoptera: Crambidae), is a serious insect pest widely distributed in Asia. Phenology and chemical constituents together with weather conditions and habitat features influenced borer damage. The borer has been found feeding on temperate, tropical, and subtropical fruit crops. Integrated approach embracing field sanitation, pruning trees, scraping tree bark, thinning fruits, and avoiding insecticidal applications enhance the activities of natural enemies and subsequently reduce the borer population without inimical effects on the nontarget organisms and the environment.

**Keywords**

Bioecology · *Conogethes* · Managements · Tropical fruits

S. Gundappa (✉)
ICAR-Central Institute for Subtropical Horticulture, Lucknow, Uttar Pradesh, India
e-mail: Gundappa@icar.gov.in

H. Khader Khan
Department of Agricultural Entomology, University of Agricultural Sciences, Gandhi Krishi Vignan Kendra (GKVK), Bengaluru, Karnataka, India

A. K. Chakravarthy
Division of Entomology and Nematology, ICAR-Indian Institute of Horticultural Research (IIHR), Bengaluru, Karnataka, India





A. K. Chakravarthy (ed.), *The Black spotted, Yellow Borer, Conogethes punctiferalis Guenée and Allied Species*,
https://doi.org/10.1007/978-981-13-0390-6_13

## 13.1 Introduction

The shoot and fruit borer, *Conogethes punctiferalis* Guenée, is a serious insect pest distributed widely across a range of habitats on many important crops (Inoue and Yamanaka 2006; Shashank et al. 2015). Studies on the different aspects of *C. punctiferalis* on tropical and subtropical fruit crops are scanty and scattered. The feeding of the borer on fruits is important economically because it directly attacks fruits and other reproductive parts that are of commercial value. Secondly, it is ecologically important as growers usually resort to insecticidal applications in order to protect fruits from the borer damage. Varieties or crops that bear fruits in clusters are preferred for feeding. Insecticides are not only undesirable and ineffective to apply on trees that are tall with well-spread-out canopies. Therefore, semiochemicals like pheromones, ovipositional deterrents, repellents, and resistant varieties are the most suited tools for managing borer pests on fruit trees. The aspects on geographical distribution, bioecology, population dynamics, and management of *C. punctiferalis* on tropical and subtropical fruit crops are dealt within this chapter.

## 13.2 Status of *Conogethes* on Fruit Crops

The shoot and fruit borer, *C. punctiferalis*, mainly occurs in tropical and subtropical countries (Pena et al. 2002) and is distributed in Asia and Australia. Among the fruit crops where this pest infestation was reported on were banana, citrus, sapota, pomegranate, mango, guava, grape, apple, and peach (Flether 1922; Konno et al. 1980; Kodoi and Kaneda 1990; Gour and Sriramulu 1992; Jarvis 1914; Butani 1978; Ram et al. 1997; Singh et al. 2002). In the 1980s and 1990s, *C. punctiferalis* was found infesting mango fruits occasionally and was considered a minor pest. But in the twenty-first century, the crambid has attained the status of a major pest, especially in Uttar Pradesh, North India.

## 13.3 Biology of *Conogethes* on Fruit Crops

Phenology of host plant influences the size, growth, and development of insect eggs. Temperature and relative humidity play an important role in egg hatching (Rajan 1965). The development of *C. punctiferalis* was investigated on apple fruits by Kadoi and Kaneda (1990) in the laboratory. The number of eggs laid on caged fruits averaged 0.25 at 26 °C and 70% relative humidity.

The growth and development of different larval instars varied with varying temperature and relative humidity. Sloan (1945) reported that larval period of *C. punctiferalis* lasts 3 weeks under normal conditions and 2–3 weeks in winters. Yang and Shaw (1962) studied the peach borer biology in China and reported that *C. punctiferalis* completes 4–5 generations in a year and larvae overwinter in the flowers, stem, and fallen leaves. The duration of the larval stage varied from 20 to 23 days in August to September at 21–35 °C to 22–26 days in October to January at 14–28 °C, and also the total larval period got extended up to 12–14 days (Jacob 1981; Kondo and Miyahara 1930; Twine 1971; Wang and Cai 1997; Xi et al. 1996). The larvae rolled the needles in twigs into a bag with silk, in which they overwinter as third or fourth instar.

Sloan (1945) reported that larval period of *C. punctiferalis* lasted 3 weeks under normal conditions and 2–3 weeks in winter. Young and Shaw (1962) studied the peach borer biology in China and reported that there were four to five generations in a year with the larvae overwintering in the flowers.

The mature larva is pinkish and speckled with minute black spots, measuring approximately 1.5 cm in length. Development of *C. punctiferalis* from the egg to the fully grown larva occupied 32 days on apples with more than 50% survival (Kodoi and Kaneda 1990). Wang and Cai (1997) reported that there were five generations a year on Younai plums in China. The larva of *C. punctiferalis* bores into the unripe mango fruit near the stalk end and feeds within. Black frass was found near the place of entrance. The attacked fruits start rotting and finally drop from the tree, and those left clinging to the tree lose market value. The fully grown larva was stout and reddish brown, with numerous short tubercles on body measuring 25–30 mm in length (Singh et al. 2002).

Pupation of *C. punctiferalis* takes place in cocoons inside, and the pupal stage lasted for 7–10 days (Patel and Gangrade 1971; Bilapate and Talati 1978; Gour and Sriramulu 1992; Kumar et al. 2017). The duration of development of pupa varies from 27 at 30 °C to 48–51 days at 20 °C. Male and female pupa differs in size, shape, and weight. There is a morphological dimorphism in the *Conogethes* pupa that helps in sex determination. Female pupae are usually larger (17.81 × 6.29 mm with 0.127 g in weight) than male pupae (14.30 × 4.26 mm with 0.108 g in weight) and are shorter and slightly narrower, and the genital opening was located in the posterior region of the ninth abdominal segment and flanked by a pair of pads. Pupal period lasts for 7.90 ± 2.80 days under laboratory conditions. Significant differences were also noticed between the sexes in terms of length and diameter (Thyagaraj et al. 2001).

The adult moth is medium sized and yellow with black spots on the body. Bilapate and Talati (1978) revealed that the adult female moths (15.80 ± 2.50 days) survived for 1 or 2 days longer than male moths (14.00 ± 3.80 days). The longevity of adults was significantly affected by adult nutritional conditions. Adults fed on water, 10% honey solution, or 10% sugar and vinegar liquid survived significantly longer than when they remain unfed. The ratio of males to females of the progeny ranged between 1:1 and 1:2. Kaneko (1978) observed a clear abdominal constriction in female *C. punctiferalis* that had already paired. This proved a reliable indication of pairing and greatly facilitated the separation of virgin and paired females. The constriction was apparent 45–60 min after pairing. Kuang et al. (2009) reported that in China, adults of *C. punctiferalis* emerged mainly between 10:00 PM and 08:00 AM. The emergence rate was 92.22%. *C. punctiferalis* males have hair pencils at the abdominal end and thus can be differentiated from the other sex. Mating occurs only in the dark after 7:30 PM (Stanley et al. 2009).

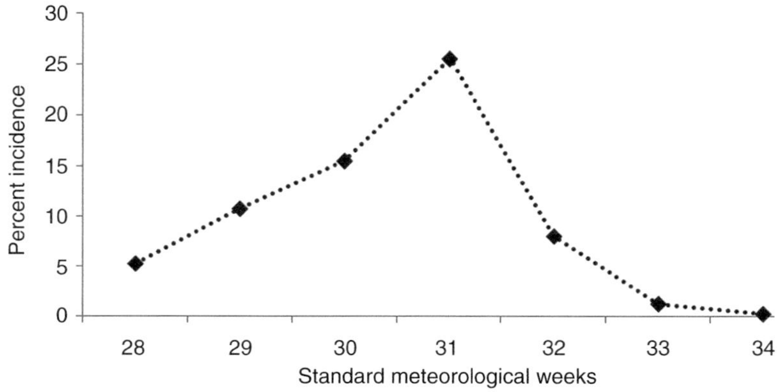


**Fig. 13.1** Seasonal incidence of guava fruit borer, *C. punctiferalis*, at Lucknow, Uttar Pradesh, India

## 13.4 Estimation of Yield Loss

Xi et al. (1996) reported that fruits of Chinese chestnut (*Castanea mollissima*) suffered 25–80% loss due to *C. punctiferalis*. A total of 120 tonnes of *C. mollissima* was damaged, and 1 million yuan RMB of revenue was lost every year due to *C. punctiferalis*, one of the main insect pests in China. Therefore, effective control was carried out on the basis of occurrence rhythm (Xu et al. 2001).

## 13.5 Seasonal Incidence

The incidence of *C. punctiferalis* has been recorded from guava-growing regions of North India. The incidence is more common in rainy season crop. Incidence of *C. punctiferalis* varied from 7.79% to 25.20%. Seasonal incidence of guava fruit borer *C. punctiferalis* has been studied under Lucknow (Uttar Pradesh, India) conditions. Its incidence started during the 28$^{th}$ standard meteorological week (SMW), and peak incidence (25.5%) was recorded during the 31$^{st}$ SMW (Fig. 13.1). Fruit borer incidence was significant, which negatively correlated with maximum relative humidity, rainfall, and evaporation and positively with sunshine hours. Akashe et al. (2015) also determined incidence of the capsule borer on castor and other crops in Maharashtra and found peak activity during September to December.

The affected fruits are generally deformed at the point of entry of larva (Fig. 13.2). Such fruit becomes weak, rots, and falls down. Larvae of this moth mainly bore fruits but it may also attack buds and tender shoots. The larvae feed on pulp and seeds of fruits resulting in premature drop of fruits. The pupa is formed in a chamber made at the basal end of the fruit. The pest completes life cycle within 25–33 days on guava.

Singh et al. (2002) observed reddish-brown larvae of *C. punctiferalis* feeding on unripe mango fruits in the Sambhalhera village of Muzaffarnagar, Uttar Pradesh. The

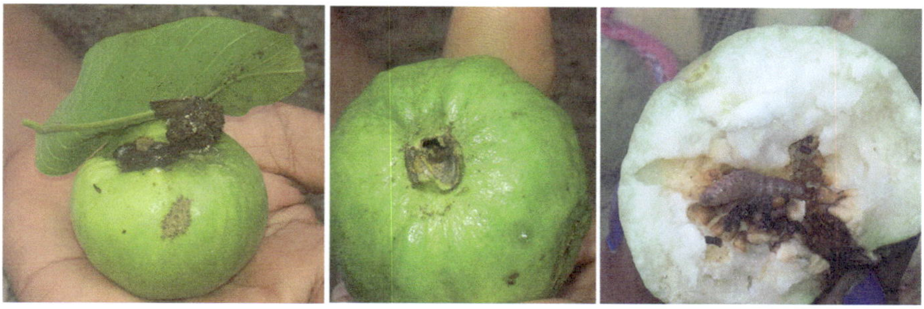

**Fig. 13.2** Symptoms of damage of *C. punctiferalis* on guava

highest incidence (20%) of *C. punctiferalis* was noticed during the 32nd standard meteorological week at Udheywala, while the other peak incidence (9%) was recorded during the 33rd standard week at Raya. A comparison of the infestation levels at Udheywala, Jammu and Kashmir, and Raya, Nepal, revealed that fruit borer infestation was higher in irrigated area compared to the rainfed area. Multiple correlation analysis of the data of both the locations showed that abiotic factors like maximum temperature, minimum temperature, relative humidity (morning and evening), and rainfall contributed up to 63.2% of the borer infestation (Virender Kaul Kesar 2003). Kannan and Rao (2007) reported that *C. punctiferalis* infested matured leaves of mango during September to October and ripened fruits during April to June.

Ram et al. (1997) recorded *C. punctiferalis* infesting grapes for the first time in Karnataka during December 1993 to February 1994. Seasonal occurrence pattern of *C. punctiferalis* was studied by Kang et al. (2002) in several persimmon orchards under different control pressures in the southern region of Korea during 2000 and 2001. The pattern showed three distinct peaks: the first one in mid to late June, the second one in mid to late August, and the third one in late September.

Yellow peach moth (*C. punctiferalis*) damage on rambutan fruits (*Nephelium lappaceum*: Sapindaceae) has been reported in Queensland, Australia. The larvae feed externally on green and mature fruits from December to March. On green, developing fruit, the larvae also feed on pedicels causing premature fruit drop. As the fruit matures, the larvae bore into fruit making it unmarketable.

## 13.6 Management

Cultural practices are important to change the microclimate of the orchard ecosystem, and thereby one can reduce the *C. punctiferalis* infestation. Apart from intercropping, removal of infested and damaged fruits from orchards reduces the pest incidence. Scraping off the fruit tree bark where the larvae of *C. punctiferalis* overwinter also helps in minimizing the pest incidence. Setting light traps and sugar vinegar traps in orchards proved best in attracting the adult moths. In China, covering the surface of the undamaged fruits with paper bags to prevent from laying eggs and boring has proved to be successful.

A significantly higher incidence of *C. punctiferalis* was recorded on cultivars/genotypes like Sardar, Arka Amulya, and Sardar Selection 4/10, while its incidence was low on Bangalore Seedling Selection 6/10, 6/12, and 6/17, Banarasi Surkha, red flesh, Baharampur seedling, and Fazilka Selection. Shillong-1, Tehsildar, and Cattley guava were found free from infestation of this pest. Parasitoids like *Trathala flavoorbitalis* (Cameron), *Brachymeria atteviae* JNJ, *Chelonus blackburni* Cameron, *Anthrocephalus decipiens* (Masi), and *Epitranus erythrogaster* Cameron were found to parasitize *C. punctiferalis* larvae.

Varadarasan and Kumaresan (1987) suggested that chemical control was unsatisfactory because it reached only the early larval instars, correct timing of pesticide application based on adult emergence was not always possible, and *C. punctiferalis* also attacked a number of other crops. The use of light traps, pheromones, and handpicking was suggested as possible alternative means of control. He (1997) showed that spraying 50% Sumithion (fenitrothion) solution, 50% omethoate emulsion, or 50% dichlorvos at fruit coloring, after summer pruning, and 3 weeks before flowering gave adequate control of *C. punctiferalis*. Wang et al. (2002) revealed that peach fruit moth was controlled by ground spraying with 50% phoxim when overwintering of larvae occurred. *C. punctiferalis* was effectively controlled using fenitrothion 30WP (3%) three times on *C. mollissima* (Xie et al. 2002). Zhu et al. (2000) reported that the major pest of chestnut was *C. punctiferalis* and can be suppressed by spraying dichlorvos 76 EC (4%) or decamethrin 2.8 EC. Studies on the combined efficacy of proven management strategies against oriental fruit fly, *Bactrocera dorsalis* (Hendel), conducted at Ludhiana revealed that the practices also significantly reduced the incidence of *C. punctiferalis*. The per cent fruit infestation ranged from 0 to 16.67 in IPM treatment compared to 76.67 in the control. Overall, the mean infestation was significantly lower in IPM (4.00%) compared to the control (19.39%) (Gill and Mann 2008). Sekiguchi (1974) reported that *C. punctiferalis* has been found feeding on sweet chestnut, peach, plum, Japanese apricot (*Prunus mume*), pomegranate, persimmon, pear, and grapevine. Experiments conducted in chestnut groves showed that PAP 0.05% emulsion, PAP (2%) dust, or DEP (4%) dust each gave good results.

## 13.7 Integrated Pest Management

- While pruning young fruits, avoid leaving fruits/pairs together.
- Collection and destruction of infested as well as fallen fruits by burning.
- Healthy fruits that hang in pairs can be separated by placing the sticks or pieces of cardboard in between.
- A combination of phytosanitary inspection and bagging of young fruits with small-holed polythene covers to prevent distribution and egg laying, respectively.
- Conserving the natural enemies is important as they actively control the pest and perpetuate under favorable conditions.

- Setting light traps (black/blue light) at 1/ha for moth catches in pomegranate and guava.
- Young caterpillars feeding on the outside can be removed using a stick or piece of wire.
- Spray insecticides like chlorpyrifos 50 EC (2 ml/l) or neem oil (3 ml/l) and acephate 75 SP (2 g/l of water).

**Acknowledgment** Authors are thankful to the director, ICAR-Central Institute for Subtropical Horticulture, Lucknow, Uttar Pradesh, and ICAR-Indian Institute of Horticultural Research, Bengaluru, Karnataka, India.

## References

Akashe VB, Indi DV, Patil SR, Jadhav JD, Pawar PB (2015) Incidence of insect pest damage in castor in relation to meteorological parameters in the scarcity zone of Maharashtra. J Agrometeorology 17(1):139–141

Bilapate GG, Talati GM (1978) Some studies on bionomics of castor short and fruit borer *Dichocrocis punctiferalis* from Gujarat. J Maharashtra Agric Univ 3(1):75–76

Butani DK (1978) Insect pests of fruit crops and their control. Pesticides 12(8):53–59

Flether TB (1922) Report of the imperial entomologist. Scientific reports of the Imperial Institute of Agricultural Research. Pusa, pp 51–67

Gill HS, Mann GS (2008) Effect of management strategies for *Bactrocera dorsalis* (Hendel) on *Dichocrocis punctiferalis* Guenee infestation in guava orchards. J Insect Sci 21:133–138

Gour TB, Sriramulu M (1992) Grapevine, *Vitis vinifera* a new host of castor shoot and capsule borer *Conogethes punctiferalis*. Trop Pest Manag 38(4):459

He RQ (1997) Study on the control of yellow peach moth. South China Fruits 26:33

Inoue H, Yamanaka H (2006) Redescription of *Conogethes punctiferalis* (Guenee) and description of two new closely allied species from Eastern Palaearctic and Oriental regions (Pyralidae, Pyraustinae). Tinea 19(2):80–91

Jacob SA (1981) Biology of *Conogethes punctiferalis* on turmeric. J Plant Crop 9(2):119–123

Jarvis E (1914) A new fruit boring caterpillar of Banana occurring at Tweed keeds. Queens Land Agric J 5:280–284

Kaneko JI (1978) Abdominal constriction in the copulated female of yellow peach moth, *Dichocrocis punctiferalis* (Guenee.) (Lepidoptera : Pyralidae). J Appl Entomol Zool 13:131–133

Kang CH, Lee KC, Park CG, Lee DW (2002) Seasonal occurrence pattern of peach pyralid moth, *Dichocrocis punctiferalis* in Fuyu persimmon orchards and fruit damage at harvesting time. Korean J Appl Entomol 41:107–112

Kannan M, Rao NV (2007) Seasonal incidence of lepidopteran pests in relation to weather parameters in mango (Mangifera indica). Crop Res 33:198–203

Kodoi M, Kaneda M (1990) Development of yellow peach mouth *Conogethes punctiferalis* on apple fruit. Res Bull Plant Prot Serv Japan 26:61–63

Kondo T, Miyahara T (1930) Fruit borer and a summary of their life histories in Knangtuny. China J Plant Prot 17:85–94

Konno Y, Honda N, Matsumoto Y (1980) Observation on the mating behaviour and biology for the six pheromone of yellow peach moth, *Dichocrocis punctiferalis*. J Appl Entomol Zool 15(3):321–322

Kuang MH, Liu SW, Ji BZ, Gao JY, Gao YG, Wang GX (2009) Investigation on the overwintering status and bionomics of needle feeding type of *Conogethes punctiferalis*. Chin Bull Entmol 46:569–573

Kumar KP, Chakravarthy AK, Khader Khan H (2017) Biology of castor shoot and capsule borer, *Conogethes punctiferalis* Guenee on castor (*Ricinus communis* L.). Int J Sci Environ Technol 6(2):1134–1139
Patel RK, Gangrade GA (1971) Note on the biology of castor capsule borer, *Dichocrocis punctiferalis* Gn. (Lepidoptera: Pyralidae). Ind J Agric Sci 41(5):443–444
Pena JE, Nadel H, Pereira MB, Smith D (2002) Pollinators and pests of *Annona* Sp. In: Pena J, Sharp J, Wysoki (eds) Tropical fruit pests and pollinators: biology, economic importance, natural enemies and control. CABI Publication, Wallingford, p 208
Rajan SV (1965) Protect your cardamom against insect pests. Ind Farm 14(11):21–24
Ram KT, Hugar PS, Somasekhar PBV (1997) Record of *Conogethes punctiferalis* (Guenee) on grapes. Pest Manag Hortic Ecosyst 3:37–39
Sekiguchi K (1974) Morphology, biology and control of the yellow peach moth, *Dichocrocis punctiferalis* Guenée (Lepidoptera: Pyralidae). Bull Ibaraki Hort Expt Stn Special Issue, pp 89
Shashank PR, Doddabasappa B, Kammar V, Chakravarthy AK, Honda H (2015) Molecular characterization and management of shoot and fruit borer *Conogethes punctiferalis* Guenee (Crambidae: Lepidoptera) populations infesting cardamom, castor and other hosts. In: New horizons in insect science: towards sustainable pest management. Springer, New Delhi, pp 207–227
Singh YP, Singh GK, Kumar S (2002) Occurrence of *Dichocrocis punctiferalis* Guen. (Lepidoptera: Pyralidae) on mango in western UP. Progress Hortic 34(1):130
Sloan WJS (1945) Insect pests of grain sorghum. Queensland Agric J 61:221–239
Stanley J, Chandrasekaran S, Preetha G (2009) *Conogethes punctiferalis* (Lepidoptera: Pyralidae) its biology and field parasitization. Ind J Agric Sci 79(11):906–909
Thyagaraj NE, Singh PK, Chakravarthy AK (2001) Sex determination of cardamom shoot and fruit borer, *Conogethes punctiferalis* Guenee pupae. Insect Environ 7:93
Twine PH (1971) Insect pests of maize. Queens Land Agric J 97(10):523–531
Varadarasan S, Kumaresan D (1987) The possibilities of integrated management of shoot or capsule borer, *Dichocrocis punctiferalis* Guen. on cardamom. J Coffee Res 17:135–136
Virender Kaul Kesar YK (2003) Incidence and management of lepidopteran fruit borers of guava (Psidium guajava L.) in Jammu, India. J Asia Pac Entmol 6:201–205
Wang YX, Cai RT (1997) The emergence and control of *Dichocrocis punctiferalis* for Younai plum variety. South China Fruits 26(3):45
Wang DX, Lou Z, Li P, Gao ZH (2002) The main diseases and pests of pomegranate in Huaiyuan area and their control. China Fruits 1:36–38
Xi K, Xu FY, Dai YS (1996) Studies on the Chinese chestnut cone life table. J Joangsu Sci Technol 23(4):6–10
Xie JX, Zhou ZF, Xu RS, Ni DW, Wang H (2002) Preliminary experiment report on pest control by fenitrothion of *Castanea mollissima* nut. J Zhejiang Sci Tech 22:28–30
Xu YC, Luo H, He JR, Xu MJ, Hong MH (2001) Occurrence rhythm and control of *Dichocrocis punctiferalis* in Fuyang city. J Zhejiang Sci Tech 21:53–55
Young HY, Shaw KY (1962) Preliminary studies on the peach borer *Dichocrocis punctiferalis*. Entomol Sinica 1(2):109–134
Zhu GL, Wu ZD, Zhu CH, Zhou PF (2000) Integrated control of the main diseases and pests of chestnut. China Fruits 3:40–41

# The Shoot and Fruit Borer, *Conogethes punctiferalis* (Guenee): An Important Pest of Tropical and Subtropical Fruit Crops

14

Sandeep Singh, Gurlaz Kaur, S. Onkara Naik, and P. V. Rami Reddy

**Abstract**

*Conogethes punctiferalis* is a polyphagous species, well distributed in the tropics and subtropics. In Punjab and Himachal Pradesh, India, *Conogethes* has become a serious pest of plum, peach, pear, litchi and pomegranate. On guava, *Conogethes* is a major pest from the beginning of the twenty-first century infesting up to 20% fruits in Jammu and Kashmir, and Allahabad Safeda is the most susceptible cultivar to infestation. One of the ideal measures of protection would be to prevent moths from ovipositing on fruits for which the work on pheromones should be stepped up. While the components of pheromones have been identified, their proportion, wax supplement from the female hair pencils and trap designs need to be refined. Mass production and release of natural enemies on community basis and cultivation of tolerant/resistant cultivars will go a long way in protecting fruit crops from borer damage.

**Keywords**

*Conogethes* · Host plants · Integrated pest management · Tropical fruits

S. Singh (✉) · G. Kaur
Department of Fruit Science, Punjab Agricultural University, Ludhiana, Punjab, India
e-mail: sandeep_pau.1974@pau.edu; gurlaz@pau.edu

S. Onkara Naik · P. V. Rami Reddy
Division of Entomology and Nematology, ICAR-Indian Institute of Horticultural Research, Bengaluru, Karnataka, India
e-mail: pvreddy@iihr.res.in





A. K. Chakravarthy (ed.), *The Black spotted, Yellow Borer, Conogethes punctiferalis Guenée and Allied Species*,
https://doi.org/10.1007/978-981-13-0390-6_14

## 14.1 Introduction

The fruit borer, *Conogethes punctiferalis* Guenee (Lepidoptera: Crambidae), mainly occurs in tropical and subtropical countries (Pena et al. 2002). The genus *Conogethes* Meyrick, 1884 (Lepidoptera: Crambidae), is a large, taxonomically complex taxon of crambids distributed throughout the Orient, Australia and tropical Asia (Shaffer et al. 1996; Inoue and Yamanaka 2006). This shoot and fruit borer is an important insect pest causing severe crop losses owing to the direct larval damage to reproductive or economically important plant parts (Chakravarthy 2015; Chakravarthy et al. 2015a). Also called yellow peach moth, castor shoot and capsule borer, castor seed caterpillar, fruit and shoot borer, durian fruit borer, durian husk borer, maize moth or peach pyralid moth, Queensland bollworm, smaller maize borer and cone moth, it is an important pest infesting a large number of cultivated and wild plants. It is polyphagous (Fletcher 1914; USDA 1957; Mishra and Teotia 1965; Singh 1970; Sekiguchi 1974; Butani 1979, 1993; Jacob 1981; Zhang 1994; Evangelista 1995; Haseeb 2007; Chakravarthy 2015; Chakravarthy et al. 2015a; Molet 2015; Singh and Kaur 2016) and attacks fruits of 30 plant species belonging to 15 families (Konno et al. 1981). It has been reported feeding on guava, castor, mulberry, pomegranate, mango, peach, loquat, citrus, apple, plum, cherry, papaya, chestnut, cocoa, durian, cotton, tamarind, hollyhock, sorghum, macadamia, etc. (Butani 1979; Mishra and Teotia 1965; Hill 1983; Fang and Zhe 2009; Chakravarthy 2015; Anonymous 2015; Chakravarthy et al. 2015a; Singh 2015; Singh and Kaur 2015a). Astridge (2001) considers this species to be a major pest of rambutan and durian in North Queensland, Australia. Cardamom is the most preferred host followed by *Hedychium* sp., *Alpinia* sp. and *Ammommum* sp. (Thyagaraj 2003). But the *Conogethes* species feeding on all Zingiberaceae plants is not *C. punctiferalis*. Currently it has been identified as *Conogethes sahyadriensis* on cardamom from South India (Shashank et al. 2018). The alternate hosts include chikoo, marang, rambutan, jackfruit and cacao in the Philippines (Evangelista 1995).

*Conogethes* and *Dichocrocis* are considered synonyms, and *punctiferalis* is often seen combined with *Dichocrocis* (Wang 1980). Also, it is a pest of quarantine category, and there is a lot of confusion about its species. Robinson et al. (1994) suggested that there may be 20 species in the genus, at least 7 of which may occur in Southeast Asia. Until now, 15 named species of genus *Conogethes* have been known from Eastern Palaearctic and Indo-Australian regions. The type locality of *C. punctiferalis* is India, and so many closely allied species may be included. However, their taxonomic revision has been neglected for a long time (Ganesha 2011). *Conogethes punctiferalis* and *Conogethes pinicolalis* Inoue and Yamanaka are the most well-known species in genera *Conogethes* as pests of agricultural and forest plants in Japan (Honda 2015).

## 14.2 Geographical Distribution and Host Range

*C. punctiferalis* is found in southern and eastern Asia, Australia, Indonesia and New Guinea (USDA 1957; Walker 2007). Honda et al. (1979) stated that *C. punctiferalis* is an important pest of chestnut and peach in Japan. It has been reported on papaya in Italy (IBPGR 1988). Evangelista (1995) reported *C. punctiferalis* as a major pest of durian fruit in the Philippines, with 6.66% to 31.82% infestation. This species is also present in Hawaii (Nishida 2002), it is a pest of quarantine importance for apple in Mexico (Anonymous 2005) and Truscott (2007) has reported *C. punctiferalis* as new record in Cornwall, England.

The distribution of *C. punctiferalis* extends from Asia to Australia (CABI 2011). In Asia, it is found in China (Zhang 1994; AQSIQ 2007), India, Indonesia, Japan, Korea, Malaysia, Taiwan, Thailand, Vietnam, Brunei Darussalam, Cambodia, Myanmar, the Philippines and Sri Lanka (Gour and Sriramulu 1992; Hang et al. 2000; Kang et al. 2002; Chakravarthy 2015; Chakravarthy et al. 2015a; Singh and Kaur 2015a, 2016). Considering suitable area for the pest, most parts of North America, Southeast Asian countries, Australia, Africa and European countries are reported climatically suitable for this pest, although it is currently problematic in Asia and Australia. In Australia, New Zealand and European Union, it is a pest of quarantine importance. Therefore, invasiveness of this borer in these regions is critically important (Chakravarthy 2015).

From Oceania, the borer has been reported from Australia and Papua New Guinea (Waterhouse 1993; Zhang 1994; Púcat 1995; EPPO 2013; Chakravarthy 2015; Molet 2015). Further, *C. punctiferalis* has been recorded as a pest of durian in Vietnam (Loc et al. 2004), and Cai et al. (2005) reported it as a leading pest of loquat in China. While much of the species' distribution is in the subtropics, *C. punctiferalis* has also been recorded from Hokkaido prefecture, north Japan (Inoue and Yamanaka 2006) and northern China (CABI 2007, 2011).

*C. punctiferalis* has been mentioned as a national threat in the USA for grapes (Mellinda and Breiter 2007) and is considered as a potential hazard in New Zealand (Reed 2009). Ikemiya et al. (2008) reported grapefruits as hosts of this borer in Osaka, Japan. *C. punctiferalis* has been recorded as a pest of durian in the Philippines with 21% to 30% infestation (Anonymous 2008). It has been included in the list of pests regulated by the Canadian Food Inspection Agency, Canada, under the *Plant Protection Act* (CFIA 2016). The simulation model predicted southern, eastern, central and northeastern parts of India climatically suitable for the successful establishment of the pest (Sridhar et al. 2015).

Serious damage of this pest was recorded both in winter and rainy seasons on guava (Ansari 1945; Butani 1979; Tandon 1988; Sandhu et al. 1979). It is a serious pest on guava and castor in Punjab (Raj 1979). Pruthi and Batra (1960) reported this borer as a sporadic pest on guava. The crop failure due to the attack of this borer is not uncommon, and it is a major pest of castor and cardamom and a minor pest of pomegranate, jackfruit and tamarind (Atwal and Dhaliwal 1997). Rambutan is the host to *C. punctiferalis* (Osman Mohd and Chettanachitara 1987). Hussain et al. (1987) surveyed mango orchards in Bangladesh and reported *C. punctiferalis* as a major pest. Singh and Kumar (1992) reported *Dichocrocis festivalis* (Swinhoe) infesting 11.36% to 33.33% of litchi fruits in the first fortnight of June during1988–1990 in Bihar, India. Larvae of the new pest made holes at the stalk end of the fruit, bored through the pulp and into the seeds, where they fed on the endocarp and cotyledons and affected fruits fall prematurely. The insect was described as being of restricted and local occurrence. He (1997) observed shoots, flower clusters, young leaves and fruits of *Fortunella* trees (citrus) attacked by *C. punctiferalis* larvae throughout the year in Renshan, Sichuan, China. The most critical stage was when fruits began to turn orange. Ram et al. (1997) recorded *C. punctiferalis* infesting grapes for the first time in Karnataka and recorded 50% reduction in yield due to this pest. At Ludhiana, 9% to 14.6% guava fruits were infested from the last week of July to the first week of October (Anonymous 2000). Huang et al. (2000) reported longan, *Dimocarpus longan*, as a host of *C. punctiferalis* in China.

Singh et al. (2002) observed the pest boring into unripe mango fruits near the stalk end in Sambhalhera village of Muzaffarnagar, Uttar Pradesh, India. It is reported to attack mango fruits during April to June in Tirupati, Andhra Pradesh (Kannan and Rao 2007). Gundappa and Kumar (2015) reported that castor capsule borer has recently become a serious problem in the mango-growing tracts of Uttar Pradesh.

The pest has also been reported on apricot, citrus, ginger, guava, jackfruit, mango, mulberry, peach, pear, plum, tamarind and turmeric (Fletcher 1914, 1922; Pruthi and Batra 1960; Butani 1979; Chakravarthy 2015; Singh and Kaur 2015a, b, 2016; Singh et al. 2016). It is also an emerging pest of cocoa in India (Alagar 2013) which feeds on rind and pods causing premature drying and shedding of flowers and fruits. Chakravarthy (2015) reported that larvae of this crambid moth are typically polyphagous attacking more than 120 wild and cultivated plants. *Punica granatum* also served as host plant for this pest (Gurmey 1918; Tryon 1920; Fletcher 1922; Ballard 1924; Clausen 1927; Veitch 1931; Hutson 1937; Swamy 1937; Sloan 1945; Twine 1971; Bilapate and Talati 1977; Cai and Mu 1993; Lu et al. 1995; Wu 1995; Konno and Shishido 1996; Peter 1996; Chakravarthy et al. 1997; Park et al. 1998). During May to June 2015, about 10% infestation of this borer was observed on litchi fruits in Hoshiarpur and Gurdaspur (Singh and Kaur 2015b). Recently, the borer was observed infesting grapes during June 2016 in the vineyards of Ludhiana, Punjab.

In China, this species has caused serious damage to peach, chestnut, sunflower and sorghum (Wei 1956; Wang 1991; Wu et al. 1999). It is the most serious insect pest of *Carica papaya* L. in Australia (Smith 1937; Chay-Prove et al. 2000) and *Durio zibethinus* in Thailand (CABI 2005). Ma and Bai (2004) reported *C. punctiferalis* as a major pest of pomegranate fruits in China.

*Nephelium lappaceum* L. is reported from Australia on banana, apple, papaya, orange, cotton, maize and sorghum, from China on pear, from Pakistan on loquat, from Java on teak trees and cocoa and from Malaya on rambutan, as reviewed by workers (Anonymous 1913a, b; David et al. 1964; Mishra and Teotia 1965). It is also reported on peach, plum, pomegranate and guava from Burma (Gosh 1940), on pear (Kondo and Miyahara 1930) and peach from China (Yang and Shaw 1961) and on mango, guava, mulberry, pomegranate, peach, pear (Fletcher 1921), loquat (Hussain 1924), orange (Pruthi and Mani 1945), jackfruit, avocado (Anonymous 1952) and guava (Ansari 1945) from India. *C. punctiferalis* has also been recorded on peach as a major pest (Konno et al. 1980; Konno et al. 1981; Kadoi and Kaneda 1990; Abe and Sanari 1992; Kimura and Honda 1999) in Japan.

Biosystematic studies of Japanese *Conogethes* sp. with special reference to host plant preference (Honda 2013a) revealed that the two *Conogethes* species – *C. punctiferalis* (CPU) and *C. pinicolalis* (CPI) – larvae showed host preference as CPU larvae were polyphagous, but their development was delayed on CPI hosts, while CPI larvae feed typically on conifer needles. Male moths of both species were cross attracted to calling females and to pheromone gland extracts because female sex pheromone systems of both are similar. Final conspecific sexual recognition in each species is accomplished by a male pheromone identified from hair pencil organs of CPU but no volatile pheromones from male CPI. Reports of *C. punctiferalis* on *Pinus* (pine), *Larix* (larch), *Cedrus* (cedar), *Abies* (fir) and other pinaceae are likely to be *C. pinicolalis*, a species that was first described in 2006 (Inoue and Yamanaka 2006).

During the last decade, the fruit borer *C. punctiferalis*, an internal tissue borer, has become an important insect pest of fruit crops in Punjab. In order to study its host plants and severity on different fruit crops, surveys were conducted during 2013–2014 and 2014–2015 in different agroclimatic zones of Punjab, i.e. south-western arid zone, central plain zone and sub-mountainous zone along with fixed plot surveys at the fruit research farm and college orchard of Punjab Agricultural University (PAU), Ludhiana. Studies revealed that *C. punctiferalis* was an active pest on eight fruit crops, i.e. *ber*, loquat, peach, mango, pomegranate, plum, pear and guava in Punjab (Singh and Kaur 2014, 2015a, b, 2016; Singh et al. 2016).

## 14.3 Damage Symptoms

**Papaya** After hatching, larva penetrates the hollow leaf stalk of papaya and, after feeding for a time on its succulent bases, bores into the crown in which it pupates (Anonymous 1913a).

**Guava** Ansari (1945) reported that larvae fed on guava fruits and passed larval and pupal stages inside them and the infested fruits dry up and drop before maturity. Sengupta and Behura (1957) observed larvae boring into young mango fruits but confined to pulp only and did not attack the stone. The presence of granular faecal

Damage on litchi fruits — *Conogethes* adult sitting on litchi fruits

Damage on peach fruit — Adult of *Conogethes* sp. resting on mango

**Fig. 14.1a** Infestation of *Conogethes* sp. on various fruit crops in Punjab

matter, entangled in the webbing, at the entrance hole was observed to be a typical symptom for identification of the damage caused by this borer (Figs. 14.1a and 14.1b) (Singh and Kaur 2016).

**Mango** Gundappa and Kumar (Gundappa and Kumar 2015) reported that on hatching, the caterpillars bore into mango fruit, bud or shoot and feed within on pulp and seeds or soft tissues. The fruit borer affects both mesocarp and the seed preferably the seeds in mango; the pest renders fruits unfit for human consumption. The larva feeds on the rinds of fruits, later bores inside and feeds on internal contents. The granular faecal pellets are seen outside the fruits. When fruits are in close proximity, it forms favourable niche for the larva to bore into fruits. Fruits damaged by this pest

**Fig. 14.1b** Infestation of *Conogethes* sp. on select fruit crops in Punjab

are exposed to secondary infection by pathogens leading to black corky appearance on the epicarp. Singh and Kaur (2016) recorded *C. punctiferalis* larvae feeding tunnel on mango inflorescence in Hoshiarpur, Punjab. Larvae were observed to bore inside the tunnel fruit by joining two fruits together. Black frass was also seen near the entrance, and attacked fruits were found to rot and drop off. One larva per fruit was observed causing about 10% damage to the trees (Figs. 14.1a and 14.1b) (Singh and Kaur 2014).

**Durian** Larva feed on floral buds, flowers and young and mature durian fruits. Infestation began on small fruits of marble size and was marked by the presence of massy faecal materials between spines on the durian fruit. Damage was more external and common on fruits that are borne in clusters, and it seldom feed on flesh or seed. Fruits damaged by *Conogethes* were sometimes exposed to secondary infection by pathogens that lead to fruit rot (Chong et al. 1991).

**Litchi** The larvae bore into the litchi fruits either from peduncle or the lateral side of the fruit. The presence of granular faecal matter entangled in the webbing, at the entrance hole, was a typical symptom for identification of damage. The larvae were found feeding on pulp as well as nut. As the larvae grew, they tunnelled in to the centre of the nut and pupated inside. The adults were observed to emerge from these tunnels through the entrance hole (Fig. 14.1b) (Singh and Kaur 2015b).

**Cocoa** Alagar (2013) reported that the caterpillars feed on rind and pods causing premature drying and shedding of flowers and fruits, resulting in heavy yield losses to cocoa, *Theobroma cacao* Linnaeus, at ICAR-Central Plantation Crop Research Institute, Regional Station, Vittal, Karnataka (see the chapter by Alagar in this book).

**Peach** Severe damage was observed on the fruits of peach in PAU, Ludhiana, during May to June 2015. The larvae bore into the fruits and feed on the pulp. The hanging of faecal material with the silken webs was a typical symptom of identification of this borer, and peak activity was recorded during April to May (Fig. 14.1b) (Singh et al. 2016).

**Pear** Severe damage and high population of larvae and adults were observed on pear fruits in Ludhiana and Hoshiarpur, Punjab, during April to July 2015. The hanging of faecal material with the silken webs was a typical symptom of identification of this borer (Fig. 14.1a) (Singh and Kaur 2016).

**Plum** Severe damage and higher numbers of larvae and adults were observed on plum in Ludhiana and Ladhowal, Punjab, during April to May 2015. Usually, damage was observed more when fruits were in a cluster and leaves were also webbed together with fruits (Singh and Kaur 2016).

**Pomegranate** The hanging of faecal material with the silken webs was a typical symptom of identification of this borer (Singh et al. 2016).

**Ber** Fruits were observed to be infested during March to April, and larvae were found in infested fruits.

**Loquat** The hanging of faecal material with the silken webs was a typical symptom of identification of this borer. The damage was recorded in April to May (Singh et al. 2016).

**Grapes** The hanging of faecal material with the silken webs was a typical symptom of identification of this borer.

The amount of loss caused by *C. punctiferalis* can be difficult to determine due to damage by other pests in the same crop and the attraction of secondary pests and diseases to existing damage (CABI 2011).

## 14.4 Population Dynamics

**Durian** The population was relatively high from August to September and March to April which coincided with the peak harvest of durian in the Philippines (Evangelista 1995).

**Longan** Huang et al. (2000) reported that in longan, *C. punctiferalis* was distributed in different spaces and growth stages. The highest insect number was in the middle and lower parts of the south aspect, followed by the lower part of the west aspect; the lowest insect number was in the middle and lower parts of the north aspect. The highest insect number was observed in the mature fruit stage, while the lowest insect number was observed in the young fruits.

**Guava** Gupta and Arora (2001) conducted preliminary survey on the incidence of lepidopteran fruit borers of guava in Jammu, India, on cultivars Allahabad Safeda, red flesh, Hybrid No. 1, Hybrid No. 2 and Lucknow-49. Results revealed severe infestation of *C. punctiferalis*. The pest infestation was first noticed during the second fortnight of October with 2.5% fruit infestation which subsequently reached a maximum of 23% in the fourth week of November and lasted up to the second week of March. Kaul and Kesar (2003) studied seasonal incidence at two locations, i.e. Udheywalla (irrigated) and Raya (rainfed) of Jammu, India, on the guava cultivar Lucknow-49. Highest incidence (20%) was noted in the 32nd standard week at Udheywalla, while peak incidence (9%) was recorded in the 33rd standard week at Raya. A comparison of the infestation levels at Udheywalla and Raya revealed that fruit borer infestation was higher in irrigated compared to the rainfed area. Multiple correlation analyses of the data at both locations showed that the abiotic factors like maximum temperature, minimum temperature, relative humidity (morning and evening) and rainfall contributed up to 63.2% ($R^2 = 0.632$) of the borer infestation. Comparative observations on the incidence of fruit borer indicated that on guava, pest incidence started in the last week of July and went on up to the second week of October (Singh 2004). Severe damage occurred on guava in Punjab during July to September. Activity on guava ranges from April to October (Anonymous 2004) with maximum incidence during the last week of August in Punjab (Singh 2004). On guava, during mid-August to end-October, 10% to 15% fruit infestation was recorded in Ludhiana, Patiala and Moga, Punjab (Singh and Kaur 2015a).

**Ber** In Punjab, *ber* (jujube) was the first crop to be infested with peak pest activity during the last week of March with 2% to 3% per fruit infestation in Ludhiana and Moga (Singh et al. 2016).

**Loquat** Peak activity period was the last week of April with 5–10% fruit infestation in Ludhiana, Punjab (Singh et al. 2016).

**Mango** On mango, borer infestation was recorded in Ludhiana, Hoshiarpur and Patiala (Punjab) during May to June with 5% to 10% fruit infestation (Singh et al. 2016).

**Pomegranate** Pomegranate fruits were found infested with yellow peach moth larvae during May and July to August with 15% to 20% infestation in Ludhiana, Jalandhar and Fazilka, Punjab (Singh et al. 2016).

**Plum** About 10–15% fruit infestation on an average was recorded during the last week of May in Ludhiana and Hoshiarpur, Punjab (Singh et al. 2016).

**Pear** Peak activity was observed in July with fruit infestation up to 10–15% in Ludhiana, Amritsar and Tarn Taran, Punjab (Singh et al. 2016).

**Litchi** During May to June 2015, about 10% infestation of this borer was recorded on litchi fruits in Hoshiarpur and Gurdaspur, Punjab (Singh and Kaur 2015b).

**Peach** Peak activity period was observed during mid-May with 10–15% damage in Ludhiana, Punjab (Singh et al. 2016). An attempt was made to study the influence of climate change on distribution of *C. punctiferalis* using the CLIMEX software to predict climatically suitable locations for *C. punctiferalis* (base year 1962–1991) and future climate change situation, i.e. 1 °C increase in global temperature. A north world expansion of suitability was predicted in North America, China and European countries, and south world expansion was predicted in South America mainly due to a decrease in cold stress (Sridhar et al. 2015).

**Biology** The biology of the pest is described in other chapters, and only select fruits are covered in this chapter.

**Peach** Yang and Shaw (1961) studied the biology of peach borer, *C. punctiferalis*, in China and reported that the larvae overwintered in the flowers, stem and fallen leaves and that they completed four to five generations in a year. Females of *C. punctiferalis* preferred young peach fruits over turnip taproots for oviposition, indicating that oviposition was induced by plant olfactory stimuli and non-host plants potentially stimulate oviposition (Luo and Honda 2015). These results indicated that not only oviposition was induced by plant olfactory stimuli but also that various non-host plants potentially stimulate oviposition. The workers further observed that even in host plants, biophysical stimuli restricted oviposition suggesting that the real host range for oviposition may be narrower than that expected from laboratory assessments based on olfactory responses of female moths.

**Pomegranate** On pomegranate, the larval population is reported to develop faster compared to other hosts, i.e. maize, brinjal, *bhindi* and bottle gourd. All other parameters including adult emergence, multiplication per generation and innate capacity of increase were higher in pomegranate than other hosts. Mean duration of generation was reported to be 35.73 days with 0.0841 as innate capacity of increase and 1.84 as weekly multiplication rate of population (Bilapate 1978).

**Durian** The husk borer, *C. punctiferalis*, completed its life cycle in 31–40 days on durian fruits in the Philippines. There were several (two to three) generations in a fruiting season (Evangelista 1995).

**Plum** Wang and Cai (1997) reported that there were five generations of this pest in a year on Younai plum variety in China.

## 14.5 Management

### 14.5.1 Resistant Varieties

**Guava** Sandhu et al. (1979) evaluated nine guava cultivars and categorized them into five distinctly separate groups, i.e. very low, low, moderate, serious and very serious, under heavy field infestation of this borer. Infestation was examined on all trees of each variety. The frass tunnel of lower end indicated confirmed borer infestation. The remaining fruits were dissected to see either pink caterpillar on tunnels plugged with frass and excreta inside. Population of *Conogethes* was the maximum on winter crop of seedless guava (50%). Apple, guava and Allahabad Safeda showed serious and moderate infestation, respectively. Guinea guava and red flesh cultivars showed lower infestation of borer, while the others were categorized as low-infested cultivars. Skin character appeared to have no relation with the incidence. This pest was found to cause 9% to 14% fruit damage in guava orchards in Punjab. Gupta and Arora (2001) evaluated guava germplasm in Jammu, India, and revealed that the cultivar Allahabad Safeda was the least susceptible with 9.50% fruit infestation at maturity, whereas red flesh was the most susceptible with 14.30% fruit infestation. Incidence was highest on cultivar Sardar (8.08%) and ranged from 2.08 (August) to 20% (September) followed by Hisar Safeda, Mridula and Hisar Surkha with significantly higher incidence of 5.42 (1–13%), 5.41 (1–12.29%) and 5.40 (1–9.33%), respectively, as compared to Arka Amulya (3.86%) and Sardar selection 4/10 (3.55%). No incidence was observed on Tehsildar, Shilong-1 and cattley guava which did not differ significantly from Behrampur seedling (0.16%), Fazilka selection (0.18%), Bangalore seedling selection 17/7 (0.23%) and Banarasi Surkha (0.47%) (Gill et al. 2005). Sharma et al. (2008) reported *C. punctiferalis* to cause damage to guava throughout the plains of India. Allahabad Safeda was found to be the most susceptible cultivar in Ludhiana (Singh and Kaur 2016).

**Litchi** Germplasm evaluation of litchi cultivars for resistance to *C. punctiferalis* revealed that the maximum damage was on the cultivar Dehradun followed by Seedless Late and Calcutta at the Fruit Research Station, Gangian, Hoshiarpur (Singh and Kaur 2016).

**Peach** Florida Prince cultivar of peach was more susceptible than Shan-i-Punjab. Pathar Nakh variety of pear was more susceptible compared to Punjab Beauty and Baggu Gosha (Singh and Kaur 2016).

**Plum** Plum variety Satluj Purple was more susceptible compared to Kala Amritsari (Singh and Kaur 2016).

**Chestnut** Du et al. (2016) reported that field infestation rates differed significantly among chestnut cultivars against *C. punctiferalis*. The highest infestation rate (15.5%) was in Huaijiu cultivar, followed by Yanhong (8.5%), Huaihuang (7.0%) and Shisheng (4.0%) cultivars.

### 14.5.2 Cultural Practices

Unprofitable host plants in the vicinity of orchard and infested fruits should be destroyed (Anonymous 1913a). Removing infested/damaged fruits from orchards and fields, scraping off the fruit tree bark in which *Conogethes* spp. larvae overwinter (Wang and Cai 1997) and pruning of infested shoots during early crop season have been found to be beneficial in reducing population of this borer (Kumar et al. 2015). In Punjab, the number of insecticidal sprays was reduced in litchi, guava and mango orchards where sanitation was maintained (Singh and Kaur 2016).

### 14.5.3 Bagging of Fruits

In China, covering the surface of undamaged fruits with paper bags to keep *C. punctiferalis* from laying eggs and boring proved successful. This also helped to remove the infested fruits bagged after egg laying (Anonymous 2004). Bagging of fruit is likely to prevent adult *C. punctiferalis* from laying eggs on the surface of fruit (Reed 2009; Sushil et al. 2015). However, bagging will be a viable mitigation option if the bags are in place during the entire growing season. Bagging can be combined with phytosanitary visual inspection (Reed 2009).

### 14.5.4 Trap Crops

Trap cropping is based on the pest's host plant preference and accomplished by providing a plant variety or species preferred over the main crop. A pest moving into an area is lightly attracted to a preferred trap plant and be diverted away from the main crop (Atwal and Dhaliwal 1997).

**Light Traps** Light trapping has been attempted in parts of Japan; however, it is not considered efficient (Konno et al. 1982). Setting light traps (some fumigant on a piece of cotton placed under a 60 W black light) and sugar-vinegar traps (containing a mixture of sugar, vinegar water and insecticide) in orchards and fields may assist in the control of adults (Anonymous 2004).

*C. punctiferalis* has been trapped in light traps previously (Korycinska 2012). Kim et al. (2014) studied the impact of insect capture in a chestnut orchard using three different light traps (A, B and C type) with lamps. The mercury lamp trap captured 125 insect species, out of which 115 were chestnut pests. The B and C type light traps, comprising a Dulux-EL white lamp, were examined for their capturing ability. The type B trap attracted Coleopteran insects (83%), while type C captured Lepidopteran insects (73%). The Dulux-EL lamp captured the highest number of *C. punctiferalis* adults (mean 10.2 adults) using the type C light trap. These results suggested that selection of the appropriate type of light traps and lamps based on the target pest species is critical in ensuring effective and eco-friendly control of the pest population. Light trap can be used to attract *C. punctiferalis* on pecan nut in Queensland, Australia (Anonymous 2016a).

### 14.5.5 Mechanical Control

Removal of infested buds and shoots (Anonymous 2004) and collection and proper disposal of infested fruits proved effective against this pest in guava orchards in India (Ansari 1945; Singh and Kaur 2015a, 2016). The infested plant parts should be collected and destroyed (Singh 1970; Singh and Kaur 2015a, 2016).

### 14.5.6 Botanical Insecticides

Lee (2009) conducted a study to develop environmentally friendly control techniques to reduce chestnut insect pests in the intensive chestnut orchards of Jinju city, Gyeongnam, province of Korea. In early- and middle-ripening cultivars of chestnut tree, the damage to the fruits by *C. punctiferalis* was significantly lower in wromstop than other treatments such as wromstop+wood vinegar and capture machine, while there was no significant difference among treatments in late-ripening cultivars. The high control effect for *C. punctiferalis* was the highest in wromstop treatment with 40.49% and 41.89% in early- and late-ripening cultivars. The diverse biological effects of neem like repellence, phago-deterrence, growth inhibition, abnormal development and oviposition deterrents could be better exploited by combining them with the trap crop. Diversification of pest by neem application results in pest congregation. And the area for treatment with biocontrol is reduced.

### 14.5.7 Microbial Control

Xu et al. (2002) conducted experiments using *Bacillus thuringiensis* for the control of *C. punctiferalis* on chestnut in China. They concluded that *B. thuringiensis* preparation Sulijing and Ludebo is safe and pollution-free to control *C. punctiferalis* on chestnut. Control efficiency amounted to 73.8% and 79.1%, respectively, for both the preparations, obviously higher than usual chemical control. The methods deserve widespread popularity as their cost is as economical as 1.3–3 yuan per 667 $m^2$. Injection of *B. thuringiensis* (*Bt*) preparation in water into the bored holes of castor capsule borer effectively checked the resurgence of the pest. The native isolates of entomopathogenic fungi (EPF), *Metarhizium* and *Beauveria bassiana*, and entomopathogenic nematode (EPN), *Heterorhabditis indica*, successfully cross-infected the larvae of *C. punctiferalis* in the laboratory. Though biocontrol agents like *Bt* and EPF are promising, certain factors such as highly specific activity limit their use on crops as they must be eaten to be effective and coverage must be thorough. Manikandan et al. (2016) screened the *cry1* gene in six indigenous isolates of *Bt* by PCR with degenerate primers showing amplification in all the Bt isolates. The toxicity analysis of *Bt* strain T27 against *C. punctiferalis* showed 100% mortality on the fifth day after treatment.

### 14.5.8 Biological Control

Rodrigo (1941) recorded three natural enemies of *C. punctiferalis* from Ceylon, namely, the braconid, *Phanerotoma hondecasiella* Cam.; the ichneumonid, *Xanthopimpla* sp.; and *Dolichurus* sp., but none of them in an appreciable number. David et al. (1964) listed *Angita* (*Dioctes*) *trochanterata* Morl., *Theronia inareolata*, *Bracon brevicornis* Wesmael and *Apanteles* sp. on larvae and *Brachymeria euploeae* West as pupal parasitoid of *C. punctiferalis*. Patel and Gangrade (1971) noticed *Microbracon hebetor* as a larval parasitoid, causing 9.61% parasitism.

Joseph et al. (1973) reported two hymenopterans *Brachymeria nosatoi* Habu and *B. lasus* Westwood parasitizing *C. punctiferalis*. Jacob (1981) reported *Myosoma* sp. and *X. australis* as parasites of *C. punctiferalis*. Evangelista (1995) reported larval-pupal parasite, braconid wasp (*Suallonius* sp.) and larval predator and earwig (*Euborellia annulata*) among natural enemies of *C. punctiferalis* on durian in the

Philippines. Huang et al. (2000) reported *Apanteles* sp., *Brachymeria lasus* and *Temelucha* sp. as natural enemies of *C. punctiferalis* on longan fruits in China. The parasite ratio was the highest in the expansion stage of fruits. Varadarasan (2001) reported *Temelucha* sp., *Agrypon* sp. and *Friona* sp. as parasites of *C. punctiferalis*. In Australia, a tachinid parasitoid *Argyrophylax proclinata* Crosskey has been reported to parasitize *C. punctiferalis* even up to 40% (Pena et al. 2002) in *Annona*. Two peak periods were noticed on the parasitization of *C. punctiferalis*, i.e. during the first week of June (55%) and first week of November (40%). The details of biocontrol are covered in another chapter.

Tachinid flies were recovered in large numbers from larvae of castor capsule borer collected from pomegranate fruits at PAU, Ludhiana, during June 2015 and 2016; from litchi fruits collected from Fruit Research Station, Gangian, Hoshiarpur, and Regional Research Station, Gurdaspur, during June 2015; from plum fruits from PAU, Ludhiana, and University Seed Farm, Ladhowal, Ludhiana, during April to May 2015; from pear fruits at PAU, Ludhiana, during April to June 2015; from peach fruits collected from PAU, Ludhiana, during May to June 2016; from mango fruits from Hoshiarpur during May to June 2015; and from guava fruits collected from PAU, Ludhiana, during July 2015 (Singh and Kaur 2016).

*Bracon hebetor* Say and *B. brevicornis* (Wesmael) (Hymenoptera: Braconidae) are highly polyphagous, gregarious and ecto-larval parasitoids of several species of lepidopteran larvae. One of the common insect pests which is the host of *Bracon* is *C. punctiferalis* (Anonymous 2016b). Several parasites, parasitoids, predators, microbial agents and entomopathogens have been recorded, but their efficacy is limited under field conditions (Chakravarthy 2015).

### 14.5.9 Semiochemicals

(a) Pheromones

Konno et al. (1982) identified the sex pheromone as (E)-10-hexadecenal. The addition of (Z)-10-hexadecenal led to an increase in trap catches. Mori et al. (1990) found that a 10:1 mixture of the (E)- and (Z)-formate was shown as attractive as a 10:1 mixture of (E)- and (Z)-10-hexadecenal (genuine pheromone) in a 100 μg dose against this pest. Liu et al. (1994) found that the most attractive blends were a mixture of 16-Ald, E10-16-Ald and Z10-16-Ald at 16:100:8 and a blend of E10-16-Ald and Z10-16-Ald at 100:8.

Cai and Mu (1993) reported that placing traps consisting of plastic bowls filled with water below a pheromone dispenser (containing 250 μg) significantly reduced damage caused by *C. punctiferalis* in two citrus orchards, at 4 traps/0.23 $hm^2$, in China. There seems to be a difference in attraction between populations as this pest seems to consist of two different populations in the Northeastern Asian region, with one group responding to the blend of 100:8-100:11 between (E)-10-hexadecenal and (Z)-10-hexadecenal and the other to that of 100:43. The first group was found in Japan and China and the second in Korea and China (Boo 1998).

Chakravarthy and Thyagaraj (1998) trapped this species using 9:1 for (E)-10-hexadecenal and (Z)-10-hexadecenal. This blend has been used for mass trapping, monitoring and mating disruption in Japan, Korea and China (Kimura and Honda 1999). In Korea, an 80:20 ratio of (E)-10-hexadecenal and (Z)-10-hexadecenal had the highest attractiveness in orchards (Jung et al. 2000). Further work by Xiao et al. (2012) found that certain hydrocarbons had a synergistic effect on responses to pheromones. (E)- and (Z)-8-tetradecenyl formate have been synthesized and tested for effectiveness in trapping *C. punctiferalis*.

Response to two main sex pheromone components was identified as E-10-hexadecenal and Z-10-hexadecenal, and their ratios tend to vary in natural conditions, and female body wax was observed to be synergistic in improving the field efficacy of sex pheromone (Sithanantham et al. 2013). Honda (2013b) separated female body wax by a column chromatography, and five monoenyl hydrocarbons showed synergistic effect thus concluding that sex pheromone set for yellow peach moth consisted of E10-16-Ald and Z10-16-Ald for a long range attraction and Z9-27-CH and Z3Z6Z9-23-CH for the final recognition of females by males in near pheromone source. A blend of (E)-10-hexadecenal (E10-16-Ald) and (Z)-10-hexadecenal (Z10-16-Ald) in 95.4:4.5 was less effective than pheromone extracts and can be improved by adding non-polar fractions (NPF). Therefore, Honda and Wei (2015) concluded that a full set of sex pheromone system is more efficient in capturing yellow peach moth males for a long range and near pheromone source attraction, and Sithanantham (2015) stated that improving trap would facilitate moth catches.

(b) Kairomones

Attraction and oviposition responses of the yellow peach moth to seven species of fungi were investigated by Honda et al. (1988) in laboratory and field cages. The gravid moths were attracted to and oviposited on fruits of codling, a host plant. The attraction was removed from a fresh codling fruit by dipping it in ether, but higher attraction reappeared subsequently when the fruit was inoculated with fungi. Similar attraction was observed on a mouldy rice cake. In contrast, a rotten codling or rice cake, infected with bacteria rather than fungi, inhibited oviposition by the moths. From the mouldy codling or rice cake, four fungi (*Penicillium* sp., *Cladosporium* sp., *Aspergillus fumigatus* and *Mucor* sp.) were isolated. When grown on Czapek's medium, the first two species showed highest attraction, while the last two species were less attracted. Among the four phytopathogenic fungi tested, *Endothia parasitica* and *Alternaria solani* were also attractive. In a field cage, the moths oviposited significantly more on the oviposition substrates baited with *Penicillium* sp. than on a fresh codling or unbaited control.

Through field surveys and the tests of behavioural response, electroantennogram (EAG) response and multiple-choice oviposition, Chen et al. (2010) studied the effects of volatiles from Nongda No. 1 chestnut (NC) and Heyuan oil chestnut (HC) on the host-selection behaviour of adult *C. punctiferalis*. Their field survey showed that the moth-eaten rate of NC fruits by adult *C. punctiferalis* was 16.1%–25.3%, while that of HC fruits was less than 5%. The volatiles from NC fruits and leaves were more attractive to female than to male moths, and the fruit volatiles were more attractive than leaf volatiles. However, the volatiles from HC fruits and leaves were not attractive to both female and male moths. The EAG response showed that female moths had significantly higher response to NC fruit volatiles than to that of HC, but male moths had no significant difference in this response. For both NC and HC, the EAG responses of female and male moths to fruit volatiles were higher than those to leaf volatiles. The number of eggs laid by female moths was much greater on NC fruits than on NC leaves and on HC fruits and leaves, but there was no significant difference in the impact on the latter three.

### 14.5.10 Radio-Frequency Treatments As Alternative Non-chemical Methods

Chemical fumigation has been widely used to control insects in postharvest chestnuts, but there are inherent risks while using fumigants. Hou et al. (2015) conducted a study to validate the application of radio-frequency (RF) treatments for disinfesting chestnuts as an alternative to chemical fumigation. A practical process protocol was developed to control insect pests in chestnuts using a 27.12 MHz free-running oscillator RF system. Mortality of fifth-instar *C. punctiferalis* increased with increasing holding time at 55 °C using RF heating and reached 100% while holding in hot air for at least 5 min. Furthermore, there was no significant quality difference in colour, fat, firmness, moisture, protein and soluble sugar of chestnuts observed between RF treatments and controls. RF treatment methods hold potential to be scaled up for industrial applications for disinfesting chestnuts.

### 14.5.11 Chemical Control

The larvae would be difficult to control by chemical treatments because they can bore into fruit and are hence more difficult to target. In addition, some of the hosts have flowers that attract honeybees and other pollinators (Korycinska 2012).

**Citrus** Spraying with 50% Sumithion (fenitrothion) solution, 50% omethoate emulsion or 50% dichlorvos at fruit colouring after summer pruning and 3 weeks before flowering gave adequate pest control on *Fortunella* trees (Citrus) in Renshan, Sichuan, China (He 1997).

**Guava** Kaul and Kesar (2003) tested six insecticides, namely, cypermethrin 10 EC (0.02%), carbaryl 50WP (0.10%), dimethoate 30EC (0.05%), neem oil 30EC (3%), endosulfan 35EC (0.07%) and polytrin C (a combination of profenofos 40EC + cypermethrin 4EC, 0.04%), at recommended dosages for management of the borer in guava orchards of Jammu, India. They reported that two applications of dimethoate 30 EC (0.05%) were the most effective in controlling the infestation followed by polytrin C, and neem oil was the least effective. In case of severe infestation, the crop can be sprayed with 1 kg Sevin 50 WP (carbaryl) in 250 litres of water/acre and can be repeated in 15 days (Anonymous 2004). Chlorpyriphos (0.05 and 0.1%) was found to be effective against *C. punctiferalis* in the guava orchards of Punjab (Singh 2004). Studies on the combined efficacy of proven management strategies against oriental fruit fly, *Bactrocera dorsalis* (Hendel) conducted at Ludhiana, Punjab, revealed that the practices also significantly reduced the incidence of fruit borer, *C. punctiferalis*. The fruit infestation ranged from 0% to 16.67% in IPM compared to 76.67% in control. Overall, mean borer infestation was significantly lower in IPM (4%) compared to control (19.39%) (Gill and Mann 2008).

**Apple** Combined cold treatment and methyl bromide fumigation treatment of *C. punctiferalis* which overwinters as mature (fifth-instar) larvae can be used. Cold treatment alone is unlikely to mitigate the risk of live larvae entering New Zealand. The USA has two treatment schedules (for different container types) against *C. punctiferalis* on apples from Japan or Korea (TQAU USDA 2008).

**Chestnut** Yuan et al. (2011) tested the effect of 2% Thiacloprid DP on *C. punctiferalis* in chestnut orchards of China. Results showed that Thiacloprid DP sprays could achieve significant results to kill the *C. punctiferalis* in all of the Lishui region, and the kill efficiency became stronger as the dosage increased. When the dosage of Thiacloprid DP was 3000 g and 4500 g, the average controlling effect exceeded 70%, while when the dosage was 1500 g, the average controlling effect exceeded 60%. The best pesticide was effective up to 80.64%.

**Durian** Spraying with lambda cyhalothrin or deltamethrin (0.00625 kg ai/ha) from 3 weeks after first spray for 2 months significantly reduced borer infestation from 47.4% to 6.66% and 8.09%, respectively, and increased yield of durian fruits in the Philippines by 6254.48 and 4704.70 kg/ha over control (Evangelista 1995).

**Castor** Dusting 5% lindane at 20–25 lb. per acre was effective in managing castor capsule borer (Rao et al. 1954). Chlorantraniliprole 18.5% SC, spinosad 0.018% and flubendiamide 480 SC were superior for the management of *C. punctiferalis* on castor. Flubendiamide, chlorantraniliprole, lambda cyhalothrin and acephate recorded less per cent damaged capsules, and spinosad also resulted in high seed yield (Narayanamma 2013). Spraying malathion and lambda cyhalothrin was found effective against *C. punctiferalis* (Kumar et al. 2015).

## 14.6 Integrated Management

Since *Conogethes* attacks diverse categories of crops in diversified habitats, information on host plant relationships is required for developing appropriate IPM practices (Sekiguchi 1974; Chakravarthy et al. 2015a, b, c). Hang et al. (2000)

recommended binding bundles of straw on the trunk to induce insects to hibernate in them and burn, hanging up sex attractants, dusting 45–60 kg lindane powder per ha in mid-May and spraying a 3000–4000 times solution of 2.5% kung fu or a 6000 times solution of 1.8% abamectin. For integrated control of *C. punctiferalis* on pomegranate, Ma and Bai (2004) recommended that besides agricultural measures (scraping trunk bark, pruning and using black lamp and sugar-vinegar mixed water), it is necessary to spray a 2000–3000 times dilution solution of 1.8% abamectin suspoemulsion 2–3 times in May to early July. Bagging the durian fruit 1 month after fruit set, monitoring fruit borer infestation at an early stage of fruit development and fruit thinning to reduce infestation are beneficial, and fallen fruits must be collected and destroyed.

The following points can be included for integrated management:

- Growing pest-free variety is the safest, cheapest and the easiest technique if available (Pruthi and Batra 1960).
- General cultivation of coloured (flesh) varieties may be advocated to avoid field infestation of fruit borer (Sandhu et al. 1979).
- Collection and destruction of infested as well as fallen fruits by burning or burying in soil (Sushil et al. 2015; Singh and Kaur 2016).
- Destroy all the infested shoots, buds and fruits in the initial stage of attack (Butani 1979).
- Pruning infested shoots during early crop season (Kumar Senthil et al. 2015).
- Intercropping and mixed cropping are important management tools.
- Mechanical methods also checked the pest to an extent (Butani 1978) and found up to 5%–10% infestation in Japanese persimmon (Tomomatsu et al. 1995).
- A combination of phytosanitary inspection and bagging of young fruits with small-holed polythene covers prevent distribution and egg laying, respectively (Korycinska 2012). Bagging at 1 month after fruit set was equally effective against *C. punctiferalis* on durian in the Philippines, reducing infestation to 9.2% (Evangelista 1995). One of the control methods used in China is to put bags over individual fruits (Korycinska 2012).
- Conserving natural enemies is an important activity as they can suppress this pest. Therefore, pesticides that kill beneficial insects can be avoided (http://thailand.ipminfo.org/pests/Durian_fruit_borer.html).
- Setting light traps (black/blue light) at 1 per ha for adult moth catch in pomegranate and guava (http://agritech.tnau.ac.in/crop_protection/pome_pest/pome_2.html).
- Young caterpillars feeding outside the fruit can be removed or dislodged using a stick or a piece of wire (http://thailand.ipminfo.org/pests/Durian_fruit_borer.html).
- Insecticides like chlorpyriphos 50 EC (2 ml/l), neem oil (3 ml/l) and acephate 75 SP (2 g/l) (Chakravarthy et al. 2015a) and malathion and lambda cyhalothrin (Kumar Senthil et al. 2015) can be applied at appropriate time on reproductive parts.

Strict quarantine measures should be taken to prevent further spread of *Conogethes* and introduction to areas where they have not yet gained entry. There is an urgent need to do more studies on its identification, host specification and management.

**Acknowledgement** The authors are thankful to the Project Coordinator, ICAR-AICRP on Fruits; Project Coordinator, ICAR-Consortium Research Platform on Borers in Network Mode; and Director of Research, PAU, Ludhiana, for providing funds and research facilities, and photographs included in this chapter are of the senior author.

## References

Abe Y, Sanari G (1992) Notes on the peach moth *Conogethes punctiferalis* attacking cynipid and aphid galls. Jpn J Entomol 60(1):108

Alagar M (2013) *Conogethes punctiferalis*: An emerging insect pest of cocoa (*Theobroma cacao* Linn.) In abstracts: Chakravarthy AK, Kumar CTA, Verghese A, Thyagaraj (eds) NE New Horizons in Insect Science: international conference on insect science, GKVK, Bangalore, Karnataka, India, February 14–17, 2013, p 31

Anonymous (1913a) Insects injurious to Pawpaw apples in Queensland. Agric J Brisbane 3:33–35

Anonymous (1913b) The cotton boll-worm. Queens Land Agric J 3:142

Anonymous (1952) List of pests of agricultural importance in Uttar Pradesh. Agric Anim Hus, UP 3:121–129

Anonymous (2000) Insect pests of fruits. In: Dhooria MS, Mann GS (eds) Annual research report (1999–2000). Department of Entomology, Punjab Agricultural University, Ludhiana, p 388

Anonymous (2004) Field problems of fruit crops. Punjab Agricultural University, Ludhiana, p 252

Anonymous (2005) Protocol of phytosanitary requirements for the apple export from China to Mexico. senasica.gob.mx/includes/asp/download.asp?iddocumento=599&idurl=21178

Anonymous (2008) Integrated pest and disease management of durian. Republic of Philippines. http://ecoenclave.com/wp-content/uploads/2015/11/IPM.pdf

Anonymous (2015) Survey protocol for pests in ginger and lentil. Integrated Environment Awareness and Community Development Centre Lazimpat, Kathmandu. http://www.npqpnepal.gov.np/downloadfile/Survey%20protocols%20of%20pest_1440996317.pdf

Anonymous (2016a) Vortex insect control. http://www.vortexics.com.au/insects.htm

Anonymous (2016b) www.agrinaturals.in

Ansari MAR (1945) *Dichocrocis punctiferalis* Gn. as a pest of guava. Short notes and exhibits Indian J Ent 7:241

AQSIQ (2007) Chinese table grape export technical information. Animal and plant quarantine service of the People's Republic of China, p 99

Astridge D (2001) Insect fauna surveys on rambutan, durian and mangosteen in North Queensland. Technical Bulletin Department of Primary Industry and Fisheries, Northern Territory of Australia, pp 75–79

Atwal AS, Dhaliwal GS (1997) Agricultural pests of South Asia and their management. Kalyanai Publishers, Ludhiana, p 487

Ballard E (1924) Entomological Note. Queens Land Agric J 11:31

Bilapate GG (1978) Life table for the castor capsule borer, *Dichocrocis punctiferalis* Guenee on different hosts. Proc Ind Acad Sci 87(9):217–220

Bilapate GG, Talati GM (1977) Effect of different food plants on the larval and pupal development of *Dichocrocis punctiferalis* Guenee. Madras Agric J 64(1):53–54

Boo KS (1998) Variation in sex pheromone composition of a few selected lepidopteran species. J Asia Pac Entomol 1(1):17–23

Butani DK (1978) Insect pests of fruit crops and their control. Pesticides 12(8):53–59

Butani DK (1979) Insects and fruits. Periodical Expert Book Agency, New Delhi, p 415
Butani DK (1993) Mango: pest problems. Periodical Expert Book Agency, Delhi, p 290
CABI (2005) Crop protection compendium. CAB International, Wallingford
CABI (2007) Crop protection compendium, 2007 edition. CAB International, Wallingford http://www.cabicompendium.org/cpc/home.asp
CABI (2011) *Conogethes punctiferalis* datasheet. In: Crop protection compendium. CAB International, Wallingford
Cai RX, Mu ZL (1993) Trap trials for *Conogethes punctiferalis* with sex pheromones in citrus orchards. China Citrus 22(1):33
Cai P, Bai L, Xiang R, Zhang L (2005) The biological characteristics of the leading pests of loquat and their control. South China Fruits (2):38–41
CFIA (2016) List of pests regulated by Canada. http://www.inspection.gc.ca/plants/plant-pests-invasive-species/pests/regulated-pests/eng/1363317115207/1363317187811
Chakravarthy AK (2015) Shoot and fruit borer *Conogethes* spp.: an overview. In abstracts: Chakravarthy AK, Krishnamoorthy A (eds) Round table on *Conogethes punctiferalis* and allied species (Series 1). IIHR, Bengaluru, May 22, 2015, p 4
Chakravarthy AK, Thyagaraj NE (1998) Evaluation of selected synthetic sex pheromone of the cardamom shoot and fruit borer *Conogethes punctiferalis* in Karnataka. Pest Manag Hortic Ecosyst 4:78–82
Chakravarthy AK, Vasudevkammar, Kumar KP (2015) Monograph on shoot and fruit borer, *Conogethes punctiferalis* and allied species. Consortia Research Platform on Borers, ICAR-Indian Institute of Horticultural Research (IIHR), Bengaluru
Chakravarthy AK, Shivanna S, Chandrappa L, Naik L (1997) Insect pests on castor in Karnataka. Insect Environ 2(4):135
Chakravarthy AK, Kammar V, Kumar KP (2015a) Monograph on shoot and fruit borer, *Conogethes punctiferalis* and allied species. ICAR-Indian Institute of Horticultural Research (IIHR) Hesaraghatta, Bengaluru, p 56
Chakravarthy AK, Kammar V, Lokeshwari D, Rani AT, Nagaraj T, Rajan VV (2015b) The black-spotted yellow shoot-and-fruit borer, *Conogethes* spp. (Crambidae: Lepidoptera) – a global perspective. Curr Sci 109(6):1016–1018
Chakravarthy AK, Shashank PR, Doddabasappa B, Kandakoor SB, Chandrashekariah (2015c) Biosystematic, molecular characterization and management of shoot and fruit borer *Conogethes spp.* (Crambidae: Lepidoptera). In: Singh B, Arora R, Gosal SS (eds) Biological and molecular approaches in pest management. Scientific Publishers, Jodhpur, pp 329–343
Chay-Prove P, Ross P, O'Hare P, Macleod N, Kernot I, Evans D, Grice K, Vawdrey L, Richards N, Blair A, Astridge D (2000) Agrilink series: your growing guide to better farming. Queensland horticulture institute and department of primary industries, Nambour, Queensland
Chen BX, Dong YZ, Liang GW, Lu H (2010) Effects of volatiles from chestnut on host preference of adult *Conogethes punctiferalis* (Lepidoptera: Pyralidae). Ying Yong Sheng Tai Xue Bao 21(2):464–469
Chong KK, Ooi PAC, Tuck HC (1991) Crop pests and their management in Malaysia. Tropical Press Sdn, Bhd, p 242
Clausen CP (1927) The citrus insects in Japan. Tech Bull USDA 15:15
David BV, Narayanaswamy PS, Murugesan M (1964) Bionomics and control of castor shoot and capsule borer, *Dichocrocis punctiferalis* Guenee (Lepidoptera: Pyralidae). Ind Oilseeds J 8:146–158
Du Y, Jiaxin Z, Zengguang Y, Yongqiang M, Mengmeng Y, Minzhao Z, Zhiyong Z, Ling Q, Cao Q (2016) Host preference and performance of the yellow peach moth (*Conogethes punctiferalis*) on chestnut cultivars. PLoS One 11(6):e0157609. https://doi.org/10.1371/journal.pone.0157609
EPPO (2013) PQR-EPPO database on quarantine pests. Available at: http://www.eppo.int
Evangelista C (1995) Biology, ecology and control of durian fruit borers, *Tonica lagaropis* Meyrick and *Conogethes punctiferalis* (Guen) (Pyralidae: Lepidoptera). The Science and Technology Information Network of the Philippines. http://scinet.dost.gov.ph

Fang ZW, Zhe QL (2009) A review on *Dichocrocis punctiferalis* Guenee. J Hebei Agric Sci 20:64–79
Fletcher TB (1914) Some South Indian insects. Superintendent, Government Press, Madras, p 433
Fletcher TB (1921) Annotated list of Indian crop pests. Bull Agric Res Inst Pusa 100:1–246
Fletcher TB (1922) Report of the imperial entomologist. Scientific reports of the Imperial Institute of Agricultural Research, Pusa, pp 51–67
Ganesha (2011) Studies on biology of castor shoot and capsule borer, *Conogethes punctiferalis* Guenee (Lepidoptera: Pyralidae) and its management. M.Sc. thesis, UAS, GKVK, Bangalore, p 120
Gill HS, Mann GS (2008) Effect of management strategies for *Bactrocera dorsalis* (Hendel) on *Dichocrocis punctiferalis* Guenee infestation in guava orchards. J Insect Sci 21:133–138
Gill HS, Mann GS, Sharma DR, Singh G (2005) Host plant resistance to insect pests of guava 1. Comparative incidence of castor capsule borer, *Dichocrocis Punctiferalis* Guenee, Abstract 1st international guava symposium 5–8 Decemeber 2005, Lucknow, India, p 92
Gosh CC (1940) Insect pests of Burma. Government Printing, Rangoon, p 126
Gour TB, Sriramulu M (1992) Grapevine, *Vitis vinifera* Linn. – a new host of castor shoot and capsule borer, *Conogethes punctiferalis* (Guenee). Trop Pest Manag 38(4):459
Gundappa, Balaji R, Kumar HK (2015) *Conogethes punctiferalis*: an emerging pest on mango in Lucknow region. In abstracts: Chakravarthy AK and Krishnamoorthy A (eds) Round table on *Conogethes punctiferalis* and allied species (Series 1). IIHR, Bengaluru, May 22, 2015, p 6
Gupta RK, Arora RK (2001) Lepidopteran fruit borers on guava in Jammu. Insect Environ 7(2):83–84
Gurmey WB (1918) The insect pests of maize. Agric Gazetteer Sydn 9:641–650
Hang H, Yan K, Sun X, Ma J (2000) Investigation on the kinds of fruit moth in the western part of Henan province and their control. China Fruits 2:44–45
Haseeb M (2007) Current status of insect pest problems in guava. In: Singh G, Kishun R, Chandra E (eds) Proceedings of the 1st international guava symposium, Lucknow, India, 5–8 December 2005. Acta Horticulturae vol 735. pp 453–467
He RQ (1997) Study on the control of yellow peach moth in China. South China Fruits 26(4):33
Hill DS (1983) Agricultural insect pests of the tropics and their control, 2nd edn. Cambridge University Press, New York, p 746
Honda H (2013a) Biosystematics of Japanese *Conogethes* spp. with special reference to host plant preference and reproductive isolation. In abstracts: Chakravarthy AK, Kumar CTA, Verghese A, Thyagaraj NE (eds) New horizons in insect science: international conference on insect science, GKVK, Bangalore, Karnataka, India, February 14–17, 2013, p 92
Honda H (2013b) Chemistry and function of unsaturated hydrocarbons as new synergistic pheromonal components of the yellow peach moth, *Conogethes punctiferalis* (Guenee). In abstracts: Chakravarthy AK, Kumar CTA, Verghese A, Thyagaraj NE (eds) New horizons in insect science: international conference on insect science, GKVK, Bangalore, Karnataka, India, February 14-17, 2013, p 91
Honda H (2015) Biosystematics of Japanese *Conogethes* spp. with special reference to host plants preference and reproductive isolation. In abstracts: Chakravarthy AK, Krishnamoorthy A (eds) Round table on *Conogethes punctiferalis* and allied species (Series 1). IIHR, Bengaluru, May 22, 2015, pp 17–18
Honda H, Wei X (2015) Chemistry and function of unsaturated hydrocarbons as new synergistic pheromonal components of the yellow peach moth, *Conogethes punctiferalis*. In abstracts: Chakravarthy AK and Krishnamoorthy A (eds) Round table on *Conogethes punctiferalis* and allied species (Series 1). IIHR, Bengaluru, May 22, 2015, p 25
Honda H, Kaneko J, Konno Y, Matsumoto Y (1979) A simple method for mass-rearing of the yellow peach moth, *Dichocrocis punctiferalis* Guenée (Lepidoptera: Pyralidae), on an artificial diet. Appl Entomol Zool 14(4):464–468
Honda H, Ishiwatari T, Matsumoto Y (1988) Fungal volatiles as oviposition attractants for the yellow peach moth, *Conogethes punctiferalis* (Guenée) (Lepidoptera: Pyralidae). J Insect Physiol 34(3):209–211

Hou L, Hou J, Li Z, Johnson JA, Shaojin W (2015) Validation of radio frequency treatments as alternative non-chemical methods for disinfesting chestnuts. J Stored Prod Res 63:75–79

http://agritech.tnau.ac.in/crop_protection/pome_pest/pome_2.html

http://thailand.ipminfo. org/pests/Durian_fruit_borer.html

Huang Y, Zhang X, Wei H, Hu Q, Zhan Z (2000) Studies on the *Dichocrocis punctiferalis* Guenee and its enemies. Acta Agric Univ Jiangxiensis 22(4):523–525

Husain M, Nargis A, Noor P (1987) Studies on the insect pest of mango (*Mangifera indica* L.) of Rajshahi. Bangladesh J Sci Ind Res 22(1–4):191–194

Hussain MA (1924) Annual report of the entomologist to govt, Punjab, Lyallpur, for the year 1922–23. Rep Dep Agric Punjab 2:177–204

Hutson JC (1937) Report on the work of the entomological division. Report of the Director of Agriculture Cyclone, pp 22–28

IBPGR (1988) Descriptors for papaya. International Board for Plant Genetic Resources, Rome

Ikemiya J, Yamaguti H, Nasu Y, Shibao M (2008) Lepidopterous insect pests of grape and fig newly recorded from Osaka prefecture. Annu Rep Kansai Plant Prot Soc 50:175–177

Inoue H, Yamanaka H (2006) Redescription of *Conogethes punctiferalis* (Guenée) and description of two new closely allied species from Eastern Palaearctic and oriental regions (Pyralidae, Pyraustinae). Tinea 19:80–91

Jacob SA (1981) Biology of *Dichocrocis punctiferalis* (Guenee.) on turmeric. J Plant Crop 9:119–123

Joseph KJ, Narendran TC, Joy PJ (1973) Studies on oriental *Brachymeria* (Chalcidoidea). A report of the work done under P.L. 480 research project, entitled Taxonomic studies of the oriental species of Brachymeria (Hymenoptera: Chalcididae). University of Calicut, Calicut

Jung J, Han K, Choi K, Boo K (2000) Sex pheromone composition for field-trapping of *Dichocrocis punctiferalis* (Lepidoptera: Pyralidae) males. Korean J Appl Entomol 39(2):105–110

Kadoi M, Kaneda M (1990) Development of yellow peach moth *Conogethes punctiferalis* on apple fruit. Res Bull Plant Prot Serv Japan 26:61–63

Kang CH, Lee KC, Park CG, Lee DW (2002) Seasonal occurrence pattern of peach pyralid moth, *Dichocrocis punctiferalis*, in Fuyu persimmon orchards and fruit damage at harvesting time. Kor J Appl Entomol 41:107–112

Kannan M, Rao NV (2007) Seasonal incidence of lepidopteran pests in relation to weather parameters in mango (*Mangifera indica*). Crop Res Hisar 33:198

Kaul V, Kesar YK (2003) Incidence and management of lepidopteran fruit borers of guava (Psidium guajava L.) in Jammu, India. J Asia Pac Entomol 6(2):201–205

Kim Y, Kim H, Kim G (2014) Effects of light trap structure and lamp type on the attraction of chestnut pests in an orchard. Kor J Appl Entomol 53(3):217–223

Kimura T, Honda N (1999) Identification and possible functions of the hair pencil scent of the yellow peach moth *Conogethes punctiferalis* in Japan. Appl Entomol Zool 34(1):147–153

Kondo T, Miyahara T (1930) Fruit borers and a summary of life histories in Kwangtung. China J Plant Prot 17:85–94

Konno Y, Shishido T (1996) Glutathione-dependent 0-alkyl and oral conjunctions for dicapthon and fertikothine in several insects. J Pestic Sci 21(4):430–433

Konno Y, Honda N, Matsumoto Y (1980) Observation on the mating behavior and biology for the sex pheromone of yellow peach moth *Dichocrocis punctiferalis*. Appl Entomol Zool 15(3):321–322

Konno Y, Honda N, Matsumoto Y (1981) Mechanism of reproductive isolation between the fruit feeding and pin ace feeding type of yellow peach moth, *Dichocrocis punctiferalis*. Jpn J Appl Entomol Zool 25(4):253–258

Konno Y, Arai K, Sekiguchi K, Matsumoto Y (1982) (E)-10-hexadecenal, a sex pheromone component of the yellow peach moth, *Dichocrocis punctiferalis* Guenée (Lepidoptera: Pyralidae). Appl Ent Zool 17(2):201–207

Korycinska A (2012) Rapid assessment of the need for a detailed pest risk analysis for *Conogethes punctiferalis* (Guenée). The Food and Environment Research Agency, p 7

Kumar Senthil CM, Jacob TK, Devasahayam S (2015) *Conogethes punctiferalis*-a major pest on ginger, turmeric and cardamom. In abstracts: Chakravarthy AK, Krishnamoorthy A (eds) Round table on *Conogethes punctiferalis* and allied species (Series 1). IIHR, Bengaluru, May 22, 2015, p 5

Lee C (2009) Study on chestnut insect pests by environmentally friendly controls in Korea. Kor J Appl Entomol 48(1):95–100

Liu M, Tian Y, Li Y (1994) Identification of minor components of the sex pheromone of yellow peach moth, *Dichocrocis punctiferalis* Guenée and field trials. Entomologia Sin 1(2):150–155

Loc H, Le Q, Nguyen V (2004) Integrated controlling fruit borer (*Conogethes punctiferalis* Guen.) on durian in The Mekong Delta. http://sofri.org.vn/FileUpload/Download/153/99201410141962.pdf

Lu ZX, Yang ZF, Wang GY, Bo WL (1995) The niche of the corn and peach borers and interspecific competition between them in corn. Acta Agric Zhajiangensis 7(1):31–34

Luo Z, Honda H (2015) Olfactory and biophysical assessment of the oviposition stimulating potential of host and non-host plants for the yellow peach moth, *Conogethes punctiferalis* (Lepidoptera: Crambidae). In abstracts: Chakravarthy AK, Krishnamoorthy A (eds) Round table on *Conogethes punctiferalis* and allied species (Series 1). IIHR, Bengaluru, May 22, 2015, p 27

Ma J, Bai H (2004) The main pests in the pomegranate producing areas of Sichuan province and their integrated control. South China Fruits 33(5):70–71

Manikandan R, Muthukumar C, Ramalakshmi A, Balasubramani V, Udayasuriyan V (2016) Screening of new isolates of *Bacillus thuringiensis* for *cry1* genes and testing of toxicity against *Dichocrocis punctiferalis* (family: Pyralidae, order: Lepidoptera). Microbiol 85(2):191–196

Mellinda S, Breiter N (2007) Grape commodity-cased survey reference. Cooperative agricultural pest survey, USA. https://www.aphis.usda.gov/plant_health/plant_pest_info/eg_moth/downloads/grape-commodityreference.pdf

Mishra SC, Teotia TPS (1965) Studies on the biology of castor shoot and capsule borer. Ind Oilseed J 9:184–187

Molet T (2015) CPHST pest datasheet for *Conogethes punctiferalis*. USDA-APHISPPQ-CPHST

Mori K, Watanabe H, Fujiwhara M, Kuwahara S (1990) (E)- and (Z)-tetradecenyl formate, potent sex pheromone mimics against the yellow peach moth. Phero Synth 122:1257–1259

Narasimha Swamy RL (1937) Staminodes in *Elettaria cardamomum*. Maton Curr Sci 5(8):427

Narayanamma VL (2013) Evaluation of newer insecticide molecules against lepidopteran pests in castor, *Ricinus communis* L. In abstracts: Chakravarthy AK, Krishnamoorthy A (eds) Round table on *Conogethes punctiferalis* and allied species (Series 1). IIHR, Bengaluru, May 22, 2015, p 40

Nishida GM (2002) Hawaiian terrestrial arthropod checklist, 4th edn. Bishop Museum Press, Honolulu, 313 pp

Osman Mohd SB, Chettanachitara C (1987) Postharvest insects and other pests of rambutan. In: Lam PF, Kosiyachinda S (eds) Rambutan: Fruit Development, Postharvest Physiology and Marketing in ASEAN. ASEAN Food Handling Bureau, Kuala Lumpur, pp 57–60

Park JD, Lec SG, Kim CS, Byun RK (1998) The bionomics of the oak nut weevil, *Mehorisur sulus* and the insect pests of the corer in Korea. J For Sci 5:151–156

Patel RK, Gangrade GA (1971) Note on the biology of castor capsule borer, *Dichocrocis punctiferalis* Gn. (Lepidoptera: Pyralidae). Ind J Agric Sci 41:443–444

Pena JE, Nadel H, Pereira MB, Smith D (2002) Pollinators and pests of *Annona* sp. In: Pena J, Sharp J, Wysoki (eds) Tropical fruit pests and pollinators: biology, economic importance, natural enemies and control. CABI Publication, Wallingford, p 208

Peter KV (1996) Major spices. In: 50 years of Crop Science Research in India. ICAR, New Delhi, pp 771–796

Pruthi HS, Batra HN (1960) Some important fruit pests of North-WEST India. Ind Counc Agric Res Bull 80:113

Pruthi HN, Mani MS (1945) Our knowledge of insect and mite pests of citrus in India and their control. Scientific monograph no. 16, The Imperial Council of Agricultural Research, p 31

Púcat A (1995) *Conogethes punctiferalis*. Yellow peach moth. Canadian Food Inspection Agency Science Branch
Hans Raj (1979) Studies on the biology and behavior of castor shoot and capsule borer, *Dichocrocis punctiferalis* (Guen.) (Pyralidae: Lepidoptera) on castor. M.Sc. thesis, Punjab Agricultural University (PAU), Ludhiana
Ram KT, Hugar PS, Somasekhar Patil BV (1997) Record of *Conogethes punctiferalis* (Guenee) on grapes. Pest Manag Hortic Ecosyst 3:37–39
Rao VT, Rao PG, Rao PVR (1954) Guide showing methods to control crop pests and diseases. Dep Agric Fish Andhra Bull 1:1–13
Reed C (2009) Import risk analysis: table grapes (*Vitis vinifera*) from China. MAF Biosecurity New Zealand
Robinson GS, Tuck KR, Shaffer M (1994) A field guide to the smaller moths of South-East Asia. Malaysian Nature Society, Kuala Lumpur, p 193
Rodrigo E (1941) Administrative Report of the Acting Director of Agriculture for Colombo:1941
Sandhu GS, Deol GS, Sohi AS (1979) Incidence of fruit-boring insects in guava cultivars. Punjab Hortic J 19:171–173
Sekiguchi K (1974) Morphology, biology and control of the yellow peach moth, *Dichocrocis punctiferalis* Guenée (Lepidoptera: Pyralidae). Bull Ibaraki Hortic Exp Station (Special Issue): 89
Sengupta GC, Behura BK (1957) Some new records of crop pests from India. Ind J Entomol 17(2):283–285
Shaffer M, Nielsen ES, Horak M (1996) Pyraloidea. In: Nielsen ES, Edwards ED, Rangsi T(eds) Checklist of the Lepidoptera of Australia. Monogr Australia Lepid vol 4, pp 164–199
Sharma DR, Bal JS, Chahill BS (2008) Assessment of reaction of guava germplasm against fruit fly and castor capsule borer. Ind J Hortic 65(2):145–151
Shashank PR, Kammar V, Mally R, Chakravarthy AK (2018) A new Indian species of shoot and capsule borer of the genus *Conogethes* (Lepidoptera: Crambidae), feeding on cardamom. Zootaxa 4374(2):215–234
Singh JP (1970) Pests of guava and their control. Labdev J Sci Technol 8(1):19–24
Singh H (2004) Management of *Bactrocera dorsalis* (Hendel) in guava orchards and its impact *on Dichocrocis punctiferalis* Guenee. Ph.D. thesis, Punjab Agricultural University, Ludhiana, pp 31–34
Singh S (2015) Damage of castor capsule borer, *Conogethes punctiferalis* (Gueneee) on pear. http://www.nbair.res.in/insectpests/Dichocrocis-punctiferalis.php
Singh S, Kaur G (2014) Diversity of pestiferous borers of mango in Punjab. Ind J Appl Entomol 28(2):125–127
Singh S, Kaur G (2015a) Current status of incidence of fruit borer, *Conogethes punctiferalis* Guenee on fruit crops in Punjab. In abstracts: Chakravarthy AK, Krishnamoorthy A (eds) Round table on *Conogethes punctiferalis* and allied species (Series 1). IIHR, Bengaluru, May 22, 2015, p 8
Singh S, Kaur G (2015b) Incidence of fruit borer, *Conogethes punctiferalis* (Guenee) on litchi in Punjab. Abstract. National Conference on entomology, Punjabi University, Patiala, October 29–30, 2015, pp 66–67
Singh S, Kaur G (2016) Annual Report (2015–16). Consortium Research Platform on Borers in Network Mode, Ludhiana Centre, p 31
Singh YP, Kumar V (1992) *Dichocrocis festivalis* Swinh. (Lepidoptera: Pyralidae)-a new pest of litchi *Litchi chinensis* Sonn. J Bombay Nat Hist Soc 89(1):137
Singh YP, Singh GK, Kumar S (2002) Occurrence of *Dichocrocis punctiferalis* Guen. (Lepidoptera:Pyralidae) on mango in western UP. Progress Hortic 34:130
Singh S, Kaur G, Bajaj K (2016) Diversity of borer insect pests on peach in Punjab. Natural resource management ecological prospectives. February 18–20, 2016. Sher-e-Kashmir University of Agricultural Sciences and Technology, Jammu, India
Sithanantham S (2015) Scope for R&D towards improved utilization of pheromone trapping system for *Conogethes punctiferalis*-overview. In abstracts: Chakravarthy AK, Krishnamoorthy A

(eds) Round table on *Conogethes punctiferalis* and allied species (Series 1). IIHR, Bengaluru, May 22, 2015, p 24

Sithanantham S, Vardarasan S, Chandrasekar SS, Thiyagarajan P, Chakravarthy AK (2013) Semiochemical research in *Conogethes punctiferalis* (Guenee) (Crambidae: Lepidoptera): global trends and Indian scenario. In abstracts: Chakravarthy AK, Kumar CTA, Verghese A, Thyagaraj NE (eds) New horizons in insect science: international conference on insect science, GKVK, Bangalore, Karnataka, India, February 14–17, 2013, p 76

Sloan WJS (1945) Insect pests of grain sorghum. Queens Land Agric J 61(4):221–239

Smith JH (1937) Insects and allied pests of the papaya. Queens Land Agric J 48(5):553–557

Sridhar V, Vinesh LS, Chakravarthy AK (2015) Estimating the potential geographical distribution of yellow peach moth, *Conogethes punctiferalis* Guenee (Crambidae: Lepidoptera) using CLIMEX. In abstracts: Chakravarthy AK, Krishnamoorthy A (eds) Round table on *Conogethes punctiferalis* and allied species (Series 1). IIHR, Bengaluru, May 22, 2015, p 40

Sushil SN, Singh IP, Nagaraju DK, Ranawat C (2015) Integrated Pest Management (IPM) in guava (*Psidium guajava*) for export purpose. Directorate of Plant Protection, Quarantine & Storage, Faridabad (Leaflet)

Tandon PL (1988) Insect pests of tropical fruit trees and their management. In: Gupta VK, Sharma NK (eds) Tree protection. Indian Society of Tree Scientists, Solan, pp 278–293

Thyagaraj NE (2003) Integrated management of some important cardamom pests of hill region of Karnataka, South India. Ph.D. thesis, Dr. B. R. Ambedkar University, Uttar Pradesh, India, p 230

Tomomatsu S, Sakaguchi T, Ogino T, Hiramatsu T, Kaukami (1995) Methyl bromide fumigator for quarantine control of persimmon fruit moth and yellow peach moth on Japanese persimmon. Res Bull Plant Prot Serv 31:67–73

TQAU USDA (2008) Treatment schedules. https://manuals.cphst.org/Tindex/treatmentSearch.cfm

Truscott L (2007) *Conogethes punctiferalis* (Guenee, 1854) (Lepidoptera: Crambidae) at light in Cornwall, newly recorded for. Europe Entomol Gaz 58:203–204

Tryon H (1920) Report of the entomologist, vegetable pathologist. Queens Land Annu Rep Dep Agric 19(16):49–63

Twine PH (1971) Insect pests of maize. Queens Land Agric J 97(10):523–531

USDA (1957) Insects not known to occur in the United States. Yellow peach moth (*Dichocrocis punctiferalis* (Guenee), pp 37–38

Varadarasan S (2001) Insect pest management in cardamom-key to reduce cost of production, spice India, pp 14–19

Veitch R (1931) Report of the chief entomologist. Annu Rep Dep Agric Queens Land Agric J: 43–46

Walker K (2007) Yellow peach moth *(Conogethes punctiferalis)*. PaDIL. http://www.padil.gov.au

Wang PY (1980) Economic insect fauna of China, vol 21. Science Press. FASC, Beijing, p 229

Wang RJ (1991) Serious damage to sorghum by yellow peach moth in southern Gansu. Plant Prot 17:9

Wang YX, Cai RT (1997) The emergence and control of *Dichocrocis punctiferalis* Guenee for Younai plum variety. South China Fruits 26(3):45

Waterhouse DF (1993) The major arthropod pests and weeds of agriculture in Southeast Asia: distribution, importance and origin. Australian Centre for International Agricultural Research, Canberra

Wei HJ (1956) A key insect pest of sunflower-yellow peach moth. Chinese Bull Entomol 2:78

Wu LM (1995) A study on the pupation pattern of three pyralid caterpillars helps in maize field. Plant Prot 21(3):24–26

Wu JD, Zhu YJ, Zhu GL, Zhu CH (1999) Occurrence and control of yellow peach moth on chestnut. Hebei Fruits 4:17–18

Xiao W, Matsuyama S, Ando T, Millar JG, Honda H (2012) Unsaturated cuticular hydrocarbons synergize responses to sex attractant pheromone in the yellow peach moth, *Conogethes punctiferalis*. J Chem Ecol: 8

Xu J, Wu L, Hong Q, Lang X, Xu Y, Mao Q (2002) Experiment about *Bacillus thuringiensis* control of *Dichocrocis punctiferalis* Guenee on chestnut. J Jiangsu For Sci Technol 29(5):31–32
Yang H, Shaw K (1961) Preliminary studies on the peach borer *Dichocrocis punctiferalis*. Acta Oecn Entomol Sci 1(2):109–134
Yuan Z, Wang M, Lu J, Li Q (2011) Efficacy of 2% Thiacloprid DP to control *Dichocrocis punctiferalis* in chestnut orchard. J Jiangsu For Sci Technol 4:21–23
Zhang BC (1994) Index of economically important Lepidoptera. CAB International, Wallingford, p 599

# Bio-ecology, Damage Potential and Management of *Conogethes punctiferalis* Guenee in Plantation Crops

15

M. Alagar

**Abstract**

The shoot and fruit borer *Conogethes punctiferalis* is emerging as a major pest of cocoa in South Karnataka, South India: an average 2–2.5% infestation on cocoa pods was recorded in 2013, and an average 30–40% pod infestation, in 2016. The crambid moth completes its life cycle on cocoa over a mean of 24 days. Increased space between plants, fruit thinning, balanced nutrition, clean cultivation, timely harvests, encouragement of natural enemies and destruction of borer-affected pods helped to reduce the pod infestation. Concerted efforts to use pheromone technology and biopesticides are in progress to suppress borers .

**Keywords**

Biopesticides · Cocoa · Gut proteases · South India

## 15.1 Introduction

*Conogethes punctiferalis* Guenee (Crambidae: Lepidoptera) is an important polyphagous pest in South and Southeast Asia and Australia (Pena et al. 2002); it has also been reported as a new pest introduced in Europe. This pest infests 36 crop plants belonging to 23 families in India (Thyagaraj et al. 2003). The main host plants include guava, mango, peaches, pomegranate, jack, avocado, mulberry, loquat, peach, plum, pear, sorghum, sunflower, cotton, cocoa, castor, tamarind, amaranthus, soapnut, hollyhocks, annona and other zingiberaceous plants. It is the most serious insect pest of *Carica papaya* L. in Australia (Chay-Prove et al. 2000); *Durio zibethinus* in Thailand, fruits and maize crops in China (CPC 2005); more than 20 fruit

M. Alagar (✉)
ICAR-Krish Vigyan Kendra, Nagapattinam, Tamil Nadu, India



A. K. Chakravarthy (ed.), *The Black spotted, Yellow Borer, Conogethes punctiferalis Guenée and Allied Species*,
https://doi.org/10.1007/978-981-13-0390-6_15

crops including *Dimocarpus longan*, *Averrhoa carambola* and *Litchi chinensis* in Korea; and *Helianthus annuus* and *Macadamia ternifolia* in New Zealand (CPC 2005).

This pest was reported on cocoa plantations in Sri Lanka, Malaysia and Indonesia (Entwistle 1972). The caterpillars damaged up to 80% of the flower cushions in about 40% of the trees. A study of cocoa pests in Thrissur, Kerala, India, in 1975 revealed that *C. punctiferalis* severely damaged cocoa flower cushions and occasionally cocoa bark (Mohanan and Kumar 1976). *C. punctiferalis* is now emerging as a major pest on cocoa in India. *C. punctiferalis* larvae bore in to the pods, feed on their contents—thereby attracting secondary infection—then emerge as adults (Alagar et al. 2013). Even though the potential damage is low (2.07%), the main concern is in the similarity of *C. punctiferalis* feeding behavior with that of the cocoa pod borer *Conophomorpha gramerella* (Snellen) (Gracillariidae: Lepidoptera), which is a serious pest in Malaysia, Indonesia, Java, East New Britain and Papua New Guinea (Ooi et al. 1987; Azhar 1995; Azhar et al. 2001), causing 20–50% yield loss (Mumford 1984). Fortunately, *C. gramerella* has not been reported in India.

*C. punctiferalis* is distributed throughout Asia (China (AQSIQ 2007), India, Indonesia, Japan, Korea, Malaysia, Taiwan, Thailand and Vietnam (Gour and Sriramulu 1992; Hang et al. 2000; Kang et al. 2004)), Australia and Papua New Guinea. *Gossypium hirsutum* L., *Zea mays* L, *Sorghum bicolor* (L.) *Moench* spp., *Psidium guajava* L., *Citrus* L., *Mangifera indica* L., *Punica granatum* L. and *Zingiber* spp. (*Curcuma longa* L., *Zingiber officinale* Roscoe and *Elettaria cardamomum* Maton) in India serve as the host plants for this pest (Gurmey 1918; Tryon 1920; Ballard 1924; Clausen 1927; Veitch 1931; Hutson 1937; Narasimha Swamy 1937; Sloan 1945; Twine 1971; Cai and Mu 1993; Lu et al. 1995; Wu 1995; Konno and Shishido 1996; Peter 1996), which is now identified as *Conogethes sahyadriensis* (Shashank et al. 2018). *Theobroma cacao* L. and *Vitis vinifera* L. were also recorded as host plants (Mohanan and Kumar 1976; Gour and Sriramulu 1992; Ram et al. 1997).

*C. punctiferalis* has also been recorded as major pest on yellow peaches in Japan and other countries (Abe and Sanari 1992; He 1997; Kimura and Honda 1999; Konno et al. 1980, 1981; Kodoi and Kaneda 1990), on bananas (Jarvis 1914), on apples (Kodoi and Kaneda 1990), on soybean (Anonymous 1944), teak and kapok in Java (Kalshoven 1922, 1928; Tryon 1920), on macadamia in Queensland (Ironside and Davis 1969), on conifer (Shiukaji 1969), on hawthorn (Sun et al. 1992) and on orange (Anonymous 1913). *Castanea mollissima* in China has been attacked by this pest (Ni 1998). Youni plum and avocado (Wang and Cai 1997); peaches, nectarines and cherries (Tryon 1920); and jackfruit (Devasahayam et al. 1998) were also damaged by this pest.

## 15.2 Bio-ecology

### 15.2.1 Biology

Eggs of *C. punctiferalis* are light yellow and measure 0.63 × 0.41 mm. After incubation for 6–7 days, the eggs become dark brown with a dark head (Jarvis 1914; Thyagaraj 2003). The host phenology influences the size, growth and development of eggs (Jacob 1981; Twine 1971). Temperature and relative humidity play an important role in egg hatching (Kondo and Miyahara 1930; Mukerji and Gage 1978; Rajan 1965), and their effects on egg characteristics have been extensively studied in the laboratory (Kalshoven 1929; Thyagaraj 2003). There was a significant differences in the percentage of eggs that hatched from 65.0 ± 0.76 to 90.5 ± 1.38 and incubation period from 4.19 ± 0.80 to 9.35 ± 1.05 days under varied temperature and relative humidity (Thyagaraj 2003; Wang and Cai 1997).

Female moths lay pinkish, oval flat eggs, singly or in groups of two or three, mostly between warts or grooves of the pods, cherelles or flower cushions of cocoa plants. Eggs were rarely observed on unopened flowers and new leaves. The incubation period for eggs is 2.9 days. The full-grown larvae are very active, 3–3.5 cm long, reddish brown with brown marks with a pinkish tinge on each segment, fine hairs on the body, a dark head and a prothoracic shield (Fig. 15.1). Study of the biology of *C. punctiferalis* revealed that the total larval period was 23.82 days. After hatching the larvae feed on flower cushions, flower buds, the rinds of cocoa cherelles and pods, and later they bore into and feed on the contents of the pods; granular faecal pellets can be seen outside the pods (Fig. 15.2). Because the pods and cherelles touch, it is easy for the larvae to damage more than one pod/cherelle. Pods damaged by *C. punctiferalis* are sometimes exposed to secondary infection by pathogens that lead to pod rot (Fig. 15.3a, b). The attacked cushions dry out and shed prematurely.

The total developmental period was 34.7 days. In cocoa, the longevity of male and female moths was 5.7 and 6.5 days, respectively. The larvae can be seen on damaged pods and cherelles under a cover of silk and frass or excreta. The full-grown larvae pupate inside or in a thin, silken cocoon outside the damaged pods/cherelles. Adults emerge in 7–10 days. The life cycle under laboratory conditions ranged from 25 to 33 days (Alagar et al. 2013).

**Fig. 15.1** Larva of *C. punctiferalis*

**Fig. 15.2** Damaged cocoa pods with larvae

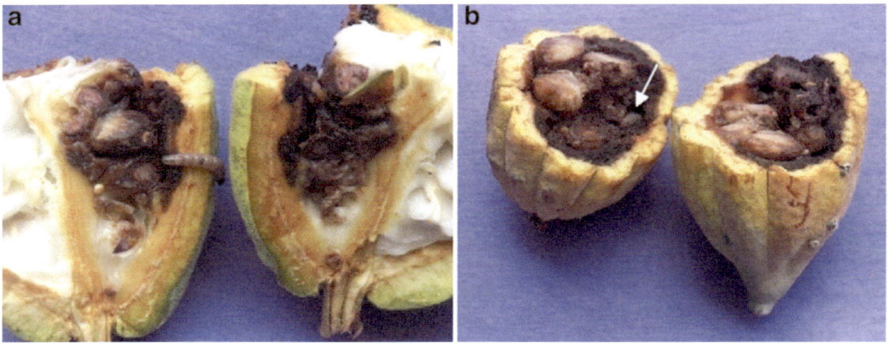

**Fig. 15.3** (**a** and **b**) Internal content of rotten cocoa pods, damaged by *C. punctiferalis*

The growth and development of different larval instars vary with temperature and relative humidity. Each larval instar lasts 3–4 days. The total larval period can extend up to 12–14 days (Jacob 1981; Kondo and Miyahara 1930; Twine 1971; Wang and Cai 1997; Xi et al. 1996). The larval period varied from 12.55 ± 2.00 to 19.59 ± 5.50 days, and the percentage survival varied from 49.6% ± 0.18% to 92.8% ± 1.39%. A temperature of 28.0 ± 1.0 °C and relative humidity at 80.0% ± 5.0% were most favorable for larval development (Thyagaraj 2003).

A clear difference can be seen in the size, shape and weight of the male and female pupae. Female pupae are larger (17.81 × 6.29 mm, weighing 0.127 g). Male pupae measure 14.30 × 4.26 mm and weigh 0.108 g (Thyagaraj 2003). The pupal period extended up to a mean of 7.90 ± 2.80 days under laboratory conditions. No significant differences in the pupal period were seen under different temperature regimes (20.0–38.0 ± 1.00 °C (Bilapate and Talati 1978; Jacob 1981; Mishra and

Teotia 1965; Wang and Cai 1997). Morphological dimorphism in the *Conogethes* pupa helps to determine sex.

### 15.2.2 Seasonal Occurrence

*C. punctiferalis* lives throughout the year on cardamom in Western Ghats in South India. Two population peaks were noticed in a year: one during April–May and the other during November–December (Thyagaraj 2003). The population coincides with periods of little or no rainfall (i.e. before and after monsoons) (Ballard 1927; Thyagaraj 2003). Damage on cocoa by *C. punctiferalis* was observed from December 2010 to May 2013. Infestation by *C. punctiferalis* started immediately after the monsoon, and peak incidence was observed during March to May (Alagar et al. 2013). Temperature and rainfall greatly influence the growth and development of this species (Moralesranous and Cote 1992; Rao 1992; Shanuowr et al. 1993). The overwintering of the needle feeding type/Pinaceae feeding type of the yellow peach moth *C. punctiferalis* was studied in China and Korea under laboratory and field conditions. Researchers observed that third or fourth instar larvae rolled needles and twigs into a bag with silk, in which they stayed during winter (Kang et al. 2004; Kuang et al. 2009).

Most insect studies deal with insects reared in a laboratory, often on artificial diets. In fact, the laboratory conditions and the diet composition are critical parameters for insect quality and yield. An artificial diet containing "soybean meal powder" and corn seeds has been developed and proposed for the maintenance and continuous rearing of *C. punctiferalis* fruit-feeding type (Honda et al. 1979; Utsumi et al. 1990). Although some efforts to rear successive generations by feeding these diets have been successful, lacunae still exist because of advances in nutritional technology.

A mean of 1.96–2.07% damage was observed on cocoa at the Central Plantation Crops Research Institute, Regional Station at Vittal (Alagar et al. 2013) Currently, 30–40% pod loss is due to *Conogethes* in a few cocoa plantations in Dakshina kannada, Karnataka, India. The damage is spreading (Chakravarthy AK, personal observation, 2016).

### 15.2.3 Gut Proteinase

Trypsin and elastase-like chymotrypsin were found to be the predominant digestive proteinases in *C. punctiferalis* infesting cardamom. The midgut proteases trypsin, chymotrypsin, elastase-like chymotrypsin and leucine-amino peptidases underwent age-related modulation. Serine protease inhibitors such as aprotinin, soybean trypsin inhibitor and phenylmethane sulfonyl fluoride inhibited peptidase activity in *C. punctiferalis*. Furthermore, aprotinin (150 mmol/L) significantly inhibited trypsin and elastase-like chymotrypsin—by 94% and 29%, respectively—under in vitro conditions (Josephrajkumar et al. 2006).

Inclusion of aprotinin (150 mm) in cardamom shoots used as feed indicated significant reductions in the body weight of larvae by day 4 compared with a control. Specific activities of two serine proteinases—namely, trypsin and elastase-like chymotrypsin—were reduced in insects fed aprotinin after 2 days of feeding. The effect of aprotinin on protein digestion and its potential to protect against the cardamom borer was established (Josephrajkumar et al. 2005). An elastase-like chymotrypsin was purified by aprotinin-agarose affinity chromatography from midgut extract of *C. punctiferalis*. The purified enzyme had a Vmax of 687.6 ± 22.1 nmol *p*Na released/min/mg protein (The rate of reaction when the enzyme is saturated with substrate is the maximum rate of reaction, this is usually expressed as the Vmax), had a Km (Michaelis constant) of 0.168 ± 0.012 mM with SAAPL*p*Na as the substrate and gave a single band on sodium dodecyl sulfide polyacrylamide gel electrophoresis, with a molecular mass of 72.1 kDa (Josephrajkumar et al. 2007a).

Among the three types of cardamom with compact tissues investigated, the Malabar type with a prostrate panicle and a thicker pseudostem tolerated damage by *C. punctiferalis*. The pseudostem diameter of cardamom variety PV-1 (2.18 cm) and Mysore-Vazhukka (2.32 cm) were significantly smaller than that of Green Gold cardamom (3.04 cm). Pseudostem girth is an important feature that confers tolerance to borers, as a hollow pseudostem accommodates the growing larvae (Josephrajkumar et al. 2002).

### 15.2.4 Molecular Identification

Molecular tools were used to confirm species identity. Intact DNA was obtained from a fifth instar larva of *C. punctiferalis* using a DNeasy kit, and it was visualized using agarose gel electrophoresis followed by ethidium bromide staining in a gel documentation system. Primers designed in this study, based on the *RPS5* and *CAD* genes. The primers were used to amplify the genomic DNA and obtain amplicons of expected sizes (Fig. 15.4). These were eluted, purified and cloned. Recombinant plasmids, extracted from clones confirmed by colony polymerase chain reaction, were sequenced. The sequences were deposited in GenBank (accession nos. KC595364 and KC595365). Sequence analysis using BLASTn revealed close homology to related moths. Phylograms constructed through the use of BLASTn analysis show the close association of *C. punctiferalis* from cocoa to *C. punctiferalis* from Hawai'i and to other related moths (Fig. 15.5a, b). The nodes were supported by high bootstrap values.

CBOL Corporation established the "All-Leps Barcodes of Life" project because Lepidoptera is the second most diverse order of insects. There are about 180,000 known species, and likely another 300,000 species await description. The initiative involves campaigns on three geographic scales: global (Geometridae, Saturniidae and Sphingidae), continental (North America and Australia) and regional (Great Smoky Mountains National Park in the United States and Area de Conservation Guanacaste in Costa Rica) (Bravo et al. 2008). Until now, 632,006 lepidopteran species have bar codes; of these, 4443 are Crambidae (International Barcode of Life 2012). Hebert et al. (2004) studied the morphological and DNA bar coding of

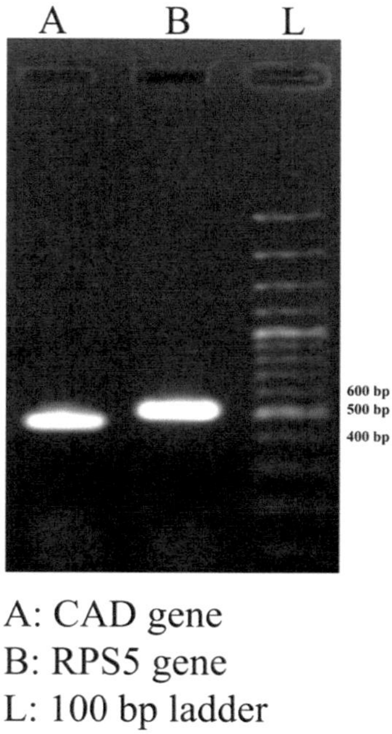


**Fig. 15.4** Amplification of DNA from *C. punctiferalis* from cocoa using gene specific primers

museum specimens of *Astraptes fulgerator* Walch, a widely distributed neotropical skipper butterfly (Lepidoptera: Hesperiidae) in northwestern Costa Rica. They showed that *A. fulgerator* is a complex of at least 10 species in Costa Rica. Largely sympatric, the caterpillars of these taxa mostly eat different plants as food, are mostly distinctive and have different ecosystem preferences, but adults differ only subtly, with no genitalic divergence.

### 15.2.5 Pest Management

Individual cultural, mechanical, biological and chemical methods are not effective in keeping this pest below economic effect thresholds. Hence, an integrated approach is required to manage this pest. Some cultural practices such as increased distance between plants have reduced pest damage (Sharma et al. 1992; Sridharan et al. 1990). Reducing nitrogenous fertilizer levels also minimizes pest damage to a large extent (Chakravarthy and Thyagaraj 1999). Clean cultivation (Sharma et al. 1995) and mechanical methods can check the pest (up to 5–10%) (Boo 1998; Butani 1978; Chakravarthy et al. 1997; Tomomatsu et al. 1995).

Some of the predators and parasites identified can be exploited to manage this pest (Abe and Sanari 1992; Anandraj and Peter 1996; Clausen 1927; Rodrigo 1940). Proven insecticides can also be used (Chakravarthy et al. 1997; He 1997; Rama 1980, Thyagaraj 2003). Individual management methods were evaluated and compared with integrated methods (Thyagaraj 2003). The golden-backed woodpecker and crow pheasant can be useful as biocontol agents on plantations. Application of

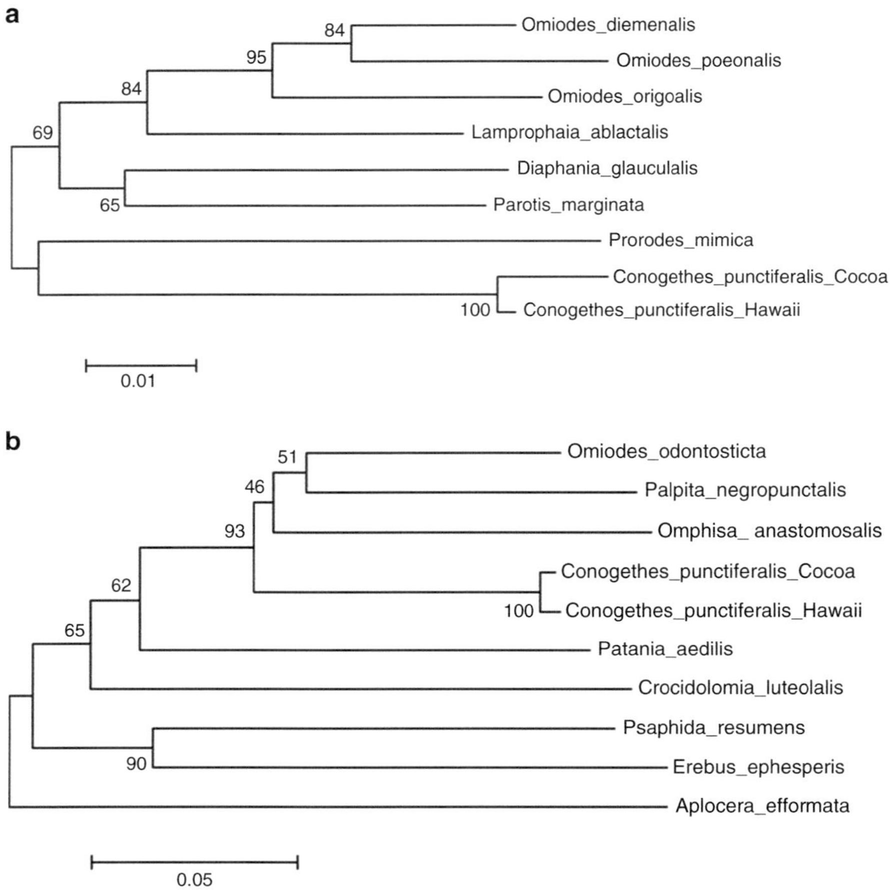


**Fig. 15.5** (**a** and **b**) Neighbor-joining phylograms of *C. punctiferalis* from cocoa using the (**a**) *RPS5* and (**b**) *CAD* genes

Steinernematid entomopathogenic nematodes or *Bacillus thuringiensis* at an appropriate and vulnerable stage of the pest (when early instar larvae are found inside the capsule or panicle or unopened leaves, i.e., within 20 days of emergence of the adult moth). Conservation of larval parasitoids (*Agrypon* sp. and *Temeluchus* sp.) at field level can suppress the pest incidence (Josephrajkumar et al. 2007b).

*C. punctiferalis* currently represents one of the most widespread and economically important lepidopteran pests, inflicting direct injury to tender vegetative and reproductive plant parts. The species is emerging as an important group because of speciation, its ability to disperse and the wide range of its host plants, which comprise more than 120 diverse species. It is emerging as a major pest on cocoa. Because cocoa is cultivated in and around evergreen tropical forest tracts in Western Ghats in India and elsewhere, suitable bioagents, cultural practices and physical methods need to be appropriately integrated.

## References


Abe Y, Sanari G (1992) Notes on the peach moth *Conogethes punctiferalis* attacking cynipid and aphid galls. Jpn J Entomol 60(1):108

Alagar M, Rachana KE, Keshava Bhat SM, Rahman S, Rajesh MK (2013) Biology, damage potential and molecular identification of *Conogethes punctiferalis* Guenee in cocoa (*Theobroma cacao* Linn). J Plant Crop 41(3):350–356

Anandraj M, Peter KV (1996) Biological control in spices. Indian Institute of Spice Research, Calicut, p 52

Anonymous (1913) Insect injurious to Papaw apple. QLD Agric J 3:33–35

Anonymous (1944) Annual report of dept. of agriculture and stock for the year. Brisbane QLD Agric J 3:88

AQSIQ (2007) Chinese table grape export technical information. Animal and plant quarantine service of the People's Republic of China, p 99

Azhar I (1995) An overview on the management of key insect pests of cocoa with major emphasis on the cocoa pod borer *Conopomorpha cramerella*. Planter 71:469–480

Azhar I, Alias A, Meriam M (2001) Research on the cocoa pod borer in Malaysia. In: Proceedings Incoped 3rd international seminar, 16th–17th October 2000, Kota Kinabalu Sabah, Malaysia, pp 105–113

Ballard E (1924) Entomological Note. QLD Agric J 11:31

Ballard E (1927) Some insects associated with cotton in Papua and in the mandated territory of New Guinea. Bull Entomol Res 27:295–300

Bilapate GG, Talati GM (1978) Some studies on bionomics of castor short and fruit borer *Dichocrocis punctiferalis* from Gujarat. J Maharashtra Agric Univ 3(1):75–76

Boo KS (1998) Variation in sex pheromone composition of a few selected lepidopteron. J Asia Pac Entomol 1(1):17–23

Bravo JP, Silva JLC, Munhoz REF, Fernandez MA (2008) DNA barcode information for the sugar cane moth borer *Diatraea saccharalis*. Genet Mol Res 7(3):741–748

Butani DK (1978) Insect pests of fruit crops and their control. Pesticides 12(8):53–59

Cai RX, Mu ZL (1993) Trap trials for *Conogethes punctiferalis* with sex pheromones in citrus orchards. China Citrus 22(1):33

Chadramohanan H, Kumar P (1975) The caster capsule borer, *Dichocrosis punctiferalis* Guen. as a pest of cocoa in India. Agricul Res Kerala 14:79–80

Chakravarthy AK, Thyagaraj NE (1999) Effect of different NPK fertilizer levels on the incidence of cardamom pests. Insect Environ 4(4):139–140

Chakravarthy AK, Shivanna S, Chandrappa L, Naik L (1997) Insect pests on castor in Karnataka. Insect Environ 2(4):135

Chay-Prove P, Ross P, O'Hare P, Macleod N, Kernot I, Evans D, Grice K, Vawdrey L, Richards N, Blair A, Astridge D (2000) Agrilink series: your growing guide to better farming. Queensland Horticulture Institute and Department of Primary Industries, Nambour

Clausen CP (1927) The citrus insects in Japan. Technol Bull USDA 15:15

CPC (2005) Crop protection compendium, Accessed in 2005. CAB International, Wallingford, UK

Devasahayam S, Koya MA, Reddy PP, Kumar WKK, Varghese A (1998) IPM in spices, challenges for the future. Advances in IPM for horticultural crops: environmental complications and thrust, Bangalore, India, October 15–17(1997), pp 157–164

Entwistle PF (1972) Pests of cacao. Longman, London, p 404

Gour TB, Sriramulu M (1992) Grapevine, *Vitis vinifera*— a new host of castor shoot and capsule borer *Conogethes punctiferalis*. Trop Pest Manag 38(4):459

Gurmey WB (1918) The insect pests of maize. Agr Gazetteer Bydney 9:641–650

Hang HL, Yan KF, Sun XH, JX MA (2000) Investigation on the kinds of fruit moth in the western part of Henan province and their control. China Fruits 2:44–45

He RQ (1997) Study on the control of yellow peach moth in China. S China Fruits 26(4):33

Hebert PDN, Penton EH, Burns JM, Janzen DH, Hallwachs W (2004) Ten species in one: DNA barcoding reveals cryptic species in the neotropical skipper butterfly *Astraptes fulgerator*. Proc Natl Acad Sci U S A 101(41):14812–14817

Honda H, Kaneka JI, Konno Y, Matsumoto Y (1979) A simple method for most rearing of the yellow peach moth *Dichocrocis punctiferalis* on an artificial diet. J Appl Entomol Zool 14(4):464–468

Hutson JC (1937) Report on the work of the Entomological Division. Report of The Director of Agriculture Ceylon, 1936, pp 22–28

Ironside DA, Davis JJ (1969) Major Macadamia pests. QLD Agric J 95(7):468–471

Jacob SA (1981) Biology of *Conogethes punctiferalis* on turmeric. J Plant Crop 9(2):119–123

Jarvis E (1914) A new fruit boring caterpillar of banana occurring at tweed keeds. QLD Agric J 5:280–284

Josephrajkumar A, Backiyarani S, Murugan M (2002) PV-1 a pseudostem borer tolerant cardamom variety. Insect Environ 7(4):155–156

Josephrajkumar A, Chakrabarty R, Thomas G (2005) Aprotinin-induced modulation of digestive proteinases and growth suppression in cardamom shoot and capsule borer, *Conogethes punctiferali,* Guenee. Entomon 30(2):175–180

Josephrajkumar A, Chakrabarty R, Thomas G (2006) Midgut proteases of the cardamom shoot and capsule borer, *Conogethes punctiferalis* (Lepidoptera: Pyralidae) and their interaction with aprotinin. Bull Entomol Res 96:91–98

Josephrajkumar A, Chakrabarty R, Thomas G (2007a) Purification of elastase-like chymotrypsin from cardamom shoot and capsule borer. Indian J Exp Biol 45:998–1002

Josephrajkumar A, Backiyarani S, Sivakumar G (2007b) Compendium of cardamom. KAU Publication, Thrissur, p 50

Kalshoven LGE (1922) Notes on forest zoology for the Dutch East India. Tectona 15:944–959

Kalshoven LGE (1928) Injuries, diseases and pests of Teak forests in Java. Tectona XXI:593–623

Kalshoven LGE (1929) Injuries, diseases and pests of Teak forests in Java. Tectona 21:593–523

Kang CH, Lee SM, Chung YJ, Choi KS, Park CG (2004) Overwintering ecology of the peach pyralid moth, *Dichocrosis punctiferalis* in southern regions of Korea. Korean J Appl Entomol 43(3):201–209

Kimura T, Honda N (1999) Identification and possible functions of the hair pencil scent of the yellow peach moth *Conogethes punctiferalis* in Japan. Appl Entomol Zool 34(1):147–153

Kodoi M, Kaneda M (1990) Development of yellow peach mouth *Conogethes punctiferalis*on apple fruit. Res Bull Plant Protect Serv Jpn 26:61–63

Kondo T, Miyahara T (1930) Fruit borer and a summary of their life histories in Knangtuny. China J Plant Protect 17:85–94

Konno Y, Shishido T (1996) Glutathione—dependent o-alkyl and oral conjunctions for dicapthon and fertikothine in several insects. J Pestic Sci 21(4):430–443

Konno Y, Honda N, Matsumoto Y (1980) Observation on the mating behaviour and biology for the six pheromone of yellow peach moth, *Dichocrocis punctiferalis*. Appl Entomol Zool 15(3):321–322

Konno Y, Honda N, Matsumoto Y (1981) Mechanism of reproductive isolation between the fruit feeding and pin ace feeding type of yellow peach moth, *Dicho crocis punctiferalis*. Jpn J Appl Entomol Zool 25(4):253–258

Kuang MH, Liu SW, Ji BZ, Gao JY, Gao YG, Wang GX (2009) Investigation on the overwintering status and bionomics of needle feeding type of *Conogethes punctiferalis*. Chin Bull Entomol 46(4):569–573

Lu Z, Yang ZF, Wang GY, Bo WL (1995) The niche of the corn and peach borers and interspecific competition between them in corn. Acta Agric Zhajiangensis 7(1):31–34

Mishra SC, Teotia TPS (1965) Studies on the biology of the castor shoot and capsule borer. Indian Oil Seeds J 9(3):184–187

Mohanan RC, Kumar RH (1976) The castor capsule borer *Dichocrocis punctiferalis* as a pest of cocoa is India. Agr Res J Kerala 14(1):79–80

Moralesranous JA, Cote JR (1992) Rate of increase and adult longevity of *Catolocus grandis* in the laboratory at four temperature. Environ Entomol 21:620–627

Mukerji MK, Gage SH (1978) A model for estimating hatch and mortality of grass hopper egg populations based on soil moisture & heat. Annu Entomol Soc Am 71(2):183–190

Mumford JD (1984) Control of the cocoa pod borer (*Acrocercops cramerella*): a critical review. In: International conference on cocoa and coconut, 15th–17th October, 1984

Narasimha Swamy RL (1937) Staminodes in *Elettaria cardamomum*. Maton Curr Sci 5(8):427

Ni DW (1998) Main pests and diseases of *Castame amollissinia* in Zhejiang and their control technique. J Zhejany For Sci Tech 6:68–71

Ooi PAC, Chan LG, Khoo KC, Teoh CH, Jusoh MM, Ho CT, Lim GS (1987) Management of the cocoa pod borer. Malaysian Plant Protection Society, Kuala Lumpur, p 192

Pena JE, Nadel H, Barbosa-Pereira M, Smith D (2002) Pollinators and pests of *Annona* species. In: Pena J, Sharp J, Wysok M (eds) Tropical fruit pests and pollinators: biology, economic importance, natural enemies and control. CABI Publication, Wallingford, pp 197–222

Peter KV (1996) Major spices, 50 years of crop science research in India. Indian Council of Agricultural Research, New Delhi, pp 771–796 (CAR)

Rajan SV (1965) Protect your cardamom against insect pests. Indian Farm 14(11):21–24

Ram KT, Hugar PS, Somasekhar Patil BV (1997) Record of *Conogethes punctiferalis* (Guenee) on grapes. Pest Manag Hortic Ecosyst 137(3):37–39

Rama CSV (1980) Plant protection in plantation crops through the soil. J Plant Crop 8(1):1–12

Rao AS (1992) Preliminary studies on the seasonal occurrence of insect pests a soap nut ( *Sapindus* sp). Indian For 118(6):432–437

Rodrigo E (1940) Administration report of the Acting Director of Agriculture for 1940. RAE 30:338

Shanuowr TG, Guttierray AP, Wightman A (1993) Effect of temperature on development rate, fecundity and longevity of ground nut leaf minor, *Aproaerema modicella* (Lepidoptera: Gelechiidae) in India. Bull Entomol Res 83:413–419

Sharma YR, Devasahayam S, Anandraj M (1992) Black Pepper and cardamom: problems and prospects. In: Proceedings of national seminar on black pepper and cardamom. May, 17–18 1992, Calicut, Kerala, India

Sharma ML, Shukla A, Reddy RK (1995) Field testing of caster cultivar for yield potential and resistance against shoot and capsule borer *Conogethes punctiferalis*. Crop Res 10(1):54–58

Shashank PR, Kammar V, Mally R, Chakravarthy AK (2018) A new Indian species of shoot and capsule borer of the genus *Conogethes* (Lepidoptera: Crambidae), feeding on cardamom. Zootaxa 4374:215

Shiukaji N (1969) Notes on the parasites attacks the fruit type of and confer type of *Dichocrosis punctiferalis*. Jpn J Appl Entomol Zool 13(2):91–93

Sloan WJS (1945) Insect pests of grain sorghum. QLD Agric J 61(4):221–239

Sridharan S, Nagarajan N, Thamburaj S, Mohidin MK (1990) Effect of planting density on the capsule damage by cardamom major pests. S India Hortic 38(2):120–121

Sun ZP, Wang W, Mung FY (1992) Survey of pests and diseases of Chinese hawthorn. Bull Agric Sci Techn 44:25–26

Thyagaraj NE (2003) Integrated management of some important cardamom pests of hill region of Karnataka, South India. Ph D thesis, Dr BR Ambedkar University, Uttar Pradesh, India

Thyagaraj NE, Singh PK, Chakravarthy AK (2003) Bioecology of cardamom shoot and fruit borer, *Conogethes punctiferalis* Guenee. Curr Res 32:3–4

Tomomatsu S, Sakaguchi T, Ogino T, Hiramatsu T, Kaukami F (1995) Methyl bromide fumigator for quarantine control of persimmon fruit moth and yellow peach moth on Japanese persimmon. Res Bull Plant Prot Serv 31:67–73

Tryon H (1920) Report of the entomologist, vegetable pathologist. Queensland annual report of department of agriculture for 1916–1917, pp 49–63

Twine PH (1971) Insect pests of maize. QLD Agr J 97(10):523–531

Utsume K, Aihara E, Watanabe N (1990) Artificial diet of yellow peach moth *Conogethes punctiferalis* eating corn seeds. Res Bull Plant Prot Serv Jpn 26:65–67

Veitch R (1931) Report of the chief entomologist, annual report of department of agriculture. QLD Agric J 1930–1931:43–46

Wang YX, Cai RT (1997) The emergence and control of *Dichocrosis punctiferalis* for Younai plum variety. S China Fruits 26(3):45

Wu LM (1995) A study on the pupation pattern of three pyralid caterpillars helps is maize field. Plant Prot 21(3):24–26

Xi K, Xu FY, Dai YS (1996) Studies on the Chinese chestnut cone life table. J Joangsu For Sci Technol 23(4):6–10

# Status of Shoot and Fruit Borer, *Conogethes sahyadriensis*, on Spice Crops

16

S. Devasahayam, C. M. Senthil Kumar, and T. K. Jacob

**Abstract**

*Conogethes sahyadriensis* (Shashank PR, Kammar V, Mally R, Chakravarthy AK, Zootaxa 4374(2):215–234, 2018) is a serious insect pest of zingiberaceous spice crops such as cardamom (*Elettaria cardamomum* Maton), ginger (*Zingiber officinale* Rosc.) and turmeric (*Curcuma longa* L.). On ginger and cardamom the borer incurs up to 25% and 80% yield losses, respectively. On cardamom the economic threshold level is fixed at 10% shoot infestation. The fecundity of female moths and the duration for completing the generations varied with the season, the spice crop and the habitat. *C. sahyadriensis* infesting spice crops in South India has alternate wild Zingiberaceae plants, viz. species of *Alpinia*, *Amomum*, *Hedychium*, *Curcuma*, *Aframomum*, etc. Over a dozen, promising parasitoids and predators have been recorded on *Conogethes* larvae and pupae. Therefore, need-based applications of selective insecticides are warranted. Use of semiochemicals for borer management should be rendered practicable.

**Keywords**

Cardamom · *Conogethes* species · Ginger · Turmeric

## 16.1 Introduction

Spices are high-value products obtained from diverse crops and are extensively used all over the world to add flavour and taste to human food besides being used in cosmetics and medicine. Many spices are also increasingly being valued as functional foods in view of their nutraceutical properties. Over 100 species of plants yield

S. Devasahayam · C. M. Senthil Kumar (✉) · T. K. Jacob
ICAR-Indian Institute of Spices Research, Kozhikode, Kerala, India





A. K. Chakravarthy (ed.), *The Black spotted, Yellow Borer, Conogethes punctiferalis Guenée and Allied Species*,
https://doi.org/10.1007/978-981-13-0390-6_16

spices, and among them zingiberaceous species such as cardamom (*Elettaria cardamomum* Maton), ginger (*Zingiber officinale* Rosc.) and turmeric (*Curcuma longa* L.) account for a major share in production and trade of spices in the world. Cardamom, known as the 'queen of spices', is the dried fruit of a perennial rhizomatous herb and is native to the evergreen forests of South India. Guatemala and India are leading producers of cardamom in the world. Ginger and turmeric are herbaceous plants native to South Asia, and their dried rhizomes yield the spice of commerce. India and China are leading producers of ginger in the world, whereas India is a leading producer of turmeric in the world. Among the insect pests infesting these crops, the shoot and fruit borer *Conogethes sahyadriensis* is considered as a serious insect pest.

## 16.2 Distribution

The borer on zingibers, now identified as *Conogethes sahyadriensis* (Shashank et al. 2018), is widely prevalent in Southeast Asia and Australia, but authentic records of the pest on cardamom, ginger and turmeric in other countries are limited. In India, the incidence of *C. sahyadriensis* on cardamom ranged from 0.4% to 7.0%, 0% to 5% and 0% to 15% on shoots, panicles and capsules, respectively, in Kodagu, Hassan and Uttara Kannada districts in Karnataka. In the nursery, the pest infestation ranged between 0.11% and 4.3% in the primary nursery and 0.39–3.0% in the secondary nursery (Rajan 1984). At Mudigere (Karnataka), the shoot damage in cardamom due to *C. sahyadriensis* infestation varied from 5% to 10% (Thyagaraj 2003).

In Nepal, infestation by *C. sahyadriensis* was observed in many of the major ginger growing districts (Ilam, Salyan Nawalparasi, Dhankuta and Syanja), though the mean infestation levels were below 3% (Gautam and Acharya 2014). In India, 23.6–25.0% of pseudostems of ginger were damaged by the pest at Kottayam and Idukki districts in Kerala (Nybe 2001). In northern districts of Karnataka, the incidence of the pest on turmeric was higher at Raibag, Chikodi (Belgaum District), Jamakhandi, Indi (Bijapur District), Bidar, Basavakalyan, and Humnabad (Bidar District) taluks, and up to 17% of plants were infested in Basavakalyan taluk (Kotikal and Kulkarni 2000).

It has been suggested that *C. punctiferalis* (earlier considered to be infesting zingiberaceous spice crops) is a combination of more than one species, especially in Southeast Asia and Australia (Konno et al. 1981; Honda 1986a, b; Honda and Matsumoto 1987; Honda and Mitsuhashi 1989; Robinson et al. 1994; Boo 1998). Recently, Shashank et al. (2014a, b) and Doddabasappa et al. (2014) have reported the occurrence of cryptic species in host-associated (castor and cardamom) populations of *C. punctiferalis* in India based on molecular, biological and behavioural studies. In a study conducted at ICAR-IISR, Kozhikode, to know if differences exist among the *Conogethes* populations infesting cardamom, ginger and turmeric collected from different locations, the partial mitochondrial CO1 region was sequenced. Phylogenetic analysis of the partial mtCO1 sequences showed close similarity among the *Conogethes* populations from ginger, turmeric and cardamom; however, they were distinctly different from the castor population (Senthil Kumar et al. 2017a). The new borer species now has been identified as *C. sahyadriensis*.

## 16.3 Damage

*C. sahyadriensis* infests cardamom plants in nurseries and plantations. The early stages of larvae bore into panicles and immature capsules and the later stages into pseudostems and feed on the internal tissues. The presence of bored holes with extruding frass on the pseudostems and capsules and the withered central shoot producing the 'dead hearts' are characteristic symptoms of the pest infestation (Fig. 16.1). In terms of capsule yield loss (dry weight basis), it varied from 6.79% to 9.18% at Mudigere (Thyagaraj 2003). Yield losses of 70–80% have been reported in severely infested plantations in India (Varadarasan 1995). The economic threshold level of pest damage was fixed at 10% shoot damage (AICRPS 1998). In Sri Lanka, the direct capsule loss due to the pest damage is 10–12% (Dharmadasa and Ranasinghe 1987).

On ginger and turmeric, the larvae bore into pseudostems and feed on the growing shoot, resulting in yellowing and drying of infested pseudostems. The presence of bored holes on the pseudostem, through which frass is extruded, and the withered central shoot is a characteristic symptom of pest infestation. Apart from boring into the pseudostem at the base, the larvae also bore into the leaf petiole to enter into the pseudostem. Sometimes, the larvae also bore into the rhizome near the base of the pseudostem.

Studies on yield loss caused by the pest on ginger in Kozhikode district, Kerala, indicated that when 50% of the pseudostems in a plant are affected, there was a significant reduction of 38 g of yield per plant (Koya et al. 1986). Yield losses of 25% have been estimated in Kottayam and Idukki districts when 23–24% of pseudostems are infested (Nybe 2001). On turmeric, 72–79 g yield loss was observed when 25–75% of shoots were infested by the pest (Devasahayam et al. 2010a).

**Fig. 16.1** Nature of damage by *C. sahyadriensis* on Cardamom: (**a**) capsule bore hole, (**b**) larva inside capsule, (**c**) excreta of larva on shoot and (d) larva inside shoot

## 16.4 Life History

The adults are medium-sized moths with a wingspan of 18–24 mm; the wings and body are pale straw yellow with minute black spots (Fig. 16.2). The adults have been described by Hampson (1856). Freshly laid eggs are elliptical, pitted on the surface, and creamy white. There are five larval instars; fully grown larvae are light greenish-brown or pale pink with sparse hairs and measure 16–35 mm in length; the pupa measures 13–15 mm in length. The dimensions of adults and larvae may vary depending on the host in which they are raised. The morphometrics of various stages when reared on turmeric has been provided by Jacob (1981). Thyagaraj et al. (2001) suggested a method for determination of sex in this species based on the size and morphology of pupae. Recently Shashank et al. (2014a) reported that male moths of this species could be identified based on the presence of anal tuft of black hairs.

Studies on the biology of the borer on cardamom have been conducted in various locations. At Appangala (Kodagu District, Karnataka), the pre-oviposition and oviposition periods lasted for 2–3 days and 4–6 days, respectively, and the fecundity of females ranged from 12 to 17 eggs. The total larval period ranged from 10 to 16 days and pupal period 10–14 days (Rajan 1983). At Thadiyankudisai (Dindigul District, Tamil Nadu), the duration of life history varied during summer (March to May) (temperature range: 16–29 °C) and winter (November to February) (temperature range: 16–25 °C). The pre-oviposition, egg and larval periods lasted for 2–3, 6–7 and 21–32 days, respectively, during summer, and 17–19, 6–8 and 40–62 days, respectively, during winter. The prepupal and pupal periods lasted for 2–3 and 10–12 days, respectively, during summer, and 4–7 and 17–27 days, respectively, during winter (Varadarasan et al. 1991). At Coimbatore, the mean duration of I–V instar larva and pupa was 2.90, 3.95, 4.20, 5.25, 6.60 and 7.90 days, respectively (Stanley et al. 2009). At Bengaluru, the egg period lasted for a mean of 3.51 ± 0.3 days on Mysore type of cardamom compared to 2.76 ± 0.2 and 2.68 ± 0.2 days on Malabar and Vazhuka types, respectively. The pre-oviposition period lasted for 5.51 ± 0.2, 4.66 ± 0.2 and 5.16 ± 0.3 days on Malabar, Mysore and Vazhuka types, respectively. The fecundity of females was 46.25 ± 3.9, 66.25 ± 0.8 and 59.33 ± 4.0 eggs on Malabar, Mysore and Vazhuka types, respectively. The mean total larval period was 30.9 ± 0.4, 34.9 ± 1.0 and 29.4 ± 1.9 days on Malabar, Mysore and Vazhuka types, respectively; the mean total pupal period was 7.95 ± 1.3, 9.94 ± 0.6

**Fig. 16.2** Adult moth of *Conogethes sahyadriensis*

and 9.61 ± 0.5 days on Malabar, Mysore and Vazhuka types, respectively (Doddabasappa et al. 2014).

The life history of *C. sahyadriensis* on ginger has been studied at Coimbatore, Tamil Nadu and the mean duration of I–V instar larva and pupa was 3.05, 4.25, 4.80, 5.15, 6.60 and 7.85 days, respectively (Stanley et al. 2009). Studies on biology of the borer on turmeric at Kasaragod (Kerala) indicated that the pre-oviposition and incubation periods lasted for 4–7 and 3–4 days, respectively. The five larval instars lasted for 3–4, 5, 3–7, 3–8, and 7–14 days, respectively. The prepupal and pupal periods lasted for 3–4 and 9–10 days, respectively. Adult females laid 30–60 eggs during its lifespan, and 6–7 generations were completed during a crop season in the field. Variations were observed in the life cycle (30 and 38 days during August to October and November to December, respectively) during different seasons (Jacob 1981).

## 16.5 Seasonal Incidence

Little information is available on the seasonal population dynamics of this borer pest on cardamom, ginger and turmeric. The population of borer is recorded in the field throughout the year on cardamom but is reported to be higher during January to February, May and September to October at Idukki (Kerala) (Varadarasan et al. 1989) and during January, March, June, August and October at Thadiyankudisai (Varadarasan et al. 1991). At Mudigere (Karnataka), it was reported to peak during September to October on the cardamom tillers (Krishnamurthy et al. 1989). On ginger, the borer is known to occur throughout the crop period in Kerala during July to December. In Kottayam and Idukki districts, the damage was reported to be higher during August, September and October (Nybe 2001).

On turmeric, the borer occurs throughout the crop period in Kerala during July to December. The pattern of distribution of the pest in a turmeric field at Peruvannamuzhi was random during July to September and became more aggregated during October to December. The symptoms of new pest infestations were higher during October to December (Devasahayam et al. 2010a).

## 16.6 Host Plants

Many of the hosts of *C. sahyadriensis* are economically important plants, and the pest infests various parts of these plants, such as buds, flowers, shoots and fruits (Devasahayam and Koya 2004, 2007). The pest has been recorded on other zingiberaceous species in India that are considered as minor spices such as *Aframomum melegueta* K. Schum. (*Melegueta pepper*), *Alpinia galanga* L. (Willd.) (galanga), *Curcuma amada* Roxb. (mango ginger) and *C. zedoaria* (Christm.) Rosc. (white turmeric). However, in view of the reports on occurrence of cryptic species in host-associated populations of *Conogethes* species, the host status of the pest is to be critically reviewed.

## 16.7 Natural Enemies

Various parasitoids have been recorded on this borer species in many crops though specific records on cardamom are limited. The parasitoids recorded on the borer infesting cardamom include *Agrypon* sp., *Eriborus trochanteratus* (Morl.), *Friona* sp., *Temecula* sp. and *Xanthopimpla australis* Kr. (Ichneumonidae) (Dubey and Rajan 1985; Varadarasan 1995). Parasitisation by *X. australis* ranged between 10.5% and 11.1% at Appangala (Dubey and Rajan 1985). At Mudigere (Karnataka), parasitisation by *Agrypon* sp. and *Temelucha* sp. ranged from 53.8% to 100.0% (AICRPS 2007). *Friona* sp. and *Agrypon* sp. parasitized 8–10% and 16–18% of larvae and pupae, respectively, at Idukki (Varadarasan 1995). Natural parasitisation by larval parasitoids ranged from 3.7% to 43.3% and pupal parasitoids from 0.2% to 1.2% at Idukki (Ali et al. 2014). The reduviid predator *Sycanus indigator* Stal. was recorded to predate larvae of *C. sahyadriensis* in cardamom fields (Nagarajan and Varadarasan 2011).

At Peruvannamuzhi (Kozhikode District, Kerala), *Hexamermis* sp. (Mermithidae), *A. taragammae* and *Bracon* sp. (Braconidae) were the common natural enemies of the borer infesting ginger in the field. Studies on seasonal incidence of natural enemies indicated that *Hexamermis* sp. was active in the field during July to November with a peak parasitisation of 72% during August and the hymenopteran parasitoids during October to December with a peak parasitisation of 28% during November (Devasahayam and Koya 1996; Devasahayam et al. 2005). Eight species of entomopathogenic nematodes (EPNs) (three species belong to genus *Steinernema*, one to *Heterorhabditis* and four to *Oscheius*) were documented from rhizosphere of ginger in India among which *S. ramanai* and *O. gingeri* were described as new species (Pervez et al. 2013, 2014a).

Eleven species of hymenopterous parasitoids including *Phanerotoma hendecasiella* Cam., *Bracon brevicornis* (Wes.), *Myosoma* sp., *Apanteles* sp. (Braconidae), *Xanthopimpla* sp., *X. australis* Kr., *Angitia trochanterata* Thomson, *Theromia inareolata* (Ichneumonidae), *Dolichorus* sp. (Sphecidae), *Brachymeria euploeae* West., *B. nosatoi* Habu (Chalcididae) and a mermithid nematode (Mermithidae) in addition to five species of predators, *Euborellia stali* Dohrn (Carcinophoridae), *Philodicus* sp., *Heligmoneura* sp. (Asilidae), *Araneus* sp. (Araneidae), *Micaria* sp. (Clubinidae) and *Thyene* sp. (Salticidae) were recorded as natural enemies of the borer on turmeric at Kasaragod (Jacob 1981, 1985).

## 16.8 Resistance

The Mysore cardamom type is highly susceptible to the shoot and fruit borer among the types of cardamom (Dubey and Rajan 1985). Josephrajkumar et al. (2002a) suggested that the tolerance of the Malabar type PV-1 to the pest was probably due to the smaller girth of pseudostem. Seventy-two cardamom accessions were evaluated for 5 years at Pampadumpara (Idukki District, Kerala), for susceptibility to *C. sahyadriensis* among which S-1 was identified as the best for inclusion in breeding programmes in view of its higher yield, moderate incidence of insect pests (including *C. sahyadriensis*) and higher quality parameters (Miniraj et al. 2000).

The reaction of various types of ginger to *C. sahyadriensis* in the field was studied by Nybe and Nair (1979), who reported that among the 25 cultivars of ginger screened, the pest infestation was minimum in Rio de Janeiro and maximum in Valluvanad, although not significant. Four hundred and ninety two accessions of ginger were screened in the field against *C. sahyadriensis* at Peruvannamuzhi, and 49 accessions were rated as moderately resistant to the pest; among the popular cultivars, Jorhat, Rio de Janeiro, Thingpui and Burdwan were rated as moderately resistant to the pest (Devasahayam et al. 2010b).

The reaction of various turmeric types to the borer in the field has been studied. Sheila et al. (1980) reported that among the 13 types of turmeric screened at Vellanikkara (Thrissur District, Kerala), Dindigam Ca-69 (an aromatic type) was the least susceptible and Amruthapani Kothapeta Cll-317 (a longa type), the most susceptible. Philip and Nair (1981) reported that among the 19 turmeric types screened, Mannuthy Local was the most tolerant. Kotikal and Kulkarni (2001) screened eight genotypes of turmeric in the field at Belgaum (Karnataka) and reported that all of them were susceptible with more than 10% damage. Velayudhan and Liji (2003) recorded the incidence of the borer on 489 accessions belonging to 21 morphotypes and the lowest incidence was observed in morphotype II with a mean score of 2 on a 0–9 scale; 22 accessions were tolerant with a score of less than 3. Nine hundred and fifteen accessions of turmeric were screened in the field at Peruvannamuzhi against *C. sahyadriensis*, and 34 accessions were rated as moderately resistant to the pest (Devasahayam et al. 2011). A study of 19 cultivars of turmeric at Mudigere (Karnataka) indicated that incidence of the borer was the lowest in the cultivar Kanti followed by PTS-24 (Shetty et al. 2015).

## 16.9 Management

(a) ***Biopesticides***: Two commercial products of *Bacillus thuringiensis*, namely, Bioasp and Dipel, were evaluated along with malathion for the management of the borer on ginger and turmeric in the field at Peruvannamuzhi. The trials indicated that spraying Dipel 0.3% at 21-day intervals during July to October was the most effective for the management of the pest (Devasahayam 2000, 2002).

The efficacies of eight native EPNs isolated from rhizosphere of ginger were tested against larvae and pupae of the borer and their multiplication was assessed. Among the tested EPNs, *Heterorhabditis* sp. (IISR 01), *Steinernema* sp. (IISR 02) and *Oscheius* sp. (IISR 07 and 08) caused 100% mortality to the larvae. *Oscheius* sp. (IISR 07) was the most virulent against pupae, causing 100% mortality, followed by *Steinernema* sp. (IISR 02) and *Oscheius* sp. (IISR 05) which killed 67% of pupae (Pervez et al. 2012). A study on the pathogenicity of four promising EPNs, *Heterorhabditis* sp. (IISR-EPN 01), *Steinernema* sp. (IISR-EPN 02), *Oscheius* sp. (IISR-EPN 08) and *O. gingeri* against *C. sahyadriensis* larva, by dose response and time exposure assay and determination of lethal dosages ($LD_{50}$) and lethal time ($LT_{50}$) indicated that *Steinernema* sp. (IISR-EPN 02) and *O. gingeri* were more promising (Pervez et al. 2014b). The infectivity of four promising EPNs, *Heterorhabditis* sp. (IISR-EPN 01), *Steinernema* sp. (IISR-EPN 02), *O. gingeri* (IISR-EPN 07) and *Oscheius* sp. (IISR-EPN 08), was tested against the borer infesting ginger and turmeric under field conditions at Peruvannamuzhi. Among the tested EPNs, *O. gingeri* (IISR-EPN 07)- and *Steinernema* sp. (IISR-EPN 02)-treated plants showed minimum shoot damage in ginger and turmeric, respectively, which was on par with malathion (IISR 2015). New artificial media for the mass production of infective juveniles of entomopathogenic nematodes effective against the borer of ginger have been developed. The media are suitable to multiply infective juveniles of *Steinernema* spp., *Heterorhabditis* spp. and *Oscheius* spp. and are cheaper than other media.

(b) ***Protease inhibitors***: Protease inhibitors cause mortality in a wide range of insects, and expression of protease inhibitors can be induced in transgenic plants for protection against pest attack. The growth and modulation of midgut proteinases of *C. sahyadriensis* larvae infesting cardamom was determined after being fed with the proteinase inhibitor, aprotinin. Inclusion of aprotinin (150 μM) in cardamom shoots used as feed resulted in a significant reduction of larval body weight by the fourth day. Specific activities of the serine proteinases trypsin and elastase-like chymotrypsin were reduced in aprotinin-fed insects after 2 days of feeding. Specific activity of leucine aminopeptidase (cytosol aminopeptidase) revealed an increase in aprotinin-fed larvae. The study indicated the effect of aprotinin in protein digestion and its potential for protecting against *C. sahyadriensis* (Josephrajkumar et al. 2005).

The midgut protease profile of the gut lumen from *C. sahyadriensis* was studied to determine the conditions for optimal protein hydrolysis and to identify potential targets for proteinaceous biopesticides, such as protease inhibitors. Optimum conditions for peptidase activity were found in 50 mM Tris-HCl, pH 10 containing 20 mM $CaCl_2$, with incubation for 30 min at 40 °C. Four synthetic substrates, namely, benzoyl-arg-*p*-nitroanilide, benzoyl-tyr-*p*-nitroanilide, succinyl-ala-ala-pro-leu-*p*-nitroanilide (SAAPL*p*NA) and leu-*p*-nitroanilide, were hydrolysed by gut proteases in Tris-HCl buffer pH 10. Trypsin and elastase-like chymotrypsin were the prominent digestive proteases, and age-related modulation of midgut proteases existed for trypsin, chymotrypsin, elastase-like chymotrypsin and leucine aminopeptidase. Serine protease inhibitors such as aprotinin, soybean trypsin inhibitor and phenyl methane sulfonyl fluoride inhibited peptidase activity. Metal ions such as $Ca^{2+}$, $Mg^{2+}$, $Pb^{2+}$ and $Co^{2+}$ enhanced BA*p*NA-ase activity, whereas others like $Mn^{2+}$, $Zn^{2+}$, $Cu^{2+}$, $Fe^{2+}$ and $Hg^{2+}$ were inhibitory at 6 mM concentration. Trypsin and elastase-like chymotrypsin were significantly inhibited by 94% and 29%, respectively, by aprotinin (150 nM) under in vitro conditions. The study indicated the potential of incorporation of protease inhibitors into transgenic plants for the management of the pest (Josephrajkumar et al. 2006).

(c) ***Plant products***: Trials conducted at Idukki with neem-based insecticide and other natural products against the borer on cardamom indicated that none of them were effective (Josephrajkumar et al. 2002a, b). Several neem-based products were evaluated at Mudigere, Karnataka, South India, against the borer on cardamom. Of all, neem oil 0.03% was the most promising (Naik et al. 2006). The bioefficacy of neem formulation (TNAU neem oil 0.03 EC), diafenthiuron and profenofos was evaluated in different IPM modules at Lower Pulneys (Dindigul District, Tamil Nadu, South India), and sequential application of neem, profenofos, diafenthiuron and profenofos was the most effective (Rajabaskar and Regupathy 2013).

In Nagaland, mulching with mahaneem (*Melia dubia* Cav.) leaves (Lalnuntluanga and Singh 2008) or spraying quinalphos 0.05% + Ozoneem 1500 ppm (3 ml/L) (Mhonchumo and Singh 2010) has been suggested for the management of the borer on ginger. Evaluation of neem oil 1% and commercial neem product (Nimbicidine 1%) in the field at Peruvannamuzhi, Kerala, South India for the management of the pest on ginger and turmeric indicated no consistent results (Devasahayam et al. 2005).

(d) ***Sex pheromones***: The presence of sex pheromones in *C. sahyadriensis* has been demonstrated and trials on the efficacy of synthetic sex pheromones in the field indicated potential on cardamom in India. A blend of E-10-hexadecenal and Z-10-hexadecenal at 90:10 ratio was the most attractive (Chakravarthy and Thyagaraj 1997, 1998; Rajabaskar and Regupathy 2012).

(e) ***Chemical control***: Several insecticides have been evaluated for the management of *C. sahyadriensis* on cardamom in the field in India. However, many of them are not approved by the Central Insecticide Board, India, at present. The insecticidal sprays should target second instar larvae that feed on tender panicles or immature capsules. In Sri Lanka, removal of infested pseudostems along with spraying thiamethoxam (0.5 g/L), chlorfluazuron (1.2 ml/L) or fenthion (2.5 ml/L) has been suggested (Dharmadasa 2000).

On ginger, Koya et al. (1988) evaluated six insecticides against *C. sahyadriensis* and found that monthly applications of malathion 0.1% during July to October was effective. Koya et al. (1986) evolved a sequential sampling strategy for monitoring pest infestation in a ginger field as guidance for undertaking control measures. Adoption of this strategy resulted in need-based application of insecticides leading to reduced management costs. An integrated strategy involving pruning and destroying freshly infested shoots during June to August and spraying insecticides such as malathion 0.1% during September to October has also been suggested (Devasahayam and Koya 1999). Recently spraying low risk insecticides such as chlorantraniliprole 0.01%, flubendiamide 0.02%, spinosad 0.0225% and cyantraniliprole 0.005% were found effective for the management of the pest (Senthil Kumar et al. 2017b).

In spite of the serious nature of damage caused by the borer on turmeric, few field trials have been conducted to suppress the pest. The insecticides generally recommended for *C. sahyadriensis* management on ginger have also been recommended against the pest on turmeric. Spraying lambda cyhalothrin 0.0125% during July to November at 21-day intervals was effective (Devasahayam et al. 2010a).

(f) ***Integrated management***: Reducing the nitrogenous fertiliser levels minimises the pest damage to a greater extent in cardamom (Chakravarthy and Thyagaraj 1999). An integrated approach adopting cultural, biological and chemical methods has been suggested as ideal for the management of the pest. Removal of infested suckers as indicated by extrusion of frass, during September to October when the infestation is less than 10%, and collection and destruction of adults which are generally observed on the undersurface of leaves and spraying quinalphos (0.075%) twice, during February to March and September to October coinciding with emergence of panicles and new shoots, have been recommended for the borer management on cardamom in Karnataka, South India (IISR 2015).

An integrated strategy involving pruning of freshly infested shoots during July to August (at fortnightly intervals) and chemical methods such as spraying insecticide (malathion 0.1%) during September to October (at monthly intervals) was effective against the borer on ginger. By adopting this integrated strategy, two insecticide sprays could be avoided (Devasahayam et al. 2005).

## References

Ali MAA, Manoharan T, Kuttalam S (2014) First report of larval parasitoids on small cardamom shoot, capsule and panicle borer, *Conogethes punctiferalis* Guenee (Lepidoptera; Crambidae) under lab and field conditions. Trends Biosci 7:3771–3773

All India Coordinated Research Project on Spices (1998) Annual report 1997–98. Indian Institute of Spices Research, Calicut

All India Coordinated Research Project on Spices (2007) Annual report 2005–06. Indian Institute of Spices Research, Calicut

Boo KS (1998) Variation in sex pheromone composition of a few selected lepidopteran species. J Asia Pac Entomol 1:17–23

Chakravarthy AK, Thyagaraj NE (1997) Response of the cardamom (*Elettaria cardamomum* Maton) shoot and fruit borer (*Conogethes punctiferalis* Guenee; Lepidoptera: Pyralidae) to different pheromone compounds. Insect Environ 2:127–128

Chakravarthy AK, Thyagaraj NE (1998) Evaluation of selected synthetic sex pheromones of the cardamom shoot and fruit borer, *Conogethes punctiferalis* Guenee (Lepidoptera: Pyralidae) in Karnataka. Pest Manage Hortic Ecosyst 4:78–82

Chakravarthy AK, Thyagaraj NE (1999) Effect of different levels of NPK on shoot-and-fruit borer, *Conogethes punctiferalis* Guenee (Pyralidae: Lepidoptera) infestation on small cardamom (*Elettaria cardamomum* Maton). Insect Environ 4:139–140

Devasahayam S (2000) Evaluation of biopesticides for the management of shoot borer (*Conogethes punctiferalis* Guen.) on ginger (*Zingiber officinale* Rosc.). In: Ramana KV, Eapen SJ, Babu KN, Krishnamurthy KS, Kumar A (eds) Spices and aromatic plants: challenges and opportunities in the new century. Indian Society for Spices, Calicut, pp 276–277

Devasahayam S (2002) Evaluation of biopesticides for the management of shoot borer (*Conogethes punctiferalis* Guen.) on turmeric. In: Rethinam P, Khan HH, Reddy VM, Mandal PK, Suresh K (eds) Plantation crops research and development in the new millennium. Coconut Development Board, Kochi, pp 489–490

Devasahayam S, Koya KMA (1996) Biological control of insect pests. In: Anandaraj M, Peter KV (eds) Biological control in spices. Indian Institute of Spices Research, Calicut, pp 33–45

Devasahayam S, Koya KMA (1999) Integrated management of insect pests of spices. Indian J Arecanut Spices Med Plant 1:19–23

Devasahayam S, Koya KMA (2004) Insect pests of ginger. In: Ravindran PN, Babu KN (eds) Ginger. The genus *Zingiber*. CRC Press, Washington, DC, pp 367–389

Devasahayam S, Koya KMA (2007) Insect pests of turmeric. In: Ravindran PN, Babu KN, Sivaraman K (eds) Turmeric. The genus *Curcuma*. CRC Press, Boca Raton, pp 169–192

Devasahayam S, Koya KMA, Zachariah TJ, Jacob TK (2005) Biological control of insect pests of spices. In: Final report of research project. Indian Institute of Spices Research, Calicut

Devasahayam S, Jacob TK, Koya KMA (2010a) Bio-ecology and integrated management of shoot borer (*Conogethes punctiferalis* Guen.) infesting turmeric. Final report of research project. Indian Institute of Spices Research, Calicut

Devasahayam S, Jacob TK, Koya KMA, Sasikumar B (2010b) Screening of ginger (*Zingiber officinale*) germplasm for resistance to shoot borer (*Conogethes punctiferalis*). J Med Aromat Plant Sci 32:137–138

Devasahayam S, Jacob TK, Koya KMA, Sasikumar B, Prasath D (2011) Screening of turmeric (Curcuma longa L.) germplasm for resistance to shoot borer (*Conogethes punctiferalis* Guen.) (Lepidoptera: Pyralidae), in Kerala, South India. Entomon 36:59–62

Dharmadasa M (2000) Management of insect pests of export agricultural crops. Department of Export Agriculture, Matale

Dharmadasa M, Ranasinghe MASK (1987) Severity of cardamom stem and capsule borer attack in Sri Lanka. In: Proceedings, 43rd annual session, Sri Lanka Association for the Advancement of Science, Colombo, p 157

Doddabasappa B, Chakravarthy AK, Thyagaraj NE (2014) Comparative biology of shoot and capsule borer, *Conogethes punctiferalis* (Guenee), (Crambidae: Lepidoptera) on castor and cardamom. Curr Biotica 8:228–245

Dubey OP, Rajan P (1985) Biology and bionomics of insect pests of cardamom. In: Annual report 1983. Central Plantation Crops Research Institute, Kasaragod, pp 201–204

Gautam J, Acharya B (2014) Survey and monitoring of insect pest and disease of ginger in Nepal. Nepalese J Agric Sci 12:118–123

Hampson GF (1856) Fauna of British India including Ceylon and Burma. Moths. Taylor and Francis, London

Honda H (1986a) EAG responses of the fruit- and Pinaceae-feeding type of yellow peach moth, *Conogethes punctiferalis* (Guenée) (Lepidoptera: Pyralidae) to monoterpene compounds. Appl Entomol Zool 21:399–404

Honda H (1986b) Post-mating reproductive isolation between fruit- and Pinaceae-feeding types of the yellow peach moth, *Conogethes punctiferalis* (Guenée) (Lepidoptera: Pyralidae). Appl Entomol Zool 21:489–491

Honda H, Matsumoto Y (1987) Larval feeding responses of the fruit- and Pinaceae-feeding type of the yellow peach moth, *Conogethes punctiferalis* (Guenée) (Lepidoptera: Pyralidae) to the host-plant extracts and their sugar components. Jpn J Appl Entomol Zool 31:28–35

Honda H, Mitsuhashi W (1989) Morphological and morphometrical differences between the fruit- and Pinaceae-feeding type of yellow peach moth, *Conogethes punctiferalis* (Guenée) (Lepidoptera: Pyralidae). Appl Entomol Zool 24:1–10

Indian Institute of Spices Research (2015) Cardamom – extension pamphlet. Indian Institute of Spices Research, Kozhikode

Jacob SA (1981) Biology of *Dichocrocis punctiferalis* Guen on turmeric. J Plant Crop 9:119–123

Jacob SA (1985) Biology and bionomics and insect pests of spices. In: Annual report for 1983. Central Plantation Crops Research Institute, Kasaragod, pp 198–200

Josephrajkumar A, Backiyarani S, Murugan M (2002a) PV-1, a pseudostem borer tolerant cardamom variety. Insect Environ 7:155–156

Josephrajkumar A, Kurian PS, Backiyarani S, Murugan M (2002b) Evaluation of biorationals against thrips *(Sciothrips cardamomi* Ramk.) and shoot and capsule borer *(Conogethes punctiferalis* Guen.) in cardamom. J Spices Aromat Crop 11:132–134

Josephrajkumar A, Chakrabarty R, Thomas G (2005) Aprotinin induced modulation of digestive proteinases and growth suppression in cardamom shoot and capsule borer, *Conogethes punctiferalis* Guenée. Entomon 30:175–180

Josephrajkumar A, Chakrabarty R, Thomas G (2006) Midgut proteases of the cardamom shoot and capsule borer *Conogethes punctiferalis* (Lepidoptera: Pyralidae) and their interaction with aprotinin. Bull Entomol Res 96:91–98

Konno Y, Honda H, Matsumoto Y (1981) Mechanisms of reproductive isolation between the fruit-feeding and the Pinaceae-feeding types of the yellow peach moth, *Dichocrocis punctiferalis* Guenee (Lepidoptera: Pyralidae). Jpn J Appl Entomol Zool 25:253–258

Kotikal YK, Kulkarni KA (2000) Incidence of insect pests of turmeric (*Curcuma longa* L.) in northern Karnataka. India J Spices Aromat Crop 9:51–54

Kotikal YK, Kulkarni KA (2001) Reaction of selected turmeric genotypes to rhizome fly and shoot borer. Karnataka J Agric Sci 14:373–377

Koya KMA, Balakrishnan R, Devasahayam S, Banerjee SK (1986) A sequential sampling strategy for the control of shoot borer (*Dichocrocis punctiferalis* Guen.) in ginger (*Zingiber officinale* Rosc.) in India. Trop Pest Manage 32:343–346

Koya KMA, Premkumar T, Gautam SSS (1988) Chemical control of shoot borer *Dichocrocis punctiferalis* Guen. on ginger *Zingiber officinale* Rosc. J Plant Crop 16:58–59

Krishnamurthy K, Khan MM, Aradhani KK, Venkatesh J, Siddaramaiah AL, Chakravarthy AK, Gurumurthy BR (1989) Three decades of cardamom research at Regional Research Station, Mudigere, Karnataka

Lalnuntluanga J, Singh HK (2008) Performance of certain chemicals and neem formulations against ginger shoot borer (*Dichocrocis punctiferalis* Guen.). Indian J Entomol 70:183–186

Mhonchumo NP, Singh HK (2010) Eco-friendly management of shoot borer and rhizome fly of ginger in Nagaland. In: Sema A, Srinivasan V, Shitri M (eds) Proceedings and recommendations, national symposium on spices and aromatic crops (SYMSAC-V), 30–31 October 2009, Medziphema, Central Institute of Horticulture, Medziphema, pp 117–123

Miniraj N, Murugan M, Joseph CR (2000) Evaluation of cardamom (*Elettaria cardamomum* Maton) germplasm. J Spices Aromat Crop 9:55–56

Nagarajan K, Varadarasan S (2011) Reduviid fauna associated with cardamom forest ecosystem. Insect Environ 17:124–126

Naik DJ, Belavadi VV, Thippesha D, Kumar MD, Madaiah D (2006) Field efficacy of neem products against thrips and capsule borer of small cardamom. Karnataka J Agric Sci 19:144–145

Nybe EV (ed) (2001) Three decades of spices research at KAU. Kerala Agricultural University, Vellanikkara

Nybe EV, Nair PCS (1979) Field tolerance of ginger types to important pests and diseases. Indian Arecanut Spices Cocoa J 2:109–111

Pervez R, Eapen SJ, Devasahayam S, Jacob TK (2012) Efficacy of some entomopathogenic nematodes against insect pests of ginger and their multiplication. Nematol Mediterr 40:39–44

Pervez R, Eapen SJ, Devasahayam S, Jacob TK (2013) A new species of entomopathogenic nematode *Oscheius gingeri* sp. n. from ginger rhizosphere. Arch Phytopathol Plant Protect 46:526–535

Pervez R, Eapen SJ, Devasahayam S, Jacob TK (2014a) Natural occurrence of entomopathogenic nematodes associated with ginger (*Zingiber officinale* Rosc.) ecosystem in India. Indian J Nematol 44:238–246

Pervez R, Devasahayam S, Eapen SJ (2014b) Determination of $LD_{50}$ and $LT_{50}$ of entomopathogenic nematodes against shoot borer (*Conogethes punctiferalis* guen.) infesting ginger. Ann Plant Prot Sci 22:169–173

Philip J, Nair PCS (1981) Field reaction of turmeric types to important pests and diseases. Indian Cocoa Arecanut Spices J 4:107–109

Rajabaskar D, Regupathy A (2012) Calling behavior and attractive response of cardamom shoot and capsule borer *Conogethes punctiferalis* Guenee (Lepidoptera: Pyralidae) to the female crude extract and synthetic blend. J Entomol 9:296–301

Rajabaskar D, Regupathy A (2013) Neem based IPM modules for control of *Sciothrips cardamomi* Ramk and *Conogethes punctiferalis* Gunee in small cardamom. Asian J Biol Sci 6:142–152

Rajan P (1983) Bionomics and control of *Dichocrocis punctiferalis* on cardamom. In: Annual report 1981. Central Plantation Crops Research Institute, Kasaragod, pp 180–181

Rajan P (1984) Biology and bionomics of insect pests of spices. In: Annual report 1982. Central Plantation Crops Research Institute, Kasaragod, p 83

Robinson GS, Tuck KR, Shaffer M (1994) A field guide to the smaller moths of South-East Asia. Malaysian Nature Society, Kuala Lumpur

Senthil Kumar CM, Jacob TK, Devasahayam S (2017a) Consortium research project in network mode. In: Final report of research project. ICARIndian Institute of Spices Research, Kozhikode

Senthil Kumar CM, Jacob TK, Devasahayam S (2017b) Evaluation of insecticides and a natural product for their efficacy against shoot borer (*Conogethes punctiferalis* Guen.) (Lepidoptera: Crambidae) infesting ginger (*Zingiber officinale* Rosc.). J Spices Aromat Crop 26:91–94

Shashank PR, Chakravarthy AK, Chandrashekharaiah R, Bhanu KRM (2014a) Behavioural studies on shoot and fruit borer, *Conogethes punctiferalis*, Guenée, (Crambidae: Lepidoptera) host-associated populations reveal occurrence of cryptic species. Entomol Gen 35:103–115

Shashank PR, Chakravarthy AK, Raju BR, Bhanu KRM (2014b) DNA barcoding reveals the occurrence of cryptic species in host-associated population of *Conogethes punctiferalis* (Lepidoptera: Crambidae). Appl Entomol Zool 49:283–295

Shashank PR, Kammar V, Mally R, Chakravarthy AK (2018) A new Indian species of shoot and capsule borer of the genus *Conogethes* (Lepidoptera: Crambidae), feeding on cardamom. Zootaxa 4374(2):215–234

Sheila MK, Abraham CC, Nair PCS (1980) Incidence of shoot borer (*Dichocrosis punctiferalis)* Guen (Lepidoptera: Pyraustidae) on different types of turmeric. Indian Cocoa Arecanut Spices J 3:59–60

Shetty GR, Hanchinamani V, Venkatesha J (2015) Correlation coefficients and occurrence of pest and diseases in turmeric (*Curcuma longa* L.) cultivars under hill zone of Karnataka. Environ Ecol 33:1201–1204

Stanley J, Chandrasekaran S, Preetha G (2009) *Conogethes punctiferalis* (Lepidoptera: Pyralidae) its biology and field parasitization. Indian J Agric Sci 79:906–909

Thyagaraj NE (2003) Integrated management of some important cardamom pests of hill region of Karnataka, South India. Ph D thesis, Dr BR Ambedkar University, Uttar Pradesh

Thyagaraj NE, Singh PK, Chakravarthy AK (2001) Sex determination of cardamom shoot and fruit borer, *Conogethes punctiferalis* (Guanee) pupae. Insect Environ 7:93

Varadarasan S (1995) Biological control of insect pests of cardamom. In: Ananthakrishnan TN (ed) Biological control of social forest and plantation crops insects. Oxford/IBH Publishing Company Private Limited, New Delhi, pp 109–119

Varadarasan S, Kumaresan D, Gopakumar B (1989) Bioecology and management of cardamom shoot/panicle/capsule borer, *Conogethes punctiferalis*. Guenee Spice India 2:15–17

Varadarasan S, Kumaresan D, Gopakumar B (1991) Dynamics of the life cycle of cardamom shoot borer, *Conogethes punctiferalis* Guen. J Plant Crop 18(Suppl):302–304

Velayudhan KC, Liji RS (2003) Preliminary screening of indigenous collections of turmeric against shoot borer (*Conogethes punctiferalis* Guen.) and scale insect (*Aspidiella hartii* Sign.). J Spices Aromat Crop 12:72–76

# Pest Risk Analysis for the Shoot and Fruit Borer, *Conogethes* spp. (Crambidae: Lepidoptera)

17

G. P. Mutturaj, S. Subhash, Sandeep Singh, and A. K. Chakravarthy

**Abstract**

The borer, *Conogethes punctiferalis*, has spread geographically through international trade. The insect has been intercepted in consignments of fruits in England, Wales, the Netherlands, Australia, New Zealand and the USA. *Conogethes* is a pest of quarantine importance and therefore should be regularly monitored for which phytosanitary measures are to be followed during harvest, processing, packing and transportation.

**Keywords**

Pest risk · Analysis · Phytosanitary · Risk rating guidelines · *Conogethes*

G. P. Mutturaj (✉)
College of Horticulture, Yalachahalli Horticulture Farm, Yelwala Hobli, Mysore, Karnataka, India

S. Subhash
Department of Agricultural Entomology, University of Agricultural Sciences, Gandhi Krishi Vignan Kendra (GKVK), Bengaluru, Karnataka, India

S. Singh
Department of Fruit Science, Punjab Agricultural University, Ludhiana, Punjab, India
e-mail: sandeep_pau.1974@pau.edu; gurlaz@pau.edu

A. K. Chakravarthy
Division of Entomology and Nematology, ICAR-Indian Institute of Horticultural Research, Bengaluru, Karnataka, India





A. K. Chakravarthy (ed.), *The Black spotted, Yellow Borer, Conogethes punctiferalis Guenée and Allied Species*,
https://doi.org/10.1007/978-981-13-0390-6_17

## 17.1 Introduction

The introduction of plants and plant-feeding arthropods of one geographical area to another is an issue of spread and worldwide concern which needs to be addressed by international agencies. One of the important organisations working in this regard is the International Plant Protection Convention (IPPC), a multilateral treaty for international cooperation in plant protection. It aims at securing common effective action to prevent the spread and introduction of pests of plants and plant products and to promote appropriate measures for their management/control. The convention also aims to protect plant health while limiting interference with international trade. The IPPC directives apply to cultivated plants, natural plants and plant products and include both direct and indirect damage not only by insects but also other arthropod pests. In addition to plants and plant products, the IPPC also extends to storage, packaging, conveyance, containers, soil and other organisms, object soil or material capable of harbouring or spreading herbivorous pests of plants. It also includes arthropods of veterinary and medical importance. The issues of pest risk analysis (PRA) with respect to the shoot and fruit borer of the *Conogethes* spp. are discussed.

## 17.2 Pest Risk Analysis

Pest risk analysis (PRA) is a science-based process providing the rationale for determining appropriate phytosanitary measures for a specified PRA area. It is a process that evaluates technical, scientific and economic evidence to determine whether an organism is a potential pest of especially cultivated plants and, if so, how should it be managed? Here the term plant pest refers to all organisms harmful to plants or plant products including other plants, bacteria, fungi, insects, mites, molluscs, nematodes and viruses. The IPPC recognizes and defines two categories of regulated pests of plants: quarantine pests and regulated non-quarantine pests. PRA assists with determining whether a pest fits either of these two categories and the strength of phytosanitary measures that can be adopted to mitigate the pest problem.

The shoot and fruit borers of the *Conogethes* spp. are potential quarantine pests of plants whose probability of introduction and spread as well as the magnitude of potential consequences is evaluated using scientific, technical and economic evidence. If the pest risk is unacceptable, the analysis may continue by suggesting management options that will reduce the pest risk to an acceptable level or eliminate the same. These pest risk management options are applied to establish phytosanitary regulations/measures.

A PRA may also consider the pest risks posed by the introduction of organisms associated with a particular pathway, such as a traded commodity. In most cases, the commodity itself does not pose a pest risk, but it may carry phytophagous pests of plants. When deliberately or inadvertently introduced and established in new areas, the organisms imported as commodities (such as plants for planting, beneficial organisms and living modified organisms (LMOs)) may pose a risk by spreading to unintended habitats causing damage to plants, and such pest risks are also analysed using the PRA process. Even flower gift bouquets may serve as carriers of organisms hitherto alien to the new landscapes.

The PRA process is not applicable to biological control agents such as parasitoids/predators/pollinators or other beneficial organisms and LMOs, pathways and processes.

## 17.3 Stages of PRA

PRA for quarantine pests follows a process in three stages or phases:

Stage 1: Initiation—This stage involves identifying the reason for the PRA and identifying the pest(s) and pathway(s) that may be considered for PRA in relation to the PRA area.

Stage 2: Pest risk assessment—In this stage, the information about the pest or pest group identified in Stage 1 is gathered and evaluated. The results are used to decide whether risk management is required. Also, the endangered area within the PRA area is identified.

Stage 3: Pest risk management—This stage determines appropriate management options to reduce the risks identified in Stage 2.

### 17.3.1 Stage 1: Initiation

#### 17.3.1.1 Name of the Pest

The shoot and fruit borer, *Conogethes punctiferalis* (Guenée 1854) (Lepidoptera, Crambidae), castor shoot and capsule borer or yellow peach moth or durian fruit borer or Queensland bollworm or maize moth or smaller maize borer, is a polyphagous pest with synonym *Dichocrocis punctiferalis* (Guenée).

**Bioecology** In general, moths lay microscopic eggs singly on reproductive plant parts. Larvae, pinkish brown with spots, undergo five instars. Average total larval period varies from 12 to 15 days and pupal period lasts for 10–12 days. On an average number of eggs per female varies from 81 to 102 eggs. Further, major mortality occurs at the pupal stage in India. Castor, tropical and temperate fruit crops are the host plants.

Females lay oval eggs on or near fruits or seeds of hosts (United State Department of Agriculture 1957). Females lay from 20 to 30 eggs in New Zealand. Once they hatch, larvae feed on or in the seeds, seed capsules and young shoots (United State Department of Agriculture 1957). Pupation usually occurs in the larval tunnels within a silk cocoon, surrounded by shelters of webbing and frass (Chong et al. 1991; MAF Biosecurity New Zealand 2009). The adult emerges about 8 days later (Chong et al. 1991). In Australia, pupation takes place from 2 to 3 weeks (in summer) to as long as 8 weeks (in winter). Adults emerge during the night; most of them are active at night and hide on the back of host leaves during the day, and the moths feed on the nectar of the larval host plant (Convention on Plant Protection 2007).

The shoot and fruit borer, *Conogethes* sp., breeds throughout the year in India and parts of Australia (USDA 1957). In Japan, this species has two to three generations per year and five to six generations in India. Full-grown larvae overwinter within thick silken cocoons spun inside the loose scales of the trunk bark or within the mummified fruit of castor or cocoa and peach or chestnut. The larvae pupate within the cocoons in spring, and moths emerge in mid-May (Púcat 1995). In laboratory conditions, *C. punctiferalis* completes its life cycle in about 28 days on castor, and *C. sahyadriensis* lasts 31 days on cardamom and 32 days on ginger. Damage caused by this species increases with relative humidity (Stanley et al. 2009). Further, under the canopy of forests in the Andaman and Nicobar islands, *Conogethes* spp. was not found infesting wild plants belonging to *Zingiber*, *Curcuma*, *Hedychium*, *Alpinia* and *Elettaria*, and borer infestation on suckers was negligible (<1–2%).

**Damage**

Since *C. punctiferalis* is a polyphagous pest, almost all plants are damaged. The first two larval instars are tiny, pinkish brown and active, and third, fourth and fifth instars larvae are light brown, with dark brown head, and larvae hang by silken thread. The duration of three instars lasts for 8–10 days. Besides, boring by *C. punctiferalis* larvae can predispose the fruits to secondary infection (Chong et al. 1991).

**Castor** Castor is the principal host plant of *C. punctiferalis* in India and neighbouring countries. Larvae bore into the tender shoot, flower clusters, young leaves and fruits. These parts are webbed, and larvae bore into the parts, staying inside a channel that is webbed with excretory pellets making it difficult for the insecticides to kill them. The larvae feed on seeds inside the capsules, incurring heavy yield losses.

**Cocoa** The larvae bore into the husk of the fruits. In cocoa, boring often starts either in the regions of pods that meet with other pods or the bark, and the pest is also often associated with severe plant bug damage (Chong et al. 1991). Tender, attacked pods wither and bored holes are plugged with excreta.

**Corn** Evidence of larval feeding includes damage to silk, grain and cob, and sometimes tunnels bored into the stalk (Púcat 1995). In the absence of preferred host plants, the larvae feed on corn.

**Cotton** Larvae bore into cotton bolls and sometimes the stems. They produce masses of webbing and excreta at the entrance of their tunnels (Cotton Catchment Communities Cooperative Research Centre 2007).

**Durian** In durian, boring usually occurs under a mass of frass, webbed together in between thorns of the fruit (Chong et al. 1991). In Malaysia and Sri Lanka and parts of Vietnam, it is the principal host plant for the pest.

**Ginger** Ginger, turmeric and cardamom form important host plants for the pest in India and neighbouring countries. Larvae begin feeding on the green contents of the leaves and later bore into the shoots, feeding on the inner core (Devasahayam et al. 2010). Larvae feeding in the centre of the stem cause the death of the "heart", which is visible when the terminal shoots yellow. Usually the larva matures before it reaches the rhizome and the stem to pupate; however, occasionally it arrives at the rhizome and damages it (Chong et al. 1991). Characteristic symptoms in ginger include the presence of bored holes on the pseudostem from which frass is extruded and the withered central shoot (Devasahayam and Koya 2005).

**Cardamom** It is another *Zingiber* plant attacked by *C. sahyadriensis*, an economically important plant that is either cultivated or grows naturally in evergreen tropical forests of the Western Ghats, South India. The borer species is yet to be identified. The pest also attacks wild *Zingiber* plants like *Hedychium*, *Alpinia*, *Curcuma*, etc.

#### 17.3.1.2 Taxonomy

*C. punctiferalis* is a complex of at least two species (CPC 2007). A polyphagous form that feeds on fruits from a number of plant families and an oligophagous form that feeds on leaves of Pinaceae have been noted in Japan (Konno et al. 1981). A similar situation has been noted in China (Chai and He 1987). On the basis of morphological differences and other evidence, Honda and Mitsuhashi (1989) concluded that the fruit- and Pinaceae-feeding types of *C. punctiferalis* are discrete taxonomic species in Japan. However, they have not mentioned which form should be named *punctiferalis* or whether a name is available for the other form (CPC 2007). Fruit-feeding and pine-feeding types are not always distinguished in the literature (FAO 2007).

#### 17.3.1.3 Geographical Distribution

*C. punctiferalis* is highly polyphagous and has been recorded on 65 host plants from 30 different families. Many hosts are economically important (Devasahayam and Koya 2005). The shoot and fruit borer, *C. punctiferalis*, occurs in Asia and is found in Bangladesh, Brunei, Cambodia, China, India, Indonesia, Iraq, Japan, Korea, Laos, Malaysia, Myanmar, the Philippines, Sri Lanka, Taiwan, Thailand and Vietnam; in North America, Hawaii (Nishida 2002); and in Oceania in Australia and Papua New Guinea (Waterhouse 1993; Zhang 1994; Púcat 1995; EPPO 2013).

**Pathway** *C. punctiferalis* can move through international trade. There are records of this species being intercepted on fruit consignments in England, Wales, the Netherlands and New Zealand (MAF Biosecurity New Zealand; Korycinska 2012).

Interceptions in England and Wales occurred on sugar apple, mango and guava (Korycinska 2012) and in New Zealand which occurred on capsicum and tomato. *C. punctiferalis* has been intercepted at the US ports of entry 748 times (742 interceptions were of larvae). The species has been intercepted and identified 104 times at the US ports of entry (all interceptions were of larvae) (AQAS 2014; queried December 9, 2014). Within Asia, the species must have spread and found new hosts from India or Sri Lanka and must have spread through plants taken from one country to another.

**Crop Loss** Currently, studies have documented the amount and type of damage caused by *C. punctiferalis*, and most reports are limited to a specific or few crops (Korycinska 2012). It has previously been described as a destructive pest of temperate fruits in China and cotton in Australia. Further, this pest has caused almost complete loss of grain sorghum in coastal Australia (USDA 1957). More recent reports refer to it as a minor and irregular pest of sorghum, corn and cotton (Korycinska 2012). Astridge (2001) considered this species to be a major pest of rambutan and durian in North Queensland, Australia.

*C. punctiferalis* has been documented as a serious pest of castor bean and fruits in tropical, sub-tropical and temperate countries (USDA 1957). Devasahayam and Koya (2005) stated that this species is the most serious pest of ginger in India, particularly in South India. Crop yield can be significantly affected when more than 45% of the shoots in a clump are damaged (Devasahayam et al. 2010). In South India, capsule yield loss in cardamom was recorded between 6.8% and 9.2%, while castor capsule damage was recorded between 11% and 27% (reviewed in Shashank et al. 2015). In Guatemala and other countries, the borer (*C. sahyadriensis*) damage on cardamom varied widely (5–30%).

Species of *Conogethes* are a major pest of *Curcuma domestica* (turmeric) and *Ricinus communis* (castor). They are also listed as a pests of *Carica papaya* (papaya), *Elettaria cardamomum* (cardamom), *Macadamia ternifolia* (macadamia), *Morus* sp. (mulberry), *Psidium guajava* (guava), *Prunus persica* (peach), *Punica granatum* (pomegranate) and *Zingiber officinale* (ginger), and most tropical and sub-tropical fruits are attacked by this pest (Hill 1983).

Studies in Kerala, India, have shown that *Conogethes* species has caused yield losses of 25% when 23–24% of ginger pseudostems were infested. Forty percent yield losses in ginger have been reported in certain parts of India (Devasahayam and Koya 2005). The shoot damage in cardamom due to borer infestation varied from 5% to 10%, and in terms of capsule yield loss (dry weight basis), it varied from 6.79% to 9.18%. Therefore, collectively the loss was estimated to be more than 20% every year (Kapadia 1996; Thyagaraj 2003). However, the crop loss due to this pest was worked out in cardamom, and economic threshold level was fixed at 10% (Anonymous 1954; Krishnamurthy et al. 1989; Ram et al. 1997). In Zingiberaceae, slight borer damage to suckers/pseudostems results in production of new suckers that may contribute to yield. Studies on castor in Salem, Tamil Nadu, South India,

**Fig. 17.1** Surveillance of pomegranate fruits to detect *Conogethes* infestation

recorded 10.80–26.70% capsule damage (Suganthy 2011). Kapadia (1996) estimated 42.30% crop loss in castor in India. Fifty percent yield reduction was recorded on grapes due to *C. punctiferalis* attack (Ram et al. 1997). *C. punctiferalis* and other species are not yet documented as vectors of disease-causing organisms (Chong et al. 1991) (Fig. 17.1).

### 17.3.2 Pest Risk Assessment

The shoot and fruit borer *C. punctiferalis* is not known to occur in Europe, South Africa, Canada and the USA. It is neither listed in the EC Plant Health Directive nor in any of the EPPO lists. However, there are records of this pest being intercepted on fruit consignments in England, the Netherlands and New Zealand (MAF Biosecurity New Zealand; Korycinska 2012), and it is a pest of concern to the NPPOs of above-mentioned countries and can cause potential hazard. Therefore, it is a pest of quarantine importance, and there is a need to rapidly assess its importance. International Standards for Phytosanitary Measures (ISPM) No. 11 (Pest Risk Analysis for Quarantine Pests Including Analysis of Environmental Risks and Living Modified Organisms, 2004) provides guidelines for further analysis of organisms that are quarantine pests.

The pest risk assessment depends on the need and economic importance of the concerned pest species. The pest risk assessment steps are documented sequentially, but it is not essential to always complete them in an order. Pest risk assessment is an iterative which requires repeated cognizance of the several factors which influence pest risk as information becomes available. The assessment of the probability of introduction and potential economic impact of introduction and spread are combined to provide an overall estimation of the risk involved due to the pest.

The pest risk assessment should support a final decision and should also provide necessary technical justification to defend decisions regarding phytosanitary measures. The pest risk assessment should be transparent based on sound science and consistent with other pest risk assessments conducted by the NPPO. It is desirable to have a national and international level that provides a standard, comprehensive framework that lays out the criteria to assess all potential risk factors.

The pest risk assessment should be comprehensive embracing several of each pest including data on spread, geographical distribution, biology and economic significance in the localities where it occurs. Experts are then employed to assess probability of the pest being introduced and its potential for establishment, spread and economic importance in the PRA area. In characterising the risk, the amount of data generated on will vary with each pest and the locality. The quality of the assessment will also vary with the resources and resource persons. For instance, one NPPO may have comprehensive pest databases and geographical information systems, while another may depend on books, printed soil maps, climate maps and expert opinion. In some cases, nil or scarce information may be available, or research data may be required. Countries where the pest is present may provide, upon request, required data for conducting the pest risk assessment in the nation of interest.

#### 17.3.2.1 Risk Rating Guidelines

The likelihood of the pest getting introduced can be graded as negligible, low, medium or high as per the NPPO guidelines. A pest's potential consequences may also range from negligible to high. Each of the ratings can be assigned a standard numeric value for the purpose of combining them. The overall risk rating for the pest is assigned by combining likelihood and consequence ratings. Following are the types of rating of risk due to introduction, according to NPPO guidelines.

**Negligible** The likelihood of introduction is negligible. No specific phytosanitary measures are necessary. However, it is felt that surveillance and monitoring are required.

**Low** No specific phytosanitary measures may be necessary. Various factors including production practices, preshipment inspection, packaging, current port-of-entry inspection, end use, season of importation, etc. are expected to provide sufficient phytosanitary security; consignment of the products may be critically examined.

**Medium** Specific phytosanitary measures may be necessary.

**High** Specific phytosanitary measures are strongly recommended. Port-of-entry inspection.

alone is not considered sufficient to provide phytosanitary security, and other factors may merit the consideration.

#### 17.3.2.2 Entry Assessment

Let us consider the case of *C. sahyadriensis* on cardamom and ginger in India. The pest is recorded in South Indian states on cardamom and in more than ten states on ginger. All life stages of the pest can be associated with these commodities.

The adults of *Conogethus* spp. have a wingspan of 25 mm and are known to feed on the nectar and fruit of host plants (CPC 2007; Kang et al. 2004). However, moths are active only at night and hide on the backs of leaves during the day. The moths can be seldom sighted in cardamom plantations. Larvae bore and tunnel into pseudostem, rhizome and cardamom capsules. Both commodities are exported from India, and, therefore, the pest is of quarantine importance.

Eggs are 2–2.5 mm long and elliptical in shape. They are laid individually on the surface of the fruit. *C. sahyadriensis* undergo between two and five generations per year, and the full-grown larvae, under favourable conditions, undergo pupation. This suggests that eggs could be present at harvest time, and owing to their small size, they may not be detected during the harvest and packing processes.

Interestingly, live larvae of *C. punctiferalis* have been intercepted twice at the New Zealand border (on capsicum in 2004 and on tomato in 2008), and both shipments were from Australia (MAFBNZ 2009). Dead larvae of *C. punctiferalis* have also been intercepted at the Canadian border on *Pyrus pyrifolia* (Lee et al. 2000). Pupae can be formed on or in the fruit. Larvae and pupae are more likely to be detected during harvest and intercepted in consignments compared to eggs. Hence, the likelihood of entry of *C. punctiferalis* larvae is considered moderate and therefore non-negligible.

In temperate fruits like plum, peaches, chestnut and pear, generally the flesh and often the skin are eaten, but the seeds and core portions are disposed off. However, whole fruit or parts of the fruit are not always consumed. In grapes, the waste material generated (pedicels, peduncle and uneaten grapes) could allow some *C. punctiferalis* larvae to disperse and find a suitable host as bunches are washed or the waste material adhering to bunches are cleaned off before the fruits are eaten.

It is likely that any *C. punctiferalis* larvae associated with imported fruit disposed of in this way would be able to pupate and emerging adults disperse. The moth is mobile and highly polyphagous. It is likely that any emerging adult would be able to find a suitable host. Hence, the likelihood of exposure is considered to be moderate and therefore non-negligible or not of any consequence.

The *C. punctiferalis* reproduces sexually, and females release sex pheromones to attract males (CPC 2007). After mating, a host plant is located to deposit eggs. The host range of *C. punctiferalis* covers plants from important fruit plantation species to arable crops like maize and sorghum. It is likely to be able to find suitable hosts in the area. Establishment also depends upon the prevailing weather conditions and top topography. Warm conditions favour the development of *C. punctiferalis* larvae by reducing the time required for development (Kang et al. 2004). However, in Asian countries like Japan and China where there are areas with temperate climates,

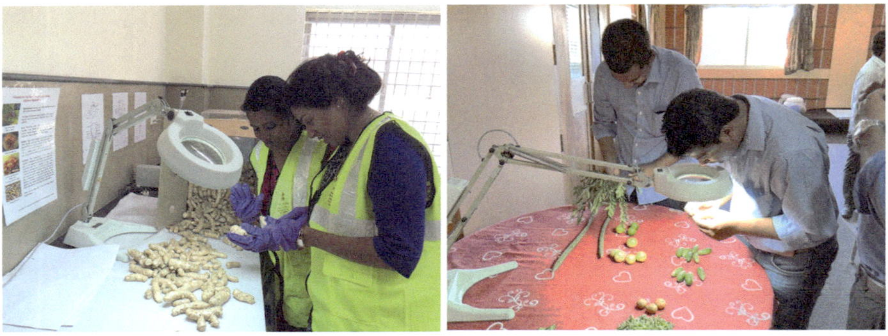

**Fig. 17.2** Physical examination of ginger and vegetable consignment for borer damage

*C. punctiferalis* overwinters in the larval stage; thus larvae hibernate once conditions become unfavourable. *C. punctiferalis* is seen as a minor and infrequent pest in Australia; it has been identified as a major and frequent pest of economic importance in the warm wet tropics of regions of North Queensland, especially on rambutan (*Nephelium lappaceum*) and durian (*Durio* spp.). It is generally more frequent in years with continuously wet summers (Astridge et al. 2005; Astridge 2006). Southeast Asian countries like Vietnam and India are characterised by incredibly diverse vegetation, habitats and end margins, and so it is likely that these countries have more than one species of *Conogethus* (Fig. 17.2).

#### 17.3.2.3 Economic Consequences

Crop yield loss caused by the target pest is an important criterion determining economic consequences concomitantly. The economic losses the pest causes in other parts of its geographical range, particularly in those areas where climatic conditions, crop production practices or other factors are similar to those in the PRA area, can be ascertained. ISPM No. 11 (2004) provides guidance on factors to consider when assessing potential economic consequences and environmental impact. It states that after obtaining information on areas where the pest currently occurs, that information should be compared with the situation in the PRA area. Case histories concerning comparable pests can also be taken into account.

*Conogethes* species can cause significant damage to stems, pseudostem, rhizomes, fruits and seeds of host plants (FAO 2007). In Australia *C. punctiferalis* is seen as a minor and infrequent pest. It is generally more frequent in years with continuously wet summers (Astridge et al. 2005). As mentioned earlier, it has been noted as a major and frequent pest in the wet tropics of North Queensland especially for rambutan and durian (Astridge et al. 2005; Astridge 2006); as also in Australia and Sri Lanka, the larvae bore into the fruit of rambutan and can destroy up to 90% of fruit clusters if not controlled (Astridge 2006). In India, infestation of grapes by *C. punctiferalis* has been reported to result in a 50% reduction in yield (Ram et al. 1997). In 2016, *C. punctiferalis* caused a loss of more than 50% to cocoa pods in coastal Karnataka, South India.

*C. punctiferalis* is an important pest of peaches and applies in southern and northern China, respectively, and contributes up to 25% of chestnut crop loss (FAO 2007). It is also a serious pest of chestnut in Korea (Kang et al. 2004). Excretions from *C. punctiferalis* have high sugar content which covers the fruit surface; based on its host range, climatic conditions, crop production practices or other factors in the PRA area, the potential economic consequences can be arrived at by accounting for the secondary insect pests and diseases that the infested fruits attract, resulting in further damage to the crop (CPC 2007).

#### 17.3.2.4 Environmental Consequences

The borer, *C. punctiferalis*, is highly polyphagous and catholic feeder of many of the plant families which includes PRA area's native members including endemic species (e.g. *Syzygium maire* in New Zealand) and genera (e.g. *Lophomyrtus* and *Neomyrtu* in New Zealand). Cardamom in South and Northeast India and castor in Central and Western India are infested by the species of *Conogethes*. The impact on native flora is currently uncertain. In addition, many exotic plant species in the same families are known hosts of *C. punctiferalis* and are found in domestic gardens, wild patches and parks in PRA area, roadsides and wastelands or are naturalised in the wild. Damage to the former might be of concern to gardeners, and colonisation of naturalised species in the wild could assist dispersal and provide reservoirs or populations for growth and damage to crops in the cultivated landscapes.

### 17.3.3 Stage 3: Risk Management

*Options* Risk management options that are relevant to this organism are listed below. Their effect in managing the risk posed by the borer is assessed.

*Pest-free area* An area can be declared as a pest-free area or a pest-free place of production, in agreement with ISPM 4 or 10, respectively (IPPC 2007). Pest freedom status is achieved through a systems approach and trapping to monitor population levels in and around orchards. Both ISPM measures rely on systems to establish freedom and phytosanitary measures and check to maintain and verify if freedom has been maintained resulting in official pest-free certification of the area or place of production. Pest-free areas can be a viable option if pest freedom is verified, but in case of *C. punctiferalis* and *C. sahyadriensis,* as both are widespread, this is not considered as a viable option.

#### 17.3.3.1 Surveillance and Management

*C. punctiferalis* produces sex pheromones which may facilitate targeted surveys of export crop areas to detect its presence. However, currently pheromone traps are not effective in detecting moths under field conditions in a majority of the areas where *Conogethes* population occurs. Detection in the surveillance programme would indicate the failure of in-field control, and any fruit from an infested area should not

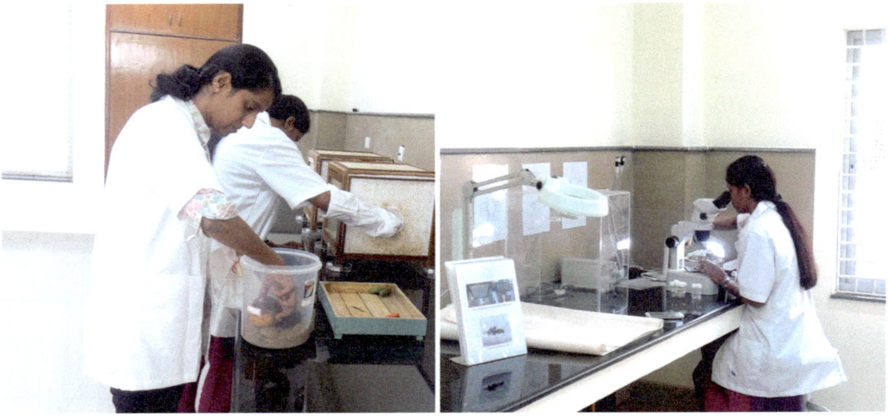

**Fig. 17.3** Rearing and identification of insect pest specimens

be permitted entry to the importing area or to areas where the pest is absent (Fig. 17.3).

#### 17.3.3.2 Bagging of Fruit

Bagging of fruits is likely to prevent adult *C. punctiferalis* from laying eggs on the surface of the fruit. However, there is a time gap of up to 4 weeks from fruit set before the fruits are bagged during which eggs could be laid. Eggs hatch in 5–18 days after laying, and larvae pupate after a further 15–18 days. It is assumed that any eggs laid on the fruit prior to the bagging would complete their development to adulthood prior to harvest of the fruit, making bagging an effective measure. However, this method in most cases may prove costly and not practically feasible.

#### 17.3.3.3 Cold Treatment

*C. punctiferalis* overwinter as mature (fifth instar) larvae in host stems or fruit or under the bark of fruit trees. Cold treatment is unlikely to mitigate the risk of live larvae entering the new area. The USA has two treatment schedules against *C. punctiferalis* on apples from Japan or Korea. They both consist of 40 days cold treatment (1.11 °C or below) followed by a methyl bromide fumigation of at least 0.5 h (USDA 2008). Since Asian pears are liable to be damaged by methyl bromide fumigation, this is unlikely to be a viable option, as also for tropical fruits like mango, guava, etc.

#### 17.3.3.4 Phytosanitary Inspection Prior to Export

The egg, larval and pupal stages are associated with plants. Eggs are white, 2.0–2.5 mm in diameter and are likely to be visible to the naked eye. Larvae are internal feeders. Since mature larvae are relatively large, it is assumed that there would be external evidence of infestation. *C. punctiferalis* adults are 12 mm long (CPC 2007). Infestations can also result in fruit drop or stunting and scorching of fruit, and such fruits are likely to be detected. *C. punctiferalis* excretions, which cover the fruit surface and have high sugar content, attract other insect pests and diseases which damage fruit, which will result in a greater likelihood of detection. Pupae occur on the surface of the fruit and are visible.

#### 17.3.3.5 Phytosanitary Measures

Many insect pests of crops are a result of transportation of the infested plant produce from one place to the new. Thus, care during the transportation of possible host plant material and fruits among countries and continents may help to limit spread of this pest (Miniraj et al. 2000). Comparative effects of gamma irradiation and methyl bromide (MeBr) fumigation were determined by Kwon et al. (2004) for fresh chestnut on mortality of *C. punctiferalis* in South Korea. The MeBr 3 irradiation caused 100% larval mortality 3 days after fumigation, and the same results are obtained in about 4 weeks by irradiation at 0.5 kGy. Considering the cumulative mortality of chestnut pests, irradiation at 0.5 kGy is recommended as one of the alternatives to MeBr fumigation for both quarantine and sprout control purposes.

Interest in the use of irradiation as a phytosanitary treatment for agricultural commodities is growing worldwide, particularly since publication of the International Plant Protection Convention (IPPC) standard that endorses and facilitates trade based on this disinfestation method. Irradiation is broadly effective against insects and mites at doses that do not compromise quality of most commodities. Unlike other disinfestation techniques, irradiation does not need to kill the pest immediately to provide quarantine security, and therefore live but sterile or non-viable insects may occur within the exported commodity making inspection for the target pests redundant. Generic irradiation treatments have been approved in the USA to control broad groups of insects in all commodities. The approved generic doses are 150 Gy for tephritid fruit flies and 400 Gy for all insects except lepidopteran pupae and adults (that require higher doses). Generic irradiation treatments will accelerate the approval of irradiation quarantine treatments for specific crops and expedite new ways in agricultural trade. The availability of generic treatments makes irradiation an attractive option (Follett et al. 2007).

For the import of papaya from Australia to New Zealand, a minimum dose of 250 Gy must be applied against *C. punctiferalis*. This treatment is considered a minimum irradiation dose for all regulated arthropod pests on the Australian papaya (Anonymous 2010). Due to reporting of damage to apples in North China (CABI 2011) and interception of the pest in fruits from Pakistan and a pest concern to many countries like New Zealand, South Africa, Canada and the USA, a rapid pest risk analysis (PRA) on *Conogethes* was conducted. Interceptions of the pest in fruits have been made both in England and Wales and in the Netherlands (FERA 2012). This species can move through international trade as it has been intercepted on fruit consignments in Europe and New Zealand (Korycinska 2012). Interceptions in England and Wales occurred on sugar apple, mango and guava. Much of the species' distribution occurs in the subtropics; however, it has also been recorded from northern Japan and China. Threshold temperatures in the Hebei province in north China were calculated at 8.4 °C for eggs, 7.3 °C for larvae and 11.3 °C for pupae (Korycinska 2012).

New Zealand Ministry for Primary Industries (NZ MPI) mandates specific pre-export phytosanitary measures for *C. punctiferalis* for the import of fresh litchi from Thailand. NZ MPI currently approves the use of infield pest control for Lepidoptera species throughout the production season or irradiation at a minimum dose rate of 250 Gy. NZ MPI approves the irradiation doses for *C. punctiferalis* with a minimum dose rate of 250 Gy (Anonymous 2014).

Interceptions in New Zealand occurred on capsicum and tomato. Species of *Conogethes* have been intercepted at US ports of entry 748 times of which 742 interceptions were of the larvae. Further, the species has been intercepted and identified 104 times at US ports of entry, and all interceptions were of larvae (AQAS 2014). The USA allows peaches into Guam and CNMI from Canada, Korea and certain parts of Japan and ginger into all ports from all countries (FAVIR 2014).

Hong et al. (2015) reported that complete control of quarantine insect pests is required for exporting domestic apples to other countries. To this end, a controlled atmosphere and heat treatment system (CATTS) has been developed as a postharvest treatment in Korea. Their study determined the CATTS conditions to completely control peach pyralid moth, *C. punctiferalis*, which exhibits different feeding behaviours. The fifth instar larvae were the most tolerant to the heat treatment. *C. punctiferalis* underwent 88% mortality under CATTS conditions with 15% $CO_2$ and 1% $O_2$ for 1 h heat treatment at 46 °C. In order to control the fifth instar larvae completely, 2 h of heat treatment is required under the same atmospheric conditions. These CATTS treatment effects were confirmed against over 3000 fifth instar larvae infesting apples. This study demonstrates that longer exposure to CATTS conditions is required for the complete disinfestation of the internal apple feeder compared to the non-internal apple feeder. The larvae of *Conogethes* spp. are cryptic internal tissue feeders owing to which low levels of infestation are not detected and the pest spreads from one country to another through infested fruits (Chakravarthy et al. 2015).

**Acknowledgement** Interaction and discussion with cardamom traders and wholesale dealers in Mudigere, Sakaleshpur, Madikeri and Calicut helped form opinions which were important for writing this chapter. Authors are thankful to authorities of the University of Agricultural Sciences, Bengaluru, Punjab Agricultural University, Ludhiana and Spice board, Calicut, Kerala for the encouragement and help. Authors are also thankful to the scientist in charge and Dr. Rashmi, Plant Quarantine Station, Hebbal, Bengaluru.

## References

Anonymous (1954) Pests of cardamom. Memories of the Department of Agriculture, Madras, pp 998–1000

Anonymous (2010) Import Health Standard Commodity Sub-class: Fresh Fruit/Vegetables Papaya, (Carica papaya) from Australia. Director, Plants, Food and Environment, New Zealand Ministry for Primary Industries, New Zealand

Anonymous (2014) Import Health Standard Commodity sub-class: Fresh fruit/vegetables Lychee (Litchi chinensis) from Thailand. Director, Plants, Food and Environment, New Zealand Ministry for Primary Industries, New Zealand

AQAS (2014) *Conogethes* sp. interceptions. Agricultural quarantine activity systems. Accessed December 9, 2014 from: https://aqas.aphis.usda.gov/aqas/HomePageInit.do#defaultAnchor

Astridge D (2001) Insect fauna surveys on rambutan, durian and mangosteen in North Queensland. In: Proceedings of the sixth workshop for tropical agricultural entomologists, Darwin, Northern Territory Department of Primary Industry and Fisheries, technical bulletin no. 288, pp 75–79

Astridge DP (2006) Rambutan – development of integrated pest management – pest monitoring and insecticide screening. RIRDC publication no. 05/187. Australian Government, Rural Industries Research and Development Corporation

Astridge D, Fay H, Elder R (2005) Yellow peach moth in rare fruit. Department of Primary Industry and Fisheries. Queensland Government. Can be accessed online at: http://www2.dpi.qld.gov.au/horticulture/5044.html

CABI (2011) Conogethes punctiferalis datasheet. In: Crop Protection Compendium. Wallingford, UK: CAB International. Wallingford, UK

Chai XM, He ZH (1987) *Dichocrocis punctiferalis* Guenee injurious to masson pine. Insect Knowl 24(2):99–100

Chakravarthy AK, Kammar V, Lokeshwari D, Rani AT, Nagaraj T, Rajan VV (2015) The black-spotted yellow shoot-and-fruit borer, Conogethes spp. (Crambidae: Lepidoptera) – a global perspective. Curr Sci 109(6):1016–1018

Chong KK, Ooi PAC, Tuck HC (1991) Crop pests and their management in Malaysia. Tropical Press Sdn.Bhd, Kuala Lumpur, p 242

Cotton Catchment Communities Cooperative Research Centre (2007) Yellow peach moth

CPC (2007) Crop protection compendium on internet, accessed in 2008. CAB International, Wallingford

Devasahayam S, Abdulla Koya KM (2005) Insect pests of ginger. In: Ravindran PN, Babu KN (eds) Ginger: the genus Zingiber. CRC Press, Washington, DC, pp 367–390

Devasahayam S, Jacob TK, Abdulla Koya KM, Sasikumar B (2010) Screening of ginger (*Zingiber officinale*) germplasm for resistance to shoot borer (*Conogethes punctiferalis*). J Med Aromat Plant Sci 32(2):137–138

EPPO (2013) PQR – EPPO database on quarantine pests. Available at: http://www.eppo.int

FAO (2007) Forest health and biosecurity working papers. Overview of forests pests. People's Republic of China. Food And Agriculture Organization of the United Nations – Forestry Department, Rome

FAVIR (2014) Fruits and Vegetables Import Requirements (FAVIR) database. USDA/APHIS

FERA (2012) http://www.fera.defra.gov.uk/plants/plantHealth/pestsDiseases/documents/conogethesPunctiferalis.pdf

Follett PA, Yang MM, Lu KU, Chen TW (2007) Irradiation for Postharvest Control of Quarantine Insects. Formosan Entomol 27:1–15

Guenée MA (1854) Histoire Naturelle des Insectes. Species Général des Lépidoptères. Roret, Paris

Hill DS (1983) Agricultural insect pests of the tropics and their control, 2nd edn. Cambridge University Press, New York, p 746

Honda H, Mitsuhashi W (1989) Morphological and morphometrical differences between the fruit- and Pinaceae-feeding type of yellow peach moth, *Conogethes punctiferalis* (Guenee) (Lepidoptera: Pyralidae). Appl Entomol Zool 24(1):1–10

Hong Y, Kwon K, Kim Y (2015) Post-harvest treatment on the peach pyralid moth and the small tea tortrix moth infesting apples using controlled atmosphere and temperature. Korean J Applied Entomol 54(1):11–17

IPPC (2007) International standards for phytosanitary measures (ISPM). https://www.ippc.int/IPP/En/default.jsp

ISPM 11 (International Standards for Phytosanitary Measures) (2004) Pest risk analysis for quarantine pests including analysis of environmental risks and living modified organisms. FAO, Rome

Kang CH, Lee SM, Chung YJ, Choi KS, Park CG (2004) Overwintering ecology of the peach pyralid moth, *Dichocrosis punctiferalis* in southern regions of Korea. Korean J Appl Entomol 43(3):201–209

Kapadia MN (1996) Estimation of losses due to pod borers in oilseed crops. J Oilseeds Res 13(1):139–140

Konno Y, Honda H, Matsumoto Y (1981) Mechanisms of reproductive isolation between the fruit-feeding and the Pinaceae-feeding types of the yellow peach moth, *Dichocrocis punctiferalis* Guenee (Lepidoptera: Pyralidae). Jpn J Appl Entomol Zool 25(4):253–258

Korycinska A (2012) Rapid assessment of the need for a detailed pest risk analysis for *Conogethes punctiferalis* (Guenée). The Food and Environment Research Agency, p 7

Krishnamurthy K, Khan MM, Avadhani KK, Venkatesh J, Siddaramaiah L, Chakravarthy AK, Gurumurthy BR (1989) Three decades of cardamom research, regional research station (1958–1988). Stn Technol Bull 2:44–68

Kwon J, Kwon Y, Byun M, Kim K (2004) Competitiveness of gamma irradiation with fumigation for chestnuts associated with quarantine and quality security. Rad Phy Chem 71(1-2):43–46 http://adsabs.harvard.edu/abs/2004RaPC7143K

Lee HK, Woo CN, Namkoong SB, Seo YS, La JH, Kim YI, Kim KG, Kim BK (2000) Quarantine pest occurrence in exporting pear fruits (*Pyrus pyrifolia* Nakai cv. Whangkeumbae) during cold storage. Korean J Hortic Sci Technol 18(1):22–27

MAF Biosecurity New Zealand (2009) Import risk analysis: table grapes (*Vitis vinifera*) from China. MAF Biosecurity New Zealand, Wellington, p 322

Miniraj N, Murugan N, Joseph CR (2000) Evaluation of Cardamom (Elettaria cardamomum Maton) germplasm. J Spices Arom Crops 9:55–56

Nishida GM (2002) Hawaiian terrestrial arthropod checklist, 4th edn. Bishop Museum Press, Honolulu, p 313

Púcat A (1995) *Conogethes punctiferalis*. Yellow peach moth. Canadian Food Inspection Agency Science Branch

Ram KT, Hugar PS, Somasekhar PBV (1997) Record of *Conogethes punctiferalis* (Guenee) on grapes. Pest Manag Hortic Ecosyst 3(1):37

Shashank PR, Doddabasappa B, Kammar V, Chakravarthy AK, Honda H (2015) Molecular characterization and management of shoot and fruit borer *Conogethes punctiferalis* Guenee (Crambidae: Lepidoptera) populations infesting cardamom, castor and other hosts. In: Chakravarthy AK (ed) New horizons in insect science: towards sustainable pest management. Springer, New Delhi, pp 207–227

Stanley J, Chandrasekaran S, Preetha G (2009) *Conogethes punctiferalis* (Lepidoptera: Pyralidae) its biology and field parasitization. Indian J Agric Sci 79(11):906–909

Suganthy M (2011) Field screening of promising castor cultures against capsule borer, *Conogethes punctiferalis* and leafhopper, *Empoasca flavescens*. Madras Agric J 98(4/6):178–179

Thyagaraj NE (2003) Integrated management of some important cardamom pests of hill region of Karnataka, South India. Ph D thesis, Dr BR Ambedkar University, Uttar Pradesh, India

USDA (1957) Insects not known to occur in the United States. Yellow peach moth (*Dichocrocis punctiferalis* (Guenee), pp 37–38

USDA (2008) Treatment schedules. https://manuals.cphst.org/Tindex/treatmentSearch.cfm

Waterhouse DF (1993) The major arthropod pests and weeds of agriculture in Southeast Asia: distribution, importance and origin. Australian Centre for International Agricultural Research, Canberra

Zhang BC (1994) Index of economically important Lepidoptera. CAB International, Wallingford, p 599

# Rearing of *Conogethes punctiferalis* Guenée (Lepidoptera: Crambidae) and Feasibility of Its Biological Control

18

Chandish R. Ballal, K. P. Kumar, P. Ambanna, A. K. Chakravarthy, Richa Varshney, and H. Khader Khan

**Abstract**

*Conogethes* spp. are economically important pests in the Orient, Australia, and North America. For conducting comprehensive and detailed studies on the fruit and shoot borer, *Conogethes punctiferalis* Guenée, efficient mass-rearing method is essential, and the correct species identification is also an important prerequisite. Provision of the perfect mating chamber, oviposition substrate, suitable larval rearing containers, and larval diet with essential nutrients are necessary for mass-rearing of *C. punctiferalis*. Host plant specific and geographic populations of *C. punctiferalis* have been recorded which may require to be reared on specific diets, and this needs to be evaluated. More than 25 natural enemies have been recorded which include species of *Bracon*, *Apanteles*, *Chelonus*, *Chrysoperla*, entomopathogenic nematodes, and entomopathogenic fungus. Attempts should be made to mass-produce the natural enemies for field utility. Bird predators too play an important role in checking *C. punctiferalis* populations in wild and cultivated ecosystems. Conservation of various promising natural enemies should be encouraged to minimize use of chemical insecticides.

C. R. Ballal (✉) · R. Varshney
ICAR-National Bureau of Agricultural Insect Resources (NBAIR), Bengaluru, Karnataka, India

K. P. Kumar
Department of Entomology, University of Agricultural Sciences, Gandhi Krishi Vignan Kendra (GKVK), Bengaluru, Karnataka, India

P. Ambanna · H. Khader Khan
Department of Agricultural Entomology, University of Agricultural Sciences, Gandhi Krishi Vignan Kendra (GKVK), Bengaluru, Karnataka, India

A. K. Chakravarthy
Division of Entomology and Nematology, ICAR-Indian Institute of Horticultural Research, Bengaluru, Karnataka, India





A. K. Chakravarthy (ed.), *The Black spotted, Yellow Borer, Conogethes punctiferalis Guenée and Allied Species*,
https://doi.org/10.1007/978-981-13-0390-6_18

**Keywords**
Shoot and fruit borer · *Conogethes punctiferalis* · Mass rearing · Meridic diets · Natural enemies

## 18.1 Introduction

The shoot and fruit borer, *Conogethes punctiferalis* Guenée, is a polyphagous moth pest of major concern in Asia, Australia, and Papua New Guinea (CAB International 2011). The economic importance and host range of the pest have been amply emphasized in this book, and it suffices to mention here that it is one of the most destructive pests of several crops. Despite efforts to manage this pest effectively (Waterhouse 1993; Chakravarthy and Thyagaraj 1999; Rajabaskar and Ragupathy 2012; Ganesh et al. 2013; Du et al. 2015), basic studies on this pest are lacking. Comprehensive studies on its physiology and toxicology are also lacking. In order to conduct the above studies, a large number of specimens of determined age, sex, and weight at a given time should be available. This step necessitates mass production of the insect (Kobayashi et al. 2008). Answers to several of the questions on bioecology and physiology of the pest will be obtained by mass rearing the species. Development of protocols for insect rearing is extremely important as this would provide reliable and affordable sources of quality insects, essential for entomological research. Several challenges are faced while developing a rearing protocol for a lepidopteran insect, which include (a) identification of optimum conditions to allow mating in caged conditions, (b) identification of a suitable oviposition substrate for the adult female, and (c) standardization of a diet that is nutritionally complete and induces feeding for the larval stage, which leads to successful pupation and adult emergence with a balanced sex ratio. Du et al. (2012, 2015) developed an artificial diet for *C. punctiferalis*. Documented publications on mass rearing of *Conogethes* species with suggestions for future research and its biological control have been discussed in this chapter. Among all orders of insects, the maximum number (258) of artificial diets has been prepared for Lepidoptera and for 12 species of crambid moths, artificial diets have been attempted (Singh 1977) (Table 18.1). From 1979 till date five teams of entomologists have attempted mass rearing of *C. punctiferalis* on semisynthetic diets (Table 18.2).

### 18.1.1 Feasibility of Rearing *C. punctiferalis*

Artificially reared phytophagous insects are important for studies on insect physiology, insecticidal action, growth, and development, effect of entomopathogens, and host plant relationships. Artificial diets can be used to synchronize insect development with the food availability and can be used to optimize fitness of insects, sometimes more than their natural host (McMorsan 1965). Bottger (1942) was the first entomologist to rear a lepidopteran pest, *Ostrinia nubilalis* (Hubn.) on artificial diet. Later, Vanderzant and Reiser (1956a, b) mass reared the notorious cotton pest, pink bollworm, *Pectinophora gossypiella* Saunders on a wheat germ-based diet. McMorsan (1965) developed a suitable diet to mass rear noctuid moths, particularly

**Table 18.1** Species of crambid moths reared on artificial/semisynthetic diets

| Crambid species | References |
|---|---|
| *Conogethes punctiferalis* (Guenée) | Honda et al. (1979), Utsumi et al. (1990), Chakravarthy et al. (1992), Ambanna (2014), Li et al. (2014), and Kumar et al. (2016) |
| *Grapholita molesta* (Busck) | Wang et al. (2011) |
| *Chilo suppressalis* (Walker) | Kamano (1965, 1971, 1973), Kamano and Yushima (1969), Hormchong et al. (1972), and Ishii (1971) |
| *Chilo agamemnon* Bleszynski | Isa (1972) |
| *Chilo auricilia* (Dudgeon) | Varma and Avasthy (1973) |
| *Chilo orichalcociliella* (=*Chilo orichalcociliellus*) Strand | Delobel (1975) |
| *Chilo partellus* (Swinhoe) | Ballal et al. (1995) |
| *Chilo zonellus* Swinhoe | Chatterji et al. (1968), Pant et al. (1960), Dang et al. (1970), and Siddiqui and Chatterji (1972) |
| *Parapediasia* (=*Crambus*) *teterrella* (Zincken) | Ward and Pass (1969) |
| *Crambus trisectus* (Walker) | Dupnik and Kamm (1970) |
| *Diatraea grandiosella* (Dyar) | Chippendale (1970), Keaster and Harrendorf (1965), and Pan and Long (1961) |
| *Diatraea saccharalis* (Fab.) | Dinther and Gossens (1970), Hensley and Hammond (1968), Miskimen (1965), and Wongsiri and Randolph (1962) |
| *Scirpophaga nivella* (Fabricius) | Wahid and Akhtar (1971) |

**Table 18.2** Mass rearing of *Conogethes punctiferalis*

| Species | References | Comments |
|---|---|---|
| *Conogethes punctiferalis (Guenée)* | Kumar et al. (2016) | Larvae reared to adult on castor and cardamom |
| | Ambanna (2014) | Reared from egg to adult |
| | Li et al. (2014) | Larvae fed on fresh corn and chestnut performed better than those fed on apple, pear, or plum |
| | Utsumi et al. (1990) | Meridic diet-fed colonies had a lower larval survival rate and a larger variation in development duration than colonies fed on natural host plant materials |
| | Honda et al. (1979) | Meridic diet-fed colonies had a lower larval survival rate and a larger variation in development duration than colonies fed on natural host plant materials |

the pest species. Over the past eight decades, entomologists have developed artificial diets for a number of insect species, and the recipes of all these diets have been compiled (Singh 1977). Like any other animal taxa, crambid moths require proteins, lipids, minerals, carbohydrates, water, vitamins, and fibers for their growth and development. For the protein source, entomologists have deployed milk proteins or casein or soy flour depending on the local availability of the protein source. Wheat germ and corn oil and leaves are provided as lipid source. Carbohydrates can be

**Table 18.3** Constituents of meridic diets for mass-rearing *Conogethes punctiferalis*

| Ingredients | Quantity (g or ml) | | | |
|---|---|---|---|---|
| | AD-I | AD-II | AD-III | AD-IV |
| Chestnut meal | 30 | 20 | 10 | 20 |
| Corn meal | 70 | 75 | 80 | 80 |
| Soybean meal | 70 | 70 | 80 | 70 |
| Yeast powder | 30 | 30 | 30 | 30 |
| Agar | 10 | 15 | 10 | 10 |
| Glucose | 9 | 9 | 9 | 9 |
| Citric acid | 2 | 2 | 2 | 2 |
| L-ascorbic acid | 3 | 3 | 3 | 3 |
| Casein | 10 | 10 | 10 | 10 |
| Methyl paraben | 3 | 3 | 3 | 3 |
| Cholesterol | 0.1 | 0.1 | 0.1 | 0.1 |
| Distilled water | 700 | 700 | 700 | 700 |

*Source*: Du et al. (2015)
*AD* artificial diet

delivered to the insects by adding sugars as glucose/sucrose/fructose into the diet. Entomologists have utilized yeast or yeast powder as a source of vitamins (Morten 1979). Commercial salts have been frequently used in diets as source of minerals, viz., Wesson salt or/and choline/ascorbic acid.

Du et al. (2015) evaluated four diets, varying in composition and sources for rearing *C. punctiferalis* (Table 18.3). The diets differed in the amounts of soybean meal, corn meal, and chestnut meal and amounts of agar used. Even slight changes in constituents and composition of a diet can have profound influence on quality and quantity of insects produced. Further, the insects should be produced efficiently at a reasonable cost. Provision of quality diets are expected to result in higher survival rates, greater pupal and larval weights, shorter life cycles, and greater fecundity rates (Du et al. 2015). Honda et al. (1979) were the first to report a meridic diet for *C. punctiferalis.* Utsumi et al. (1990) also proposed a meridic diet for *C. punctiferalis*, but both the diets had lower larval survival rate and developmental rate. However, these attempts provided useful baseline data for developing a well-defined diet for *C. punctiferalis*.

Earlier studies have indicated that each constituent in the diet has a role in the growth and development of *C. punctiferalis*. For instance, larvae of *C. punctiferalis* reared on chestnuts developed faster and had a higher survival rate than larvae reared on peach, cypress, or persimmon (Choi et al. 2006; Honda et al. 1979). This indicated that chestnut is a useful ingredient in the meridic diet for *C. punctiferalis*. Li et al. (2014) from China reported that *C. punctiferalis* larvae reared on fresh corn and chestnut performed better than those reared on temperate fruits like peach, plum, pear, and apple. Du et al. (2015) reported that diets having sorbic acid or formaldehyde as microbial inhibitors were unsuitable for rearing as they repelled the insects, while the use of methyl paraben was observed to favor growth and development of *C. punctiferalis.* Some workers like Du et al. (2015) derived the meridic diet of *C. punctiferalis* from that of a related genera the Oriental fruit moth, *Grapholita molesta* (Busck) (Lepidoptera: Tortricidae) (Wang et al. 2011).

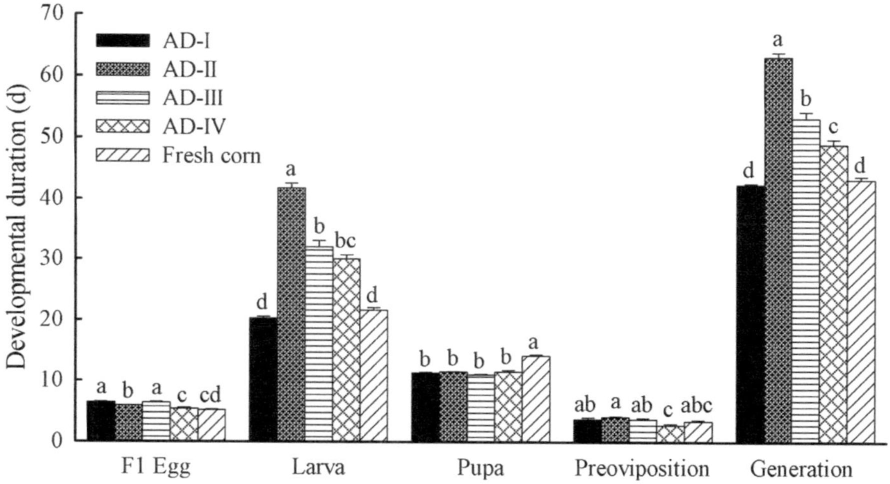


**Fig. 18.1** Developmental duration of *Conogethes punctiferalis* reared on meridic diets and fresh corn. Bars represent means ± SE; significant differences among six host plants are indicated by letters over each bar (Bonferroni test, $P=0.05$). (*Source*: Du et al. 2015)

Du et al. (2015) compared four meridic diets with different ratios of chestnut meal, corn meal, and soybean meal and reported that among the four diets tested, the diet containing 30 g chestnut meal, 70 g corn meal, and 70 g soybean meal proved the best resulting in enhanced survival rate, shortened developmental duration, increased pupal weight, and higher fecundity (Fig. 18.1). Very few attempts have been made to mass rear *Conogethes*, and hence researchers have tried to adopt the mass-rearing techniques used for rearing other species of Crambidae. *Conogethes* and other genera like *Chilo* and *Maruca* are closely related and come under the superfamily Pyraloidea. *Chilo suppressalis* Walker on rice and *Chilo partellus* Swinhoe on maize have received considerable attention with respect to mass rearing. These species were successfully reared in the laboratory, and the larval stage was reared on artificial/semisynthetic diet (Taneja and Nwanze 1990). According to Chambers (1977) mass rearing should result in production of ten thousands to one million times the native productivity of the insect population. The semisynthetic diet for *C. partellus* included glucose, salt mixture, casein, yeast, choline chloride, cholesterol, cellulose, leaf factor, agar, water, and methyl paraben (Pant et al. 1960). Subsequently Chatterji et al. (1968) reared *C. partellus* on wheat germ-based diet. Dang et al. (1970) for the first time introduced kabuli gram-based diet which is used successfully by researchers in India for rearing several species of lepidopterans including pyralids. At NBAIR, Bangalore, a healthy culture of *C. partellus* is maintained throughout the year on a semisynthetic diet (Ballal et al. 1995).

Chi et al. (2004) evaluated 12 artificial diets varying in proportions of constituents for mass rearing *Maruca vitrata* Fab. The 30–50% cowpea diets and 30–70% azuki diets were the most suitable for mass rearing considering their higher reproductive rates. Onyango and Ochieng'-Odero (1993) reared *M. vitrata* in the laboratory for ten generations on soybean flour and cowpea flower powder-based diets.

One liter of diet could yield an average of 400 pupae with adults recording an average fecundity of 200 eggs/female.

At the ICAR-National Bureau of Agricultural Insect Resources (NBAIR), attempts were made for the first time in India to rear *C. punctiferalis* on a semisynthetic diet (Ballal et al. 2015). Larvae of capsule borer were collected from infested castor plants, and the capsules were used for rearing them till pupation. The adults which emerged were placed in wooden cages with bouquets of castor capsules. Provision of red light falling on the adult cages led to successful mating and laying of fertile eggs. Earlier researchers have used apple or peach covered with cheese cloth as oviposition substrates. However, it was observed that fertile eggs were laid on castor capsules, on cloth, and also on the moist cotton plugs, which were used to make bouquets. The eggs laid on the cotton plugs could be easily collected, while those laid on castor capsules could not be separated without damaging the eggs. A semisynthetic diet, with kabuli gram flour as the base ingredient, originally developed for *C. partellus* (Ballal et al. 1995) was evaluated with and without modifications for rearing the first laboratory generation of *C. punctiferalis.* It was observed that the diet (with and without leaf powder of cardamom, castor, or maize) could be successfully utilized for rearing *C. punctiferalis* with 79–87% pupation and 90–95% emergence of healthy adults, while on natural diet (castor capsules) the corresponding values were 95% and 96%, respectively. Female and male pupal weights when reared on the different semisynthetic diets were on par with those on natural diet, the weight ranging between 0.04 and 0.05 gm. Pupation, emergence rates, and total developmental period of *C. punctiferalis* when reared on the test diets were on par with the natural diet batch. However, considering the ease of rearing, utilization of a semisynthetic diet would definitely be the best option. *C. punctiferalis* laid more eggs (54.7%) on cotton plug followed by castor capsules (31.5%), water swab (10.3%), and honey swab (3.5%).

### 18.1.2 Diet Containers

Different kinds of rearing containers for larval rearing of *C. punctiferalis* have been tried. Chakravarthy et al. (1992) utilized 12 different containers for rearing larvae of fruit-feeding (FFT) and stalk-feeding types (SFT) of *Conogethes*. For FFT, polyvinyl cup (9.6 × 6.5 cm) proved to be the most suitable, while for SFT, plastic rectangular boxes (21 × 11 cm) served as the most suitable. It may be noted here that species infesting castor was identified as *C. punctiferalis* but that attacking cardamom as *Conogethes* sp. (Chakravarthy et al. 2015). The types of containers with relative net precision and efficacy and efficiency of larval rearing methods are as presented in Tables 18.4 and 18.5. The study revealed that different species of *Conogethes* require to be reared in different rearing containers due to differences in larval behavior and host plant relationships (Chakravarthy et al. 1992).

Ambanna (2014) and Ballal et al. (1995) used sterilized glass vials (7 × 2.5 cm) and plastic containers of various sizes with different test diets (Figs. 18.2 and 18.3). Honda et al. (1979) developed a simple mass-rearing method for the fruit-feeding type yellow peach moth, *Dichocrocis punctiferalis* (now *Conogethes punctiferalis*) with an oviposition device and an artificial diet. The oviposition device was a

**Table 18.4** Mean density(x), standard deviation (SD), relative variance (RV), and relative net precision (RNP) for comparing mass-rearing methods of *Conogethes* larvae

| | Fruit-feeding type | | | Stalk-feeding type | | |
|---|---|---|---|---|---|---|
| Container | Mean ± SD | RV (%) | RNP | Mean ± SD | RV (%) | RNP |
| Plastic cup | 67.80 ± 26.37 | 12.27 | 3.45 | 62.60 ± 3.87 | 1.95 | 20.41 |
| Plastic tray | 80.82 ± 9.31 | 3.65 | 12.98 | 65.60 ± 2.32 | 1.11 | 11.24 |
| Cylindrical container | 66.53 ± 6.53 | 3.16 | 4.76 | 89.75 ± 4.45 | 1.57 | 7.82 |
| Pot-shaped ice-cream cum | 42.65 ± 0.28 | 0.21 | 178.56 | 59.90 ± 0.93 | 0.48 | 58.82 |
| Polyvinyl cup | 90.50 ± 1.02 | 0.36 | 90.90 | 57.15 ± 0.72 | 0.38 | 58.82 |
| Rectangular box | 70.35 ± 0.72 | 0.31 | 66.66 | 78.45 ± 0.79 | 0.31 | 62.50 |
| Cylindrical jar (large) | 58.32 ± 1.10 | 0.60 | 15.87 | 24.50 ± 0.33 | 0.42 | 29.74 |
| Cylindrical jar (small) | 64.10 ± 28.10 | 13.86 | 0.41 | 47.45 ± 1.69 | 1.12 | 6.02 |
| Petri dish | 17.95 ± 0.68 | 1.23 | 0.41 | 11.70 ± 1.17 | 3.16 | 9.43 |
| Tin box | 45.15 ± 42.50 | 29.76 | 1.92 | -------Not suitable--------- | | |
| Egg tray | 30.70 ± 0.84 | 0.88 | 16.95 | -------Not suitable--------- | | |
| Insect cage | 580.74 ± 2.29 | 0.92 | 44.89 | 86.10 ± 1.27 | 0.92 | 74.39 |

*Source*: Chakravarthy et al. (1992)

ball-type tea strainer (dia. 8 cm) (Fig. 18.4) which contained a small green apple or peach fruit as an odor source and was wrapped with cheesecloth. The artificial diet was composed of meal powder (for mouse), soybean meal powder, ascorbic acid, water, wood powder, and agar. The average pupal yield was 39.40%, which was quite low.

Han et al. (2012) used sterilized glass tubes (diameter by height, 2.4 by 7 cm) for rearing *Chilo suppressalis* Walker larvae. The prepared artificial diet was sliced into small and thin pieces and placed in glass tubes, covered with sterilized cotton wool plug to allow air exchange but to prevent drying of the diet. Pupae were extracted from rearing tubes, sexed, and kept separately in plastic boxes for emergence (Han et al. 2012).

Depending on the behavior and ecology of the *Conogethes* species, researchers have used varying designs, materials, and shape of the rearing containers (Honda et al. 1979). Proper hygiene should be maintained to avoid *Aspergillus* contamination which is a major concern in mass rearing of insects. Honda et al. (1979) found after attempting laboratory trials with several antimicrobial agents that wood powder could reduce infection of microorganisms, except in the initial stage of larval growth. The suppression of microorganisms might have been due to appropriate water content in the artificial diet and also antimicrobial activity of terpene compounds present in wood powder.

## 18.2 Diet Formulation

Besides the diet being nutritional, it should induce feeding in insects. The diet color, shape, and texture can influence feeding in insects. To initiate feeding, most species of phytophagous species require phagostimulants. This is facilitated in the diet by

**Table 18.5** Efficacy and efficiency parameters[a] for mass-rearing methods of *Conogethes* larvae

| Container | Fruit-feeding type | | | | | | Stalk-feeding type | | | | | | Remarks |
|---|---|---|---|---|---|---|---|---|---|---|---|---|---|
| | Pupal weight (mg) | Food required (mg) | Time required to clean (min) | Time required to prepare food (S) | Mean larval survival | Larvae reared (no.) | Pupal weight (mg) | Food required (mg) | Time required to clean (min) | Time required to prepare food (S) | Mean larval survival | Larvae reared (no.) | |
| Plastic cup | 66 | 150 | 1.62 | 0.75 | 72.5 | 8 | 85.5 | 4 | 1 | 1.5 | 62.5 | 2 | Moisture accumulation |
| Plastic tray | 90 | 600 | 1.1 | 1 | 81.5 | 260 | 110 | 1.05 | 5 | 3 | 67 | 28 | Easy to handle |
| Cylindrical container | 54 | 165 | 6 | 0.6 | 62.7 | 74 | 106 | 250 | 3.1 | 5 | 90 | 10 | Everyday changing food is cumbersome. Larvae lose weight due to frequent handling |
| Pot-shaped ice-cream cum | 60 | 85 | 1.7 | 1 | 42.5 | 25 | 110 | 120 | 1.7 | 1.75 | 60 | 8 | |
| Polyvinyl cup | 66 | 50 | 2 | 1 | 90 | 20 | 101 | 90 | 1 | 3.5 | 57 | 5 | Good quality animals obtained |
| Rectangular box | 60 | 300 | 3 | 1.7 | 70 | 135 | 97.5 | 330 | 2 | 3.25 | 78 | 8 | Easy to handle |
| Cylindrical jar (large) | 56 | 450 | 10 | 7.7 | 67.5 | 200 | 108 | 600 | 3.1 | 12 | 48.1 | 15 | Cumbersome and heavy cumbersome |
| Cylindrical jar (small) | 55 | 350 | 7 | 5.5 | 52.5 | 170 | 99 | 500 | 2.7 | 8 | 47.2 | 9 | Cumbersome and heavy cumbersome |
| Petri dish | 57 | 45 | 1.66 | 0.7 | 18 | 15 | 71.5 | 8.5 | 0.7 | 2.6 | 12 | 4 | Due to high humidity and temperature, a number of larvae were lost, delicate |
| Tin box | 55 | 45 | 1 | 0.75 | 45 | 120 | 86 | 235 | 1.2 | 1.5 | 24.5 | 12 | |
| Egg tray | 59 | 150 | 3.5 | 3.2 | 30 | 42 | – | – | – | – | – | – | |
| | | | | | | | (Larvae move out of cavities and many larvae lost) | | | | | | |
| Insect cage | 70 | 3600 | 13 | 0.45 | 71.4 | 1300 | 110 | 5.2 | 6 | 6.5 | 92 | 100 | |

*Source*: Chakravarthy et al. (1992)

[a]$\bar{x}$ of ten replications, ten larvae/replication

**Fig. 18.2** Rearing *C. punctiferalis* larvae in semisynthetic diet in rectangular and round containers

adding plant material preferred by the test insect usually in the form of powder (of leaves, fruits, stems, etc.).

The structure and texture of the diet should be appropriate for insects to feed and grow. The diet should neither be too hard nor soft. It must have optimum quantity of water so that it is acceptable for the insect and mouthparts are not physically damaged. Generally gelatin or agar is added to bring the diet to a semisolid form. In some refined or synthetic diets, a known quantity of cellulose is added to provide texture, bulk, and structure.

## 18.3 Diet Performance

Diets used for mass rearing are to be evaluated by measuring biological parameters of the insects reared on them. The survival rate of *Conogethes* larvae from egg hatch to pupation was 94.50% on diet utilizing chestnut meal and corn meal in 30:70 ratios (Du et al. 2015), while the survival rates of larvae and adults were significantly lower in other test diets.

Kumar and Khader Khan (2017) reported that *C. punctiferalis* could complete its life cycle in 31.48 ± 1.50 days when reared on semisynthetic diet with castor capsule powder which was significantly lower compared to natural diet (38.78 ± 2.34 days) (Fig. 18.5). For a balanced diet and optimum yield of quality insects, the artificial or semisynthetic diet should be inclusive of both essential and

**Fig. 18.3** Rearing *C. punctiferalis* larvae in semisynthetic diet in vials: (**a**) larval rearing on castor leaf-based diet, (**b**) larval rearing on cardamom leaf-based diet, (**c**) larva feeding on artificial diet in vials, and (**d**) pupae kept for emergence. (*Source*: Ambanna 2014)

nonessential ingredients. Further, the texture and structure of the diet should be attractive, so that the insect is stimulated to initiate feeding.

Artificial diets with a stable isotope can have industrial uses. As the biological materials are labeled with a stable isotope, the three-dimensional configurations of structures can be determined by nuclear magnetic resonance (NMR). Further mass spectrometry can be utilized to determine qualitative and quantitative analyses. The artificial diet developed by Kobayashi et al. (2008) comprises a carbohydrate, protein, and stable isotope. The protein originates from a microorganism that is subjected to a defatting treatment and a step with hydrothermal solution. The artificial diet also includes a vitamin and a mineral salt.

Unmole (2009) used mung bean sprouts (*Vigna mungo* (L.) Hepper) as one of the components in the diet for rearing of *M. vitrata*. Larval survival was 87% in two generations, and results were encouraging. One kilogram of mung bean yielded 1200 pupae. This indicated that locally available, preferred host plant may be suitable as an ingredient in the diet. Blanco et al. (2009) worked on a diet with soybean and wheat germ flour for the tobacco budworm, *Heliothis virescens* (F.), and found

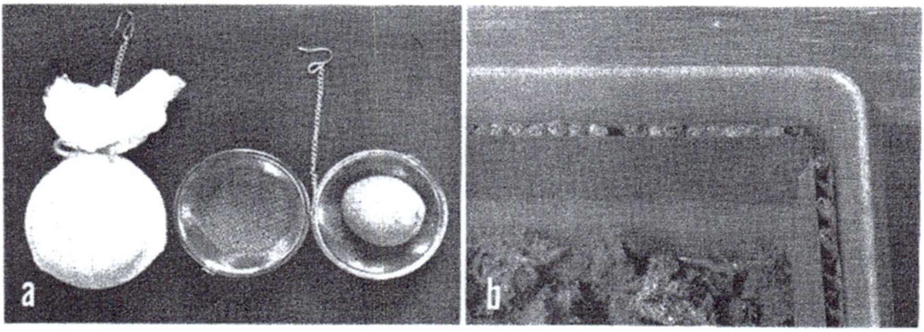


**Fig. 18.4** (**a**) Artificial oviposition device for *Dichocrocis* (now *Conogethes*) *punctiferalis*, (**b**) larval rearing on the artificial diet and pupation in the carton-board cells. (*Source*: Honda et al. 1979)

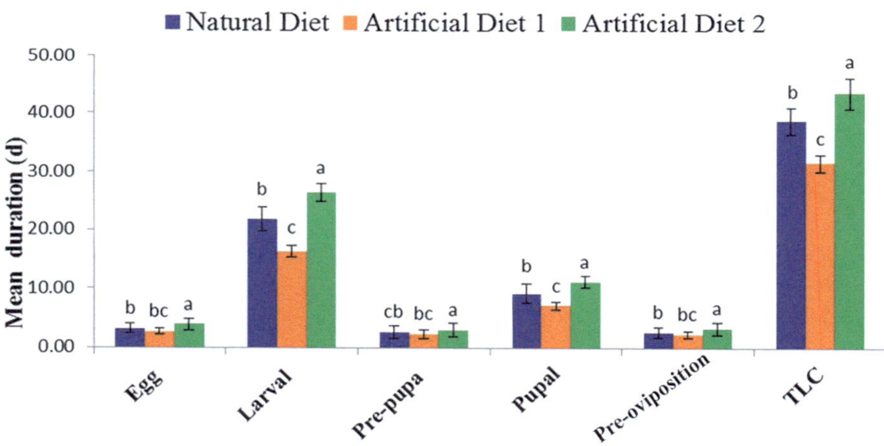


**Fig. 18.5** Duration of developmental stages (egg to adult, $N = 30$) of *Conogethes punctiferalis* reared on semisynthetic diets (1 and 2) and natural host. TLC, total life cycle. Bars represent means ± SE; significant differences among three host plants are indicated by letters over each bar ($P > 0.05$, LSD test). (*Source*: Kumar and Khader Khan 2017)

that changes in the protein content and nutrients can have negative effects on the growth and development of the noctuid moth.

### 18.3.1 Feasibility of Biological Control for Management of *Conogethes punctiferalis*

It is important to record the potential indigenous parasitoids and predators which naturally control the pests in the field so that attempts can be made to either conserve them or to multiply them and augment them in the field situations. Various natural enemies have been recorded on *C. punctiferalis* infesting ginger and

**Table 18.6** Common parasitoids recorded on the shoot and fruit borer, *Conogethes* sp.

| Sl. no. | Parasitoid/predator | Family | Life stages | Reference(s) |
|---|---|---|---|---|
| 1. | *Xanthopimpla australis* Krieger | Ichneumonidae | Pupae | Devasahayam and Abdullakoya (2007) |
| 2. | *Trathala flavo-orbitalis* Cameron | Ichneumonidae | Larvae | Stanley et al. (2009) |
| 3. | *Eriborus trochanteratus* (Morley) | Ichneumonidae | Larvae | Devasahayam (2008) |
| 4. | *Friona* sp. Cameron | Ichneumonidae | Larvae | Ali et al. (2014) |
| 5. | *Agrypon* sp. Foerster | Ichneumonidae | Larvae + pupae | Ali et al. (2014) |
| 6. | *Apanteles taragamae* Viereck | Braconidae | Larvae | Devasahayam (2000) |
| 7. | *Myosoma* sp. Brulle | Braconidae | Larvae | Stanley et al. (2009) |
| 8. | *Chelonus blackburni* Cameron | Braconidae | Larvae | Stanley et al. (2009) |
| 9. | *Brassus* sp. Fabricius | Braconidae | Larvae | Stanley et al. (2009) |
| 10. | *Bracon* sp. Fabricius | Braconidae | Larvae | Stanley et al. (2009) |
| 11. | *Glyptapanteles* sp. Ashmead | Braconidae | Larvae | Stanley et al. (2009) |
| 12. | *Brachymeria atteviae* JNJ | Chalcididae | Larvae | Stanley et al. (2009) |
| 13. | *Anthrocephalus decipiens* (Masi) | Chalcididae | Larvae | Stanley et al. (2009) |
| 14. | *Epitranus erythrogaster* Cameron | Chalcididae | Larvae | Stanley et al. (2009) |
| 15. | *Trichogramma pretiosum* Riley | Trichogrammatidae | Egg | Krishnamurthy) (Pers. observation) |
| 16. | *Brachymeria nosatoi* Habu | Chalcididae | Larvae | Joseph et al. (1973) |
| 17. | *Microbracon hebetor* (Say) | Braconidae | Larvae | Patel and Gangrade (1971) |

turmeric which include *Hexamermis* sp. (mermithid nematode), hymenopterous parasitoids (*Xanthopimpla australis Krieger*, *Apanteles taragamae*, and *Myosoma* sp.), and general predators like spiders, earwigs, and assassin flies. A list of selected major parasitoids is presented in Table 18.6.

## 18.3.2 Parasitoids

The observations at the Indian Institute of Horticulture Research (IIHR), Bangalore, India, during 2014–2016 revealed that *C. punctiferalis* infested guava fruits, pomegranate, and castor capsules. Parasitoids belonging to Bethylidae, Ichneumonidae, and Braconidae were recovered from pupae of *C. punctiferalis* during 2015 and 2016 (Kavitha SJ, Pers Observations). *Eurytoma* sp. (Eurytomidae) was recovered from larvae feeding on sapota in November–December 2016 (Kumar KP, pers. Observations). The parasitoids were observed to parasitize when the infestation level was high and sufficient crop loss had occurred. Stanley et al. (2009) reported *Trathala flavo-orbitalis* Cameron, *Brachymeria atteviae* JNJ, *Chelonus blackburni* Cameron, *Bassus* sp., *Bracon* sp., *Anthrocephalus decipiens* (Masi), *Epitranus erythrogaster* Cameron, and a phorid parasitizing *C. punctiferalis* larvae infesting castor crop. They recorded 47.3% parasitization by *T. flavo-orbitalis*, 14.8% by *B. atteviae*, 11.5% by the phorid, and less than 5% by others. *Bassus* sp. was found to parasitize *C. punctiferalis* rarely. Ali et al. (2014) reported *A. taragamae* and *Glyptapanteles* sp. (Hymenoptera; Braconidae) as indigenous larval parasitoids of *C. punctiferalis* (recently identified as *C. sahyadriensis* (Shashank et al. 2018)) in small cardamom plantations in South India. They recorded 20 species of hymenopterous parasitoids on the cardamom shoot and capsule borer among which *Eriborus trochanteratus* (Morl.), *X. australis*, *Friona* sp., and *Agrypon* sp. (Ichneumonidae) are important.

A number of natural enemies have been documented on the shoot borer on different host plants in India. On *Zingiber* crops like cardamom, ginger, and turmeric, about 15 species of hymenopterous parasitoids and 5 species of predators have been recorded on the borer, *C. sahyadriensis*.

**Castor** Parasitoids of shoot borer infesting castor are *Bracon brevicornis* West., *Apanteles* sp., *Brachymeria euploeae* West., *Angitia trochanterata* Morl., (David et al. 1964). Patel and Gangrade (1971) reported *Microbracon hebetor* Say and Joseph et al. (1973) documented *Brachymeria nosatoi* Habu and *B. lasus* (Walker) as natural enemies of the pest infesting castor.

**Cardamom** Nearly 20 species of parasitoids have been documented on the shoot borer infesting cardamom (CPCRI 1985), and they include *Palexorista parachrysops* Bazzi (Tachinidae), *Agrypon* sp., *E. trochanteratus*, *Friona* sp., *Gotra* sp., *Nythobia* sp., *Temelucha* sp., *X. australis*, *X. kandiensis* Cram., *B. brevicornis* West., *M. hebator*, *Apanteles* sp., *Phanerotoma hendecasisella* Cam. (Braconidae), *Synopiensis* sp., *Brachymeria* sp. nr. *australis* Kr., and *B. obscurata* (Chalcidae) (Varadarasan 1995; Devasahayam and Abdulla Koya 2007). Parasitization by *E. trochanteratus* and *X. australis* to an extent of 10.5–11.1% was reported at Appangala (Karnataka) (CPCRI 1985; Devashayam 2008), while 8–10% and 16–18% larval/pupal parasitization was reported by *Friona* sp. and *Agrypon* sp., respectively, at Idukki, Kerala (Varadarasan 1995; Devashayam 2008).

**Ginger** *Hexamermis* sp. (Mermithidae) and *A. taragamme* (Braconidae) were documented on the shoot borer infesting ginger in Kerala (Devasahayam and Abdulla Koya 2007).

**Turmeric** Jacob (1981) reported predators such as dermapteran (*Euborellia stali* Dohrn (Carcinophoridae), asilid flies (*Philodicus* sp. and *Heligmoneura* sp.) (Asilidae), and spiders (*Araneus* sp., *Micaria* sp., and *Thyene* sp.) feeding on the shoot borer infesting turmeric in Kerala. He also reported a mermithid nematode and parasitoids (*Myosoma* sp., *X. australis* Kr.).

A tachinid parasitoid *Argyrophylax proclinata* Crosskey was reported to cause nearly 40% parasitization to *C. punctiferalis* in Australia (Pena et al. 2002). From Sri Lanka, parasitoids, viz., *Dolichurus* sp., *Xanthopimpla* sp., and *P. hendecasisella*, were recorded (Rodrigo 1941; Devasahayam and Abdulla Koya 2007). From China several parasitoids of longan infesting shoot borer were recorded by Huang et al. (2000) which include *Apanteles* sp., *B. lasus*, *Temelucha* sp. (Devasahayam and Abdulla Koya 2007), *T. flavo-orbitalis*, and *Brachymeria obscurata* Wlk. (CABI 2002). *Apechthis scapulifera*, *Scambus persimilis*, and *B. obscurata* were reported from Japan (CABI 2002).

### 18.3.3 Bird Predators

Field observations in cardamom plantations revealed that woodpeckers predate on cardamom borer, *C. sahyadriensis* (Chakravarthy 1988). The woodpecker with its strong bill chiseled out shoot peelings to locate the larvae of the borer inside the stem (Fig. 18.6a and 18.6b). The bird was observed systematically examining the clumps in a row and gulping the larvae down. Sixty cardamom plantations were surveyed from 1984 to 1986 in Chikmagalur district, Karnataka, South India. The predatory activity of the bird occurred in plantations where pesticides were not used and indigenous old trees of species *Artocarpus*, *Terminalia*, *Albizzia*, *Acacia*, *Bombax*, *Sapindus*, *Alstonia*, *Dipterocarpus*, etc., served as shade trees.

The bird predatory activity was not found in plantations where only one species of shade tree (e.g., *Erythrina lithosperma*) was used. This could be due to absence of suitable sites for nesting and shelter for the golden-backed woodpecker (*Dinopium benghalense* Linn.) in such plantations. Similar observations were recorded in organic cardamom plantations in Kerala, South India (Pers Comm Prakash KV). It is important to encourage the predatory activity of woodpecker in cardamom plantations by minimizing chemical applications and maintaining tree diversity.

### 18.3.4 Microbials

As *Conogethes* spp. feed on a wide spectrum of plants and almost all parts of the plants, microbes could be utilized for their management. Microbes could be very effective as biopesticides against *Conogethus spp*. Devasahayam (2000) recorded *Bt* as an effective bioagent to target *C. punctiferalis*. The entomopathogenic nematode *Steinernema glaseri* Steiner (Steinernematidae) has been recorded on larvae of *C. punctiferalis* (CABI 2002). For management of *C. punctiferalis* (recently identified

***Trichogramma pretiosum* Riley**

***Apanteles taragamae* Viereck**

***Epitranus* Walker**

***Eriborus trochanteratus* (Morley)**

***Trichospilus pupivorus* Ferrière**

***Chelonus blackburni* Cameron**

**Fig. 18.6a** Parasitoids of *Conogethes punctiferalis* Guenée. (*Source*: Poorani et al. 2017) © (http://www.nbair.res.in/Featured_insects)

as *C. sahyadriensis* on small cardamom), steinernematid entomopathogenic nematodes at 100 IJ/larvae or *Bacillus thuringiensis* when early-instar larvae are found in capsule or panicle or unopened leaf buds, i.e., within 20 days of adult moth emergence can be recommended (Nybe et al. 2009). In Japan, an entomopathogenic fungus was observed to be pathogenic to *Conogethes* eggs. Application of microbial suspensions on emerging fruiting bodies is known to avert the need for application of chemical insecticides (Shashank et al. 2015). Pervez et al. (2015) reported that temperature had an effect on infectivity of entomopathogenic nematodes against

**Fig. 18.6b** Predator of *Conogethes punctiferalis* Guenée. (*Source*: Chakravarthy 1988)

larvae of *C. sahyadriensis* infesting ginger. The above findings indicate the potential of EPN as a biological control agent against this borer of ginger in India. The study revealed that temperature affected host recognition, penetration, infectivity, and multiplication of EPNs. At lower temperatures nematode activity was observed to be significantly reduced. Out of all EPN tested, *Heterorhabditis* sp. (IISR-EPN-01), *Steinernema* sp. (IISR-EPN-03), and *Oscheius gingeri* were effective at 30 °C in the laboratory against *C. sahyadriensis* larvae. Strains of fungi and entomopathogenic nematodes developed by Indian Institute of Spice Research (IISR), Calicut, Kerala were identified as promising agents against *C. sahyadriensis* infesting spices. Microbes such as *Bacillus* spp., *Pseudomonas* spp., and *Alternaria* spp. and toxins from *Metarhizium anisopliae* can be used against *Conogethes* spp. (Ganga Visalakshy et al. 2015). The native isolates of EPF-*Metarhizium*, *Beauveria bassiana*, and EPN-*Heterorhabditis indica* could successfully infect *C. punctiferalis* larvae under laboratory conditions. However, for field conditions crop canopy should be thoroughly covered with the microbial preparations. Larvae of *Conogethes* sp. should be able to consume the microbial insecticide-treated plant parts (Kavitha et al. 2015). Murphy et al. (1995) recorded Nuclear Polyhedrosis virus (NPV) for *Dichocrocis* (now *Conogethes*) *punctiferalis*. Microbial preparations are known to combine well with pheromones, biopesticides, and mechanical and cultural methods of pest suppression, and hence the promising microbial biopesticides should be field evaluated.

**Conservation** Since cardamom is cultivated in evergreen tropical forest tracts in South India, during off-season (January–June), the parasitoids need to survive on natural, alternate hosts. In and around wild and cultivated patches in cardamom plantations, species of wild plants in the form of *Curcuma, Zingiber, Hedychium, Alpinia*, etc., serve as hosts for *C. sahyadriensis*. These plants harbor *Conogethes* larvae that serve as hosts for a number of parasitoids and predators. Castor fields are generally cultivated in dry or arid areas where vegetation or plants serving as refugia will not be a part of the cultivated ecosystem. In such areas, plants/vegetation at the borders of the plot may serve as refugia, and efficacy of egg parasitoids such as *Trichogramma chilonis* Ishii or *Trichogramma pretiosum* Riley may be explored by releasing them in field. In orchards too, in addition to wild plants, a diversity of fruit, shade, and border plants may assist in sustaining populations of beneficials. Inoculative releases of parasitoids and predators like *B. hebetor*, *A. taragamae*, *C. blackburni*, *Apanteles congoensis De Saeger*, and *Chrysoperla* sp. could facilitate pest population reduction and obviate the need to spray chemical insecticides.

In India, we have succeeded in rearing *C. punctiferalis* on a semisynthetic diet; however, future thrust should be on evaluating the effect of rearing this insect continuously on semisynthetic diet on its biological parameters. Another aspect to be investigated would be to verify if the populations of *C. punctiferalis* from different host plants are equally amenable to rearing on the semisynthetic diets. Efforts should be made to field evaluate the effectiveness of biocontrol agents, viz., macrobials such as *Trichogramma* spp. *C. blackburni*, braconids, and microbials such as entomopathogenic fungus, entomopathogenic nematode, bacteria, and viruses, for managing *C. punctiferalis* and other related *Conogethes* spp.

**Acknowledgment** Authors sincerely thank the authorities of the University of Agricultural Sciences, GKVK, Bengaluru, ICAR-IIHR, Bengaluru, ICAR-NBAIR, Bengaluru, and ICAR, New Delhi, India, for encouragement and facilities.

## References

(CABI) CAB International (2002) Crop protection compendium. CAB International, Wallingford

(CPCRI) Central Plantation Crops Research Institute (1985) Annual report for 1983. Central Plantation Crops Research Institute, Kasaragod

Ali ansar MA, Manoharan T, Kuttalam S (2014) First report of larval parasitoids on small cardamom shoot, capsule and panicle borer, *Conogethes punctiferalis* Guenee' (Lepidoptera; Crambidae) under lab and field conditions. Trends Biosci 7(22):3771–3377

Ambanna P (2014) Development of a rearing protocol for *Conogethes punctiferalis* Guenee (Lepidoptera: Crambidae). M Sc (Agri.) thesis, University of Agricultural Sciences, Bangalore-6, India

Ballal CR, Kumar P, Ramani S (1995) Laboratory evaluation, storability and economics of an artificial diet for rearing *Chilo partellus* (Swinhoe) (Lepidoptera: Pyralidae). J Ent Res 19:135–141

Ballal CR, Paneknavar A, Chakravarthy AK, Thyagaraj NE (2015) Feasibility of rearing *Conogethes punctiferalis*. In Abstracts: Chakravarthy AK, Krishnamoorthy A (eds) Round table on *Conogethes punctiferalis* and allied species (Series 1). IIHR, Bengaluru, May 22, 2015, pp 14–15

Blanco CA, Portilla M, Abel CA, Winters H, Ford R, Streett D (2009) Soybean flour and wheat germ proportions in artificial diet and their effect on the growth rates of the tobacco budworm. J Insect Sci 9:59

Botteger GT (1942) Development of synthetic food media for use in nutrition studies of the European corn borer. J Agric Res 65:493–500

CAB International (2011) *Conogethes punctiferalis* datasheet. In: Crop protection compendium. CAB International, Wallingford

Chakravarthy AK (1988) Predation of Goldenbacked woodpecker, *Dinopium benghalense* (Linn.) on cardamom shoot and fruit borer, *Dichocrocis punctiferalis* (Guene). J Bombay Nat Hist Soc 85(2):427

Chakravarthy AK, Thyagaraj NE (1999) Effect of different NPK fertilizer levels on the incidence of cardamom pests. Insect Environ 4(4):139–140

Chakravarthy AK, Honda H, Thyagaraj NE (1992) Comparison of containers for larval rearing in stalk and fruit feeding type of *Conogethes punctiferalis* (Guen.) (Lepidoptera: Pyralidae). Placrosym 9:127–131

Chakravarthy AK, Kammar V, Lokeshwari D, Rani AT, Nagaraj T, Varun Rajan V (2015) The black spotted yellow shoot and fruit borer, *Conogethes* spp. (Crambidae: Lepidoptera): a global perspective. Curr Sci 109(6):1016–1018

Chambers DL (1977) Quality control in mass rearing. Annu Rev Entomol 22:289–308

Chatterji SM, Siddiqui KH, Panwar VPS, Sharma GC, Young WR (1968) Rearing of the maize stem borer, *Chilo zonellus* Swinhoe on artificial diet. Indian J Entomol 30:8–12

Chi YC, Sakamaki Y, Tsuda K, Kusigemati K (2004) Simple artificial diets for mass reproduction of the legume pod borer, *Maruca vitrata* (Lepidoptera: Crambidae). Jpn J Appl Entomol Zool 48(4):315–320

Chippendale GM (1970) Morphometric changes in fat body proteins of the southwestern corn borer. *Diatraea grandiosella*. J Insect Physiol 16:1057–1068

Choi KS, Han KS, Park IK, Hong JI, Kim CS, Chung YJ, Chin SC (2006) Difference of developmental time, survival rate and sex ratio of *Dichocrocis punctiferalis* (Lepidoptera: Pyralidae) on three hosts. J Korean For Soc 95:174–176

Dang K, Mohini A, Jotwani MG (1970) A simple improved diet for mass rearing of sorghum stem borer, *Chilo zonellus* (Swinhoe). Indian J Entomol 32:130–133

David BV, Narayanaswami PS, Murugesan M (1964) Bionomics and control of the castor shoot and capsule borer *Dichocrocis punctiferalis* Guen. in Madras state. Indian Oil Seeds J 8:146–158

Delobel A (1975) *Chilo orichalcociliellus* Strand (Lepidoptera, Pyralidae), foreur des tiges du sorgho et du mais a Madagascar. I. Caracteristiques morphologiques. Cah ORSTOM Ser Biol 10:3–9

Devasahayam S (2000) Evaluation of biopesticides for the management of shoot borer (*Conogethes punctiferalis* Guen.) on ginger (*Zingiber officinale* Rosc.). Spices and aromatic plants: challenges and opportunities in the new century contributory papers centennial conference on spices and aromatic plants, Calicut, Kerala, India, 20–23 September, 2000, pp 276–277

Devasahayam S, Abdulla Koya KM (2007) Insect pests of turmeric. In: Ravindran PN, Nirmal Babu K, Siva Raman K (eds) Turmeric the genus curcuma. CRC press, Boca Raton, pp 169–192

Devashayam S (2008) Biological control of insect pests. In: Parthasarathy VA, Kandiannan K, Srinivasan V (eds) Organic spices. New India Publishing Agency, New Delhi, pp 133–152

Dinther JBM Van, Gossens PA (1970) Rearing of *Diatraea saccharalis* on diets in Surinam. Entomol Exr Appl 13:320–326

Du YL, Guo HM, Sun SL, Zhang MZ, Zhang AH, Wang JB, Qin L (2012) Effects of temperature on the development and reproduction of yellow peach moth, *Conogethes punctiferalis* (Guene'e). Acta Entomol Sin 55:561–569

Du YL, Ai PP, Li-Xin L, Min-Zhao Z, Shu-Ling S (2015) Development of an artificial diet for rearing *Conogethes punctiferalis* (Lepidoptera: Crambidae). J Entomol Sci 50(2):89–98

Dupink TD, Kamm JA (1970) Development of an artificial diet for *Crambus trisectus*. J Econ Entomol 63:1578–1581

Ganesha, Chakravarthy AK, Naik MI, Basavaraj K, Manja Naik C (2013) Biology of castor shoot and capsule borer, *Conogethes punctiferalis* Guenee (Lepidoptera: Pyralidae). Curr Biot 7(3):188–195

Ganga Visalakshy PN, Darshana CN, Krishnamoorthy A, Swathi C, Pillai KG (2015) Microbes for management of *Conogethes punctiferalis*-future scope. In: Chakravarthy AK, Krishnamoorthy A (eds) *Conogethes punctiferalis* and allied soecies. IIHR, Bangalore, p 44

Han L, Li S, Liu P, Peng Y, Hou M (2012) New artificial diet for continuous rearing of *Chilo suppressalis* (Lepidoptera: Crambidae). Ann Entomol Soc Am 105:253–258

Hensley SD, Hammond AM (1968) Laboratory techniques for rearing the sugarcane borer on an artificial diet. J Econ Entomol 61:1742–1743

Honda H, Kaneka JI, Konno Y, Matsumoto Y (1979) A simple method for most rearing of the yellow peach moth *Dichocrocis punctiferalis* on an artificial diet. J Appl Entomol Zool 14(4):464–468

Hormchong T, Srithunya S, Podhisritong P (1972) Mass rearing of rice stem borer. Jpn Pestic Inf 10:126–127

Huang YQ, Zhang XJ, Wei H, Hu QY, Zhan ZX (2000) Studies on the *Dichocrocis punctiferalis* Guenee and its enemies. Acta Agric Univ Jiangxiensis 22:523–525

Isa AL (1972) A standard technique for rearing *Chilo agamemnon* Blesz. on an artificial diet. Agric Res Rev 50:61–64

Ishii S (1971) Nutritional studies of the rice stem borer, *Chilo suppressalis* Walker, and its mass rearing. Entomophaga 16:165–173

Jacob SA (1981) Biology of *Dichocrocis punctiferalis* Guen. on turmeric. J Plant Crop 9:119–123

Joseph KJ, Narendran TC, Joy PJ (1973) Taxonomic studies of the oriental species of *Brachymeria* (Hymenoptera: Chalcididae). Report, PL 480 research project, University of Calicut, Calicut

Kamano S (1965) Rearing of Papilio xuthus larvae on artificial diets. Jpn J Apll Entomol Zool 9:133–135

Kamano S (1971) Studies on artificial diets of the rice stem borer, *Chilo suppressalis* Walker. Bull Nut Inst Agric Sci Tokyo Ser C25:1–45

Kamano S (1973) Studies on artificial diets and laboratory methods suitable for successive generations of the rice stem borer, *Chilo suppressalis* Walker. Bull Nut Inst Agric Sci Tokyo Ser C27:1–51

Kamano S, Yushima T (1969) Rearing method of insects under aseptic and semi-aseptic conditions. Plant Prot 23:353–356

Kavitha SJ, Krishnamoorthy A, Gangavisalakshy PN (2015) Bio-control aspects of shoot and fruit borer, *Conogethus pun*ctiferalis. In: Chakravarthy AK, Krishnamoorthy A (eds) *Conogethes punctiferalis* and allied soecies. IIHR, Bangalore

Keaster AJ, Harrendorf K (1965) Laboratory rearing of the southwestern corn borer, *Zeadiatraea grandiosella* on a wheat germ medium. J Econ Entomol 58:923–924

Kobayashi C, Fukasawa Y, Hirose D, Kato M (2008) Contribution of symbiotic mycangial fungi to larval nutrition of a leaf-rolling weevil. Evol Ecol 22:711–722

Kumar KP, Khader Khan H (2017) Artificial diet for rearing of *Conogethes punctiferalis* Guenee. (Lepidoptera: Crambidae). Mysore J Agric Sci 51(2):244–249

Kumar KP, Nitin KS, Vasudev Kammar, Chakravarthy AK, Khader Khan H (2016) Artificial diet for rearing shoot and fruit borer, *Conogethes punctiferalis* Guenee (Lepidoptera: Crambidae). Souvenir, 3rd national meet of entomologists, 7–8th October 2016, ICAR- Indian Institute of Horticultural Research, p 44

Li DY, Ai PP, Du YL, Sun SL, Zhang MM (2014) Effects of different host plants on the development and reproduction of yellow peach moth, *Conogethes punctiferalis* (Lepidoptera: Crambidae). Austral Entomol 54:149. https://doi.org/10.1111/aen.12105

McMorran A (1965) A synthetic diet for the spruce budworm, *Choristoneura fumiferana* (Clem.) (Lepidoptera: Tortricidae). Can Entomol 97:58–62

Miskimen GW (1965) Non-aseptic laboratory rearing of the sugarcane borer, *Diatraea saccharalis*. Ann Entomol Soc Am 58:820–823

Morton AC (1979) Rearing butterflies on artificial diets. J Res Lepidop 18(4):221–227

Murphy FA, Fauquet CM, Mayo MA, Jarvis AW, Ghabrial SA, Summers MD, Martelli GP, DHL B (eds) (1995) Sixth report of the international committee on taxonomy of viruses, archives of virology. Springer, New York

Nybe EV, Rajan S, Mini Raj N (2009) Plantation crops and spices. In: Alexander D, Rajan S, Rajamony L, Ushakumari K, Kurien S (eds) The adhoc package of practices recommendations for organic farming. Director of Extension Kerala Agricultural University, Mannuthy, Thrissur, pp 81–150

Onyango FO, Ochieng'-Odero JPR (1993) Laboratory rearing of the legume pod-borer *Maruca testulalis* Geyer (Lepidoptera: Pyralidae) on a semi-synthetic diet. Insect Sci Appl 14:719–722

Pan YS, Long WH (1961) Diets for rearing the sugarcane borer. J Econ Entomol 54:257–261

Pant NC, Gupta P, Nayar JK (1960) Physiological studies of *Chilo zonellus* Swinh. A pest on maize crop. I. growth on artificial diets. Proc Nat Inst Sci India 26B:379–383

Patel RK, Gangrade GA (1971) Note on the biology of castor capsule borer, *Dichocrocis punctiferalis*. Indian J Agric Sci 41:443–444

Pena JE, Nadel H, Barbosa-Pereira M, Smith D (2002) Pollinators and pests of *Annona* species. In: Pena J, Sharp J, Wysoki M (eds) Tropical fruit pests and pollinators: biology, economic importance, natural enemies and control. CABI Publication, Wallingford, p 208

Pervez R, Santosh JE, Devasahayam S, Jacob TV (2015) Effect of temperature on the infectivity of entomopathogenic nematodes against shoot borer (*Conogethes punctiferalis* Guenee) infesting ginger. J Biol Control 29(4):187–193

Poorni J, Ramesh kumar A, Gupta A (2017) Factsheets on agriculturally important insects. Electronic Publication. http://www.nbair.res.in/Featured_insects

Rajabaskar D, Regupathy A (2012) Calling behavior and attractive response of cardamom shoot and capsule borer *Conogethes punctiferalis* Guenee (Lepidoptera: Pyralidae) to the female crude extract and synthetic blend. J Entomol 9:296–301

Rodrigo E (1941) Administration report 1940. Director of Agriculture, Colombo, pp 1–18

Shashank PR, Doddabasappa B, Kammar V, Chakravarthy AK, Honda H (2015) Molecular characterization and management of shoot and fruit borer *Conogethes punctiferalis* Guenee (Crambidae: Lepidoptera) populations infesting cardamom, castor and other hosts. In: Chakravarthy AK (ed) New horizons in insect science: towards sustainable pest management. Springer, New Delhi, pp 207–227

Shashank PR, Vasudev K, Mally R, Chakravarthy AK (2018) A new Indian species of shoot and capsule borer of the genus *Conogethes* (Lepidoptera: Crambidae) feeding on cardamom. Zootaxa 4372(2):215–234

Siddiqui KH, Chatterji SM (1972) Laboratory rearing of the maize stem borer, *Chilo zonellus* Swinhoe (Crambidae: Lepidoptera) on a semi-synthetic diet using indigenous ingredients. Indian J Enromol 34:183–185

Singh P (1977) In: IFI/Plenum Data Company (ed) Artificial diets for insects, mites, and spiders. Springer, New York

Stanley J, Chandrasekaran S, Preetha G (2009) *Conogethes punctiferalis* (Lepidoptera: Pyralidae) its biology and field parasitization. Indian J Agric Sci 79(11):906–909

Taneja SL, Nwanze KF (1990) Mass rearing of Chilo spp. on artificial diets and its use in resistance screening. Insect Sci Appl 3:605–616

Unmole L (2009) Laboratory rearing of the legume pod borer, *Maruca vitrata* Fabricius (Lepidoptera: Pyralidae) on mung bean sprouts. Univ Mauritius Res J 15:577–584

Utsumi K, Aihara E, Watanabe N (1990) Artificial diet of yellow peach moth *Conogethes punctiferalis* eating corn seeds. Res Bull Plant Protect Serv Jpn 26:65–67

Vanderzant ES, Reiser R (1956a) Aseptic rearing of the pink bollworm on synthetic media. J Econ Entomol 49:7–10

Vanderzant ES, Reiser R (1956b) Studies on the nutrition of the pink bollworm using purified casein media. J Econ Entomol 49:454–458

Varadarasan S (1995) Biological control of insect pests of cardamom. In: Ananthakrishnan TN (ed) Biological control of social forest and plantation crop insects. Oxford and IBH Publishing Company Private Limited, New Delhi, pp 109–111

Varma A, Avasthy PN (1973) An artificial diet for the rearing of stalk borer, *Chilo auricilius* Dudg. Experirnria 29:1161–1162

Wahid A, Akhtar M (1971) Nutritional requirements of sugarcane top borer, *Scirpophaga novella* (Fabricius). Pak J Zool 3:15–21

Wang YR, Huang M, Liu XW, Wu JX, Zhao D (2011) Artificial diet for larvae of the oriental fruit moth *Grapholita molesta* (Busck) (Lepidoptera: Tortricidae). J Northwest Sci Tech Univ Agric For Univ (Nat Sci Ed) 39:161–166

Ward AG, Pass BC (1969) Rearing sod webworms on artificial diets. J Econ Entomol 62:510–511

Waterhouse DF (1993) The major arthropod pests and weeds of agriculture in Southeast Asia: distribution, importance and origin, ACIAR monograph no. 21. Australian Centre for International Agricultural Research (ACIAR), Canberra

Wongsiri T, Randolph NM (1962) A comparison of the biology of the sugarcane borer on artificial and natural diets. J Econ Entomol 55:472–473

# Reproduction in the Shoot and Fruit Borer, *Conogethes* spp. (Crambidae: Lepidoptera): Strategizing Survival?

19

M. A. Rashmi, A. K. Chakravarthy, and S. R. Kulkarni

**Abstract**

The neurohormonal regulatory process in *Conogethes* reproductive system is poorly understood. This has impeded research on bioecology and management of pest species of *Conogethes*. The reproduction in *Conogethes* is physiologically complex embracing from other factors the influence of light, sound, plant odors, and hormones. The process is quite elaborate and requires in-depth, comprehensive studies.

**Keywords**

*Conogethes* spp. · Reproduction · Calling behavior · Mating · Oviposition

## 19.1 Introduction

Physiology of reproduction is a composite science as it requires information on endocrinology, histology, physiology, morphology, anatomy, genetics, and molecular biology. The ultimate goal of reproduction is perpetuation or procreation. The process of reproduction varies in complexity and mechanisms (Nalnabdov 1970). The male and female sexes involved in reproduction are specialized with well-regulated steps and events. The reproductive patterns in general may be similar but may vary slightly from one species of *Conogethes* to another. Reproductive events

M. A. Rashmi · A. K. Chakravarthy (✉)
Division of Entomology and Nematology, ICAR-Indian Institute of Horticultural Research, Bengaluru, Karnataka, India

S. R. Kulkarni
Department of Entomology, Mahatma Phule Krishi Vidyapeeth (MPKV), Rahuri, Ahmednagar, Maharashtra, India





A. K. Chakravarthy (ed.), *The Black spotted, Yellow Borer, Conogethes punctiferalis Guenée and Allied Species*,
https://doi.org/10.1007/978-981-13-0390-6_19

are processed by a complex of interconnecting hormones, which in turn are connected to a neural system which is not well understood in insects in general, including species of *Conogethes*. The neuroendocrine systems and the morphology, anatomy, and physiology of the sex regulatory mechanisms are controlled and regulated by neurohormonal system. The efficiency of reproduction may decline due to dysfunction in the hormonal, neural, and humoral links, disrupting the mechanisms of synchronization that underlies the reproduction system. When reproduction gets impaired, sterility results. This is important for management of *Conogethes* sp. Reproduction has either to be reduced or impaired. So the basic function of reproduction in *Conogethes* is to be well understood. Hence, understanding reproduction of *Conogethes* is basic and crucial for pest management.

*Conogethes* species represents a cryptic species complex. Up to 2006, descriptions and new descriptions of species of *Conogethes* have been documented (Inone and Yamanaka 2006). The populations of *Conogethes* in the Orient, Eastern Palaearctic, and Australian regions are undergoing speciation. The speciation in *Conogethes* appears to be host related and in turn location dependent. Speciation is an adaptive process involving intrinsic barriers to gene flow among closely related populations by development of reproductive isolating mechanisms (Bush 1975). The genetics and evolutionary process leading to reproductive isolating mechanisms are yet to be established in *Conogethes* populations. This and sex pheromone aspects of reproduction have been covered in separate chapters in this volume.

## 19.2 Reproductive Cycles

### 19.2.1 Reproductive System in *Conogethes*

The reproductive system in *Conogethes* is composed of paired ovaries, a pair of lateral oviducts, one common oviduct, a genital chamber, the spermatheca along with its gland, and accessory glands. The ovaries are secured within the abdomen by fine tracheae and fat bodies, and normally loop back and forth two or three times within the abdomen. Each ovary is composed of four ovarioles, with an average length of 10.05 mm, which connects to the calyx of each lateral oviduct. The ovarioles fuse at their apex, contact closely throughout their length, and each of them is composed of a thin membranous tunica propria containing oocytes from 7 to 10, with newly emerged female containing the greater number. The common oviduct branches anteriorly into two lateral oviducts, with an average length of 0.90 mm, and slightly constricts before joining the genital chamber. The lateral oviduct expands into the base of ovarioles. The seminal duct and the spermathecal duct open into the anterior end of the genital chamber, while the common duct of the accessory glands opens into the dorsum of the posterior region of the genital chamber slightly anterior of the opening to the rectum. The genital chamber terminates into a fleshy, telescopic ovipositor equipped with numerous sensory hairs and an ovipore. The common duct of the accessory gland attaches to the genital chamber and distally expands into an indistinctive common accessory gland reservoir (Wang et al. 2015) (Fig. 19.1).

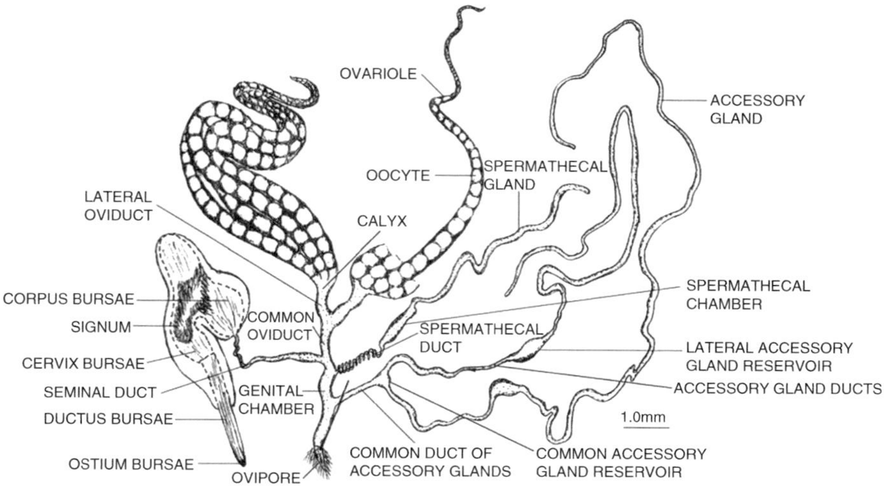


**Fig. 19.1** Female reproductive system in *Conogethes* (Wang et al. 2015)

### 19.2.2 Male Reproductive System in *Conogethes*

The internal male reproductive organs of *Conogethes* include two testes, a pair of vasa deferentia, the ductus ejaculatorius duplex, a ductus ejaculatorius simplex, and a pair of accessory glands. The two testes are encased by a scrotum which appears as a thin, transparent membrane and form a complex testis. The paired vasa deferentia arise from the ventral surface of the testis and join to the ductus ejaculatorius duplex, respectively. The ductus ejaculatorius duplex consists of two tubular organs, fusing at their posterior terminals to form the ductus ejaculatorius simplex and giving rise to the accessory glands at their anterior terminals. Each branch of the ductus ejaculatorius duplex receives one vas deferens, and the junctions are near the accessory glands and locate at the one third of ductus ejaculatorius duplex. The accessory glands are uniform, remain in close contact throughout their length, and are intertwined among other organs within the abdominal cavity (Wang et al. 2015) (Fig. 19.2).

## 19.3 Male and Female Genitalia

### 19.3.1 Male Genitalia

In *Conogethes* bred on *Ricinus communis*, the anterior margin of tegumen was completely sclerotized, and lateral arms are narrow. The uncus was narrow, slender, ventrally curved, apical one third swollen, and evenly covered with setae dorsally, and its apical tip with two small thin lobeless processes bearing a few bristles ventrally (looks like cattle egret *Bubulcus ibis* head). The vinculum is U shaped with the apex produced into conical point. The valve was broad at the apex and narrow at the

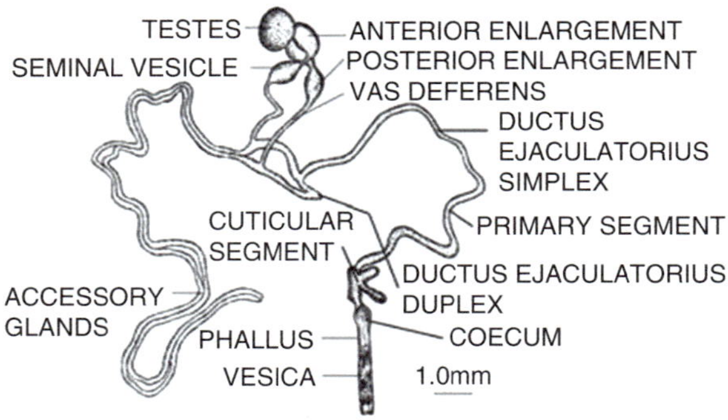


**Fig. 19.2** Male reproductive system in *Conogethes* (Wang et al. 2015)

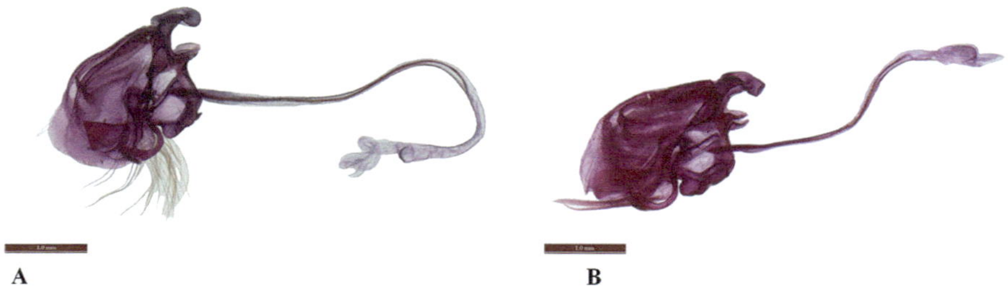


**Fig. 19.3** Male genitalia of *Conogethes* species breeding on (**a**) castor and (**b**) cardamom. (*Source*: Shashank 2012)

base, somewhat broadly oval, and saccular margin weakly protruded. The cucullus had tuft of long hairs. The clasper was short, sclerotized, decurved with blunt tip, and placed near mesoventral side of valve. The sacculus was rather narrow, sclerotized, and tapered with fringed hairs. The transtilla are somewhat broad, platelike elements weakly sclerotized and each termen of element slightly indented, and two elements closely approached but fused medially. The juxta was elongated and expanded basally. The saccus was short and tapered and anterior margin evenly rounded. The aedeagus was very long, slender, and strongly curved near base. The cornuti composed of ringlike structure at the proximal end (Shashank 2012) (Fig. 19.3).

## 19.3.2 Difference Between the Fruit- and Pinaceae-Feeding Type of *Conogethes* sp.

There is a definite difference in three characters of the male genitalia between FFT and PFT although the specimens from various host plants and localities showed a little individual variation. The first discriminating character between the males of the two types was the angle of mesal projection of valva (APV) against costa (Figs. 19.4 and 19.5, Table 19.1). APV was distinctly larger in PFT than in FFT, but no difference in this trait was observed among the populations from various host plants in each type.

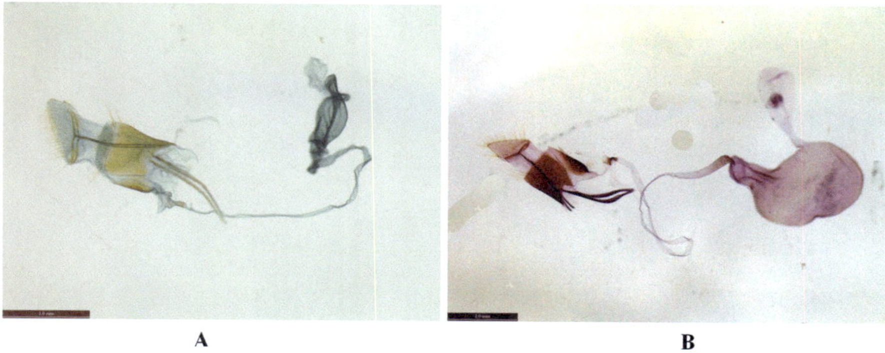

**Fig. 19.4** Female genitalia of *Conogethes* species breeding on (**a**) castor and (**b**) cardamom. (*Source*: Shashank 2012)

### 19.3.3 Female Genitalia

The ovipositor was triangular and covered with mixture of long and short setae. The apophysis anterioris was about as long as length of posterioris and anterioris more or less thicker than in posterioris. The ostium was narrow, membranous, and funnel shaped. The antrum was a sclerotized collar. The ductus bursae were narrow and very long. The corpus bursa was ovate but irregular in shape and size. The signum was absent in female CBR Shashank (2012). A comparision was made between posterior apophysis and papilla analis on female ovipositor of the yellow peach moth from Japan and the cardamom shoot borer from India. The length and width of carpus bursa of *Conogethes* breeding on castor were 1.52 ± 0.040 mm and 0.88 ± 0.018 mm long, respectively. The length of ductus bursae of CBR measured 4.14 ± 0.110 mm. The length of posterior apophyses of CBR measured 0.99 ± 0.015 mm. The length of anterior apophyses measured 0.86 ± 0.010 mm of CBR. The length of ventral side (F), dorsal side (G), and anterior side (H) of ovipositor castor moths (males) were 0.53 ± 0.014 mm, 0.30 ± 0.011 mm, and 0.36 ± 0.010 mm, respectively. The diameter of accessory gland in the case of *Conogethes* breeding on castor was 0.50 ± 0.004 mm (n = 10).

## 19.4 Calling Behavior

Insects have evolved methods of communication, using conspicuous sounds. However, the risk of eavesdropping by competitors, predators, and parasitoids is high (Greenfield 2002; Zuk and Kolluru 1998). To counteract this risk, various signalers (e.g., crickets, katydids, and moths) have modified/changed the temporal and spectral characteristics of their signals and the timing of signaling (Jia et al. 2001; Zuk and Kolluru 1998).

Mating in moths generally depends on the expression of a series of behavioral patterns (Hou and Sheng 2000). In females, these behaviors include production of

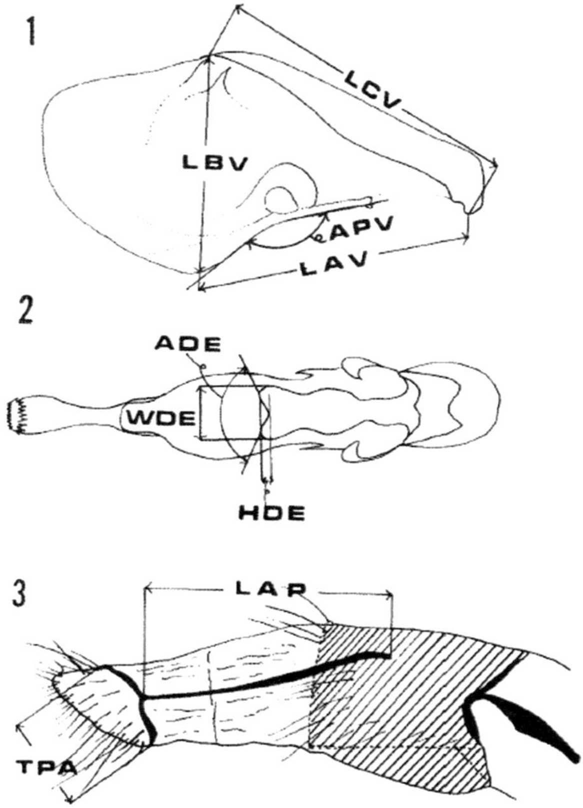


**Fig. 19.5** Morphometric characters of the two types of yellow peach moth. (1) Valva of male genitalia, (2) tegumen of male genitalia, (3) ovipositor of female. (*Source*: Honda and Mitsuhashi 1989)

**Table 19.1** Comparison of posterior apophysis and papilla analis on female ovipositor of the yellow peach moth from Japan and the cardamom shoot borer from India

| Species or type | No. of insects | LAP (mm) | TPA (mm) | TPA/ LPA | MPA |
|---|---|---|---|---|---|
| FFT of yellow peach moth *C. punctiferalis* | 24 | 0.142 | 1.392 | 3.38 | Absent |
| PFT of yellow peach moth *C. punctiferalis* | 27 | 0.496 | 1.088 | 2.20 | Present |
| Cardamom shoot borer *C. sahayadriensis Conogethes* sp. | 3 | 0.530 | 1.104 | 2.08 | Present |

Fruit-feeding type (FFT), the conifer type, the Pinaceae-feeding type (PFT); LAP, TPA (see Fig. 19.1); *MPA* microtrichia on papilla analis. (*Source*: Honda and Mitsuhashi 1989)

volatile sex pheromones; emission of sex pheromones via calling behavior, which leads to attraction of potential mates; and the receptivity to males that attempt mating (Kingan et al. 1993).

The calling positions of female are either posing the abdomen in a curved manner or maintain the abdomen straight in a horizontal position with wing buzzing occasionally. During calling position, ovipositor gets extruded beyond the abdominal tip. Simultaneously the females move their antennae frequently up and down and with little walking. Calling behavior in *Conogethes* is observed from 1 to 7 days after emergence of female moths. The onset and ending of calling behavior will not differ significantly with the age but with species, for instance, *Conogethes punctiferalis* on castor (*Ricinus communis*) and *Conogethes* sp. on cardamom (*Elettaria cardamomum*). *Conogethes* moths reared on castor are smaller than moths reared on cardamom. Moths from cardamom exhibit more elaborate behavior than moths from castor. The entire reproduction repertoire appears to last for longer periods in moths from cardamom than from castor.

Calling behavior, regardless of female age, was recorded from the 1st to 11th h of scotophase, with the peak calling at 6th to 8th h after light off ($F_{6,66} = 3.57$, $P = 0.003$). Percent calling females also increased significantly with age ($F_{11,66} = 20.66$, $P < 0.001$). The mean duration of calling differed significantly with age in castor female ($F_{6,240} = 34.7$, $P < 0.00$). Duration of calling was appreciably low in 1-day-old females in ($P > 0.05$) compared to others. Length of calling increased considerably on second day. But on 3rd, 4th and 5th day females, even though increased length of calling evident but not significant. As days advanced (6th and 7th day), length of calling decreased significantly ($P > 0.05$) but, still, significantly higher than the 1st- and 2nd-day-old females (Shashank et al. 2014). Konno (1986) reported that calling in *C. punctiferalis* started from 5 h after light off, reached a maximum 7.5 after light off, and then decreased. The calling behavior was influenced by the age of the female ($F_{11,66} = 23.20$, $P < 0.001$). There was an increase in percent calling by females throughout the scotophase and it decreased thereafter ($F_{6,66} = 9.92$, $P < 0.001$). Peak calling was observed on 5th to 7th h after light off. In cardamom female, the length calling ($F_{6,210} = 43.00$, $P < 0.001$) varied significantly with age. The calling pattern of cardamom female was similar to that of castor female (Shashank et al. 2014) (Fig. 19.6).

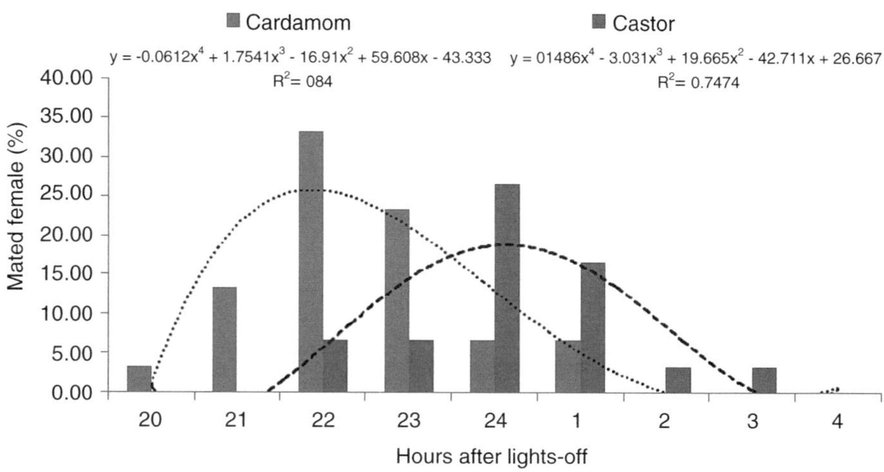


**Fig. 19.6** Temporal distribution of mating throughout the scotophase of *Conogethes* species reared on castor and cardamom ($n = 30$ pairs). (*Source*: Shashank et al. 2014)

## 19.5 Influence of Host in Calling Behavior

Shashank (2012) reported that *Conogethes* moths bred on castor began calling from 1 day onward. The daily calling patterns of 1–7-day-old females were similar. However, there were minor differences in calling behavior among females of 1–7 days old. After lights off, 1- to 2-day-old female moths started calling at 5 hours and extended up to 11 h. In the case of 3–7 days old, calling behavior began within 4 h after lights off, and percentage of female calling was maximal from 5 h to 8 h after lights off and then decreased gradually. A few moths performed calling behavior near the end of lights off on 3rd to 5th d. The temporal calling pattern shifted with age and the peak calling period was observed on 3–4-day-old moths bred both on castor and cardamom.

In 2-day-old moths, peak calling percentage (38.88%) was recorded at 7 h after lights off. In 3-day-old moth, the calling was first observed from 3 h after lights off and pronounced up to 8 h and then gradually decreased. No calling was observed 11 h after lights off. The same trend was observed in 4–7-day-old moths with peak calling at 4–6 h after lights off.

Rajabaskar and Regupathy (2012) studied the calling behavior of cardamom shoot and capsule borer, *Conogethes* species, in laboratory. The workers collected larvae from cardamom fields and reared on ginger *(Zingiber officinale)* in laboratory. The results revealed that the females start calling 3.45 h after start of scotophase and peak calling was observed 3 h prior to end of scotophase (Fig. 19.7).

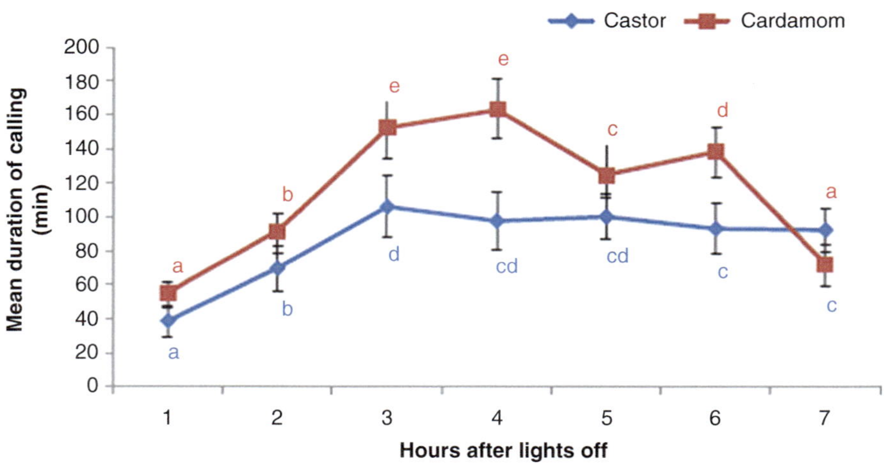


**Fig. 19.7** Comparative mean duration of calling behavior (mean of 7 days) response of *Conogethes* species reared on castor and cardamom under laboratory conditions (values are means ± SD). Values followed by the same letters in each component are not significantly different (Tukey-Kramer's test, $p < 0.05$). (*Source*: Shashank 2012)

## 19.6 Mating Behavior

For mating both sexes need to spend considerable energy, and costs are usually high and should therefore only occur if it is successful. Ultrasound production during mating behavior has been reported for an increasing number of moth species in the last three decades (Spangler 1988; Sanderford and Conner 1995; Simmons and Conner 1996; Sanderford et al. 1998; Conner 1987; Nakano et al. 2006, 2009a; Takanashi et al. 2010). Ultrasound produced by moths for sexual communication is classified into two types on the basis of its role calling song (Surlykke and Gogala 1986; Spangler 1988; Alcock et al. 1989; Heller and Achmann 1993; Heller and Krahe 1994; Sanderford and Conner 1995) and courtship song (Conner 1987; Simmons and Conner 1996; Sanderford et al. 1998; Nakano et al. 2006, 2009a, 2010; Takanashi et al. 2010). The females of most moths emit a sex pheromone, a calling signal to males, at the first stage of mating behavior (Ando et al. 2004). In other words, species of which the male first emits a calling signal to the female are rare (examples include the lesser wax moth *Achroia grisella* Stephens (Pyralidae) (Spangler 1988) and Australian whistling moths *Hecatesia* spp. (Noctuidae) (Alcock et al. 1989).

Females of *Conogethes* species release sex pheromones to attract males, and the attracted male produces courtship ultrasound near the female before attempting copulation. However, males of many moths produce a courtship song. This commonness of male courtship ultrasound in moth intraspecific sexual communication implies that acoustic signals are important for mating; however, the effect of the courtship ultrasound on mating has only been verified by behavioral experiments for a limited number of species (Conner 1987; Simmons and Conner 1996; Nakano et al. 2006, 2008, 2010; Takanashi et al. 2010).

In *Conogethes* within one scotophase (dark period), male moths can only produce one single spermatophore, which is transferred to the female during mating. Remating within the same scotophase would thus be unsuccessful. Premating period ranges from 2.40 to 2.95 days with an average of 2.67 ± 0.19 days. The mating behavior of the moths is most frequently observed for 2–4-day-old females and 3–4-day-old males (Konno et al. 1980, 1982; Kimura and Honda 1999, 2002). The high incidence of copulation in the late scotophase in *Conogethes punctiferalis* seems to be attributable to the release of sex pheromone by females in this period in Japan (Konno et al. 1982; Konno 1986).

The male courtship ultrasound of the yellow peach moth was of extremely high intensity (103 dB peSPL at 1 cm). The ultrasound also had a crucial effect on mating success: no copulation is accomplished when the female was deafened, and deaf females did not assume the wing-raising posture. Because ultrasound-producing males did not attempt to land near the deaf females, it was concluded that the wing-raising behavior of the female *Conogethes* is a mate-acceptance signal, which elicits the male's landing (Nakano et al. 2012).

The male ultrasound in *C. punctiferalis* consists of pulses and a burst. The minimum unit for the pulse and burst is a click, suggesting that males have tymbal organs whose sclerotized and striated or corrugated cuticle generates clicks, which last for 100 ms duration, by inward/outward buckling generated by inner muscles (Blest et al. 1963).

A study was conducted in Japan by Konno et al. (1981) on two types of the yellow peach moth, *D. punctiferalis*. They are a polyphagous (fruit-feeder) type that attacks the fruits of 30 species in 15 families of Angiospermae and an oligophagous (Pinaceae-feeder) type that only attacks the young foliage of *Pinus* and *Cedrus*. Although the mating behavior and the calling posture of the two types were almost identical, the calling frequency of the oligophagous type was much less frequent than that of the polyphagous type. When the two types were kept under a photoperiod of LD 15:9, the peak mating occurred 1.5 h earlier in the oligophagous than in the polyphagous type. Mating was significantly delayed in the presence of both sexes of the other type. Seol et al. (1986) observed the calling and mating behavior of the lesser mulberry pyralid, *Glyphodes pyloalis* Walker, as the mean mating time preceded the mean calling time by 3.5 to 5 h on day 2 to day 4 after emergence.

Konno and Tanaka (1996) reported that in Japan, *Chilo suppressalis* (Walker) infesting rice and the water oat (*Zizania latifolia* (Griseb.)) are two different strains, viz., rice-feeding and water oat-feeding strains. The mating time between these two strains was significantly different. The peak mating time of the rice feeder was 5 h earlier than that of the water oat feeder, which implies two strains are reproductively isolated.

Mating is preceded by wing vibration in the male prior to and during walking approach. Mating occurred only during the scotophase period between the 4th and 10th h of scotophase, with the peak occurring at the 7th h. Newly emerged *Neoleucinodes elegantalis* (Guenée) (Crambidae) couples rarely mated (2.8%), whereas 48- and 96-h-old couples mated 26.3% and 27.5%, respectively. Gland extracts from abdominal tips of 48–72 h virgin female moths evaluated in a wind tunnel were more attractive than virgin females.

The courtship behavior begins when males flew near a calling female (within 10 cm radius) and walked toward her while rapidly vibrating his wings. Occasionally, a female would fly off when approached. The beginning of copulation always occurred between 4 and 7 h, corresponding to the interval from 180 to 360 min after dusk. The mean copulation duration was 35 min. *C. punctiferalis* reared on castor and cardamom had single mating. The maximum percentage of moth mated reared on cardamom was between 4 and 6 h after lights off with the peak activity in 22.50 h (y = – 0.0612x4 + 1.7541x3 – 16.91x2 + 59.608x – 43.333, R2 = 0.84).

This was much earlier (peak mating in 24.50 h) than observed in moths reared on castor (6–9 h) (y = 0.1486x4 – 3.031x3 + 19.665x2 – 42.711x + 26.667, R2 = 0.7474). But there was no statistical difference in their peak mating time between castor and cardamom females. The percent mated were lower (maximum 26.67%) in *Conogethes* on castor compared to that on cardamom (maximum 33.33%). This observation suggests that maximally 33.33% of moths mated under laboratory conditions when reared on castor and cardamom.

Kaneko (1978) observed a clear constriction in abdomen of *C. punctiferalis* female which had already paired. This showed a consistent indication of pairing and critically helped the separation of virgin and paired females in the course of mass-rearing. The constriction was apparent 45–60 min after pairing.

## 19.7 Courtship Songs

Insects have developed diverse sexual communication systems exploiting acoustic and other signals to ensure successful mating. In insects that communicate acoustically, the males produce calling songs to attract and courtship song for mate acceptance. Calling songs are produced at the early phase of mating behavior, whereas courtship songs are produced at late phases in connection with the copulation attempt (Conner 1987; Krasnoff and Yager 1988; Simmons and Conner 1996; Conner 1999). Females, in their response, often show a preference for a song possessing particular acoustic characteristics (dominant frequency, temporal structure, and sound level) that may be relevant to the quality or size of the males (Forrest 1983; Jang and Greenfield 1996; Reinhold et al. 1998; Ryder and Siva-Jothy 2000). As a theory on the evolution of male signal and female response, the "receiver bias model" has been accepted widely (Ryan 1998; Greenfield 2002). In the case of moths, this theory assumes that sensory bias of the receivers to the ultrasonic echolocation calls of predatory bats drives the evolution of acoustic communication (Ryan 1998; Greenfield and Weber 2000). Actually, ultrasonic calling/courtship songs are employed for successful mating in an increasing number of moths (Conner 1999; Nakano et al. 2009a).

Males of the Asian corn borer moth *Ostrinia furnacalis* (Guenée) show typical courtship behavior after landing near a female releasing sex pheromone (i.e., a mating dance with wing fanning and subsequent copulation attempts) (Nakano et al. 2006, 2008). These male species produce ultrasonic courtship songs of extremely low intensity during copulation attempts (Nakano et al. 2006, 2008, 2009b). These songs increase the mating success of the males (Nakano et al. 2008).

Males produce louder songs if the first copulation attempt fails, suggesting that the males increase their sound levels to achieve successful copulation. This shows that the ultrasonic songs of the male makes the females motionless, which is the same response as that to ultrasonic bat calls (Nakano et al. 2010).

Male yellow peach moth *C. punctiferalis* produce loud courtship songs close to a female. After approaching a female releasing sex pheromones, the male hovers around her generating a series of brief pulses and then one long loud (83 dB SPL at 10 cm) pulse (Fig. 19.8a) (Nakano et al. 2012), causing the female to raise her wings upright and accept copulation (Fig. 19.8b). The wing-raising reaction is important for copulation eliciting the male's landing and attempting genital coupling. Wing-raising is evoked by any "long" ultrasonic pulse with duration of >200 ms (Nakano et al. 2014) (Fig. 19.8c). The series of short pulses (duration 28 ms, inter-pulse interval 26 ms) emitted before the long pulse did not directly influence female's mate acceptance. However, the time-based pattern is similar to the approach calls of

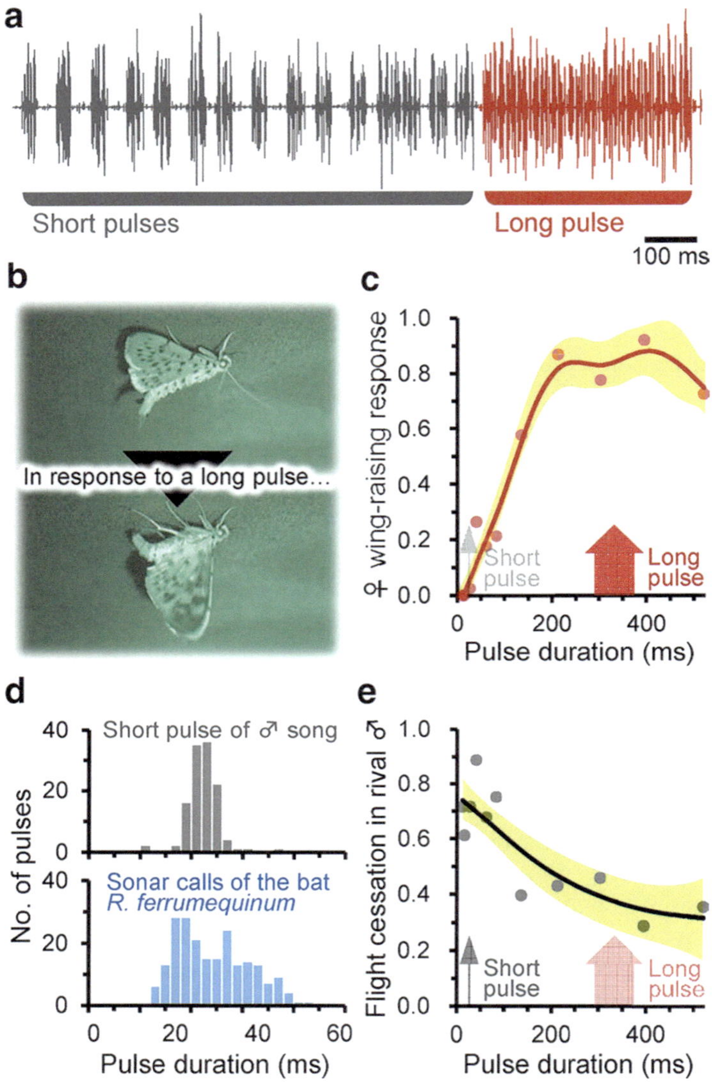


**Fig. 19.8** Communication with loud sounds. Acoustic communication with loud courtship song in a crambid moth *Conogethes punctiferalis*. (**a**) Oscillogram of male courtship song composed of short pulses in the early phase (black) and a long pulse in the late phase (red). (**b**) Females raise their wings in a mate-acceptance posture in response to a long (>200 ms) ultrasonic pulse. (**c**) The proportion of virgin receptive females raising their wings increases with duration of male courtship pulse. The response curve (solid line) is estimated from averages of binary data (circles) with 95% confidence intervals (CI) indicated by the *yellow band*. The width of two arrows on the *x*-axis denotes the 95% CI of the mean durations of short pulses (gray; 26.8–28.8 ms) and long pulse (red; 297.0–380.9 ms), respectively. (**d**) *Histograms* of pulse duration of short pulses emitted by males of *C. punctiferalis* (top, black) and that of approach-phase echolocation calls in the greater horseshoe bats *Rhinolophus ferrumequinum* (bottom, blue). (**e**) Flight suppression effect of male courtship pulses on males flying toward a female. Symbols are same as c. (Adapted from Nakano et al. 2012, 2014)

horseshoe bats *Rhinolophus* spp. which is a moth-foraging bat species, which elicits flight cessation in moths. Thus, a possible function of the short pulses reduce the flight of rival males (Fig. 19.8d, e), indicating a dual function of male courtship songs in *C. punctiferalis* which is used to fend off rivals with the short pulses and make the female accept mating with the long pulse. The evidence from both *A. sociella* and *C. punctiferalis* suggests that the high sound pressure of the male courtship song is developed for communication with rival males, whereas for communicating with the females, the loud sounds are produced using the same sound-producing mechanism for rival songs and courtship songs.

## 19.8 Scotophase

The dark phase in a cycle of light and darkness, especially artificially induced.

## 19.9 Photophase

The period of light during a day-night cycle.

## 19.10 Role of Light in Reproduction

Ecological impacts on the sensory framework assume an essential part in the support and maintenance of genetic variation (Ryan 1990, 2007; Endler 1993a). Signal traits and receptor organs are selected depending on environmental conditions, not only maximizing signal attractiveness but also minimizing negative effects of factors such as predation (Gamble et al. 2003; Fuller 2002; Bailey and Zuk 2008). Light is a highly variable factor, because of two important characteristics – spectrum and intensity (Endler 1993b) – which exhibit both spatial (among microenvironments) and temporal (diurnal and seasonal) variations. Thus, variation in visual communication through differences in light environment can favor directional selection.

For example, in *Drosophila melanogaster*, the development of the optic lobes and mushroom body depends on the light environment; optic lobe size in flies reared in dark conditions was smaller than that of flies reared in full light condition (Barth et al. 1997b). Additionally, when *D. melanogaster* are reared in light environments, photoreceptor voltage responses to light contrast changes are amplified, favoring signaling performance (Wolfram and Juusola 2004).

It has been demonstrated that light environment during early adulthood favors assortative mating and profoundly affects mating success in *D. melanogaster* (Hirsch and Tompkins 1994; Hirsch et al. 1995; Barth et al. 1997a). Also, acoustic experience at an early age can modulate female choice and shape alternative mating tactics in field crickets *Teleogryllus oceanicus* (Bailey and Zuk 2008; Bailey et al. 2010). Thus, rearing environment could be crucial in modeling mating behavior, especially in species that exhibit elaborated courtship displays. Male fruit flies reared in red, blue, and shaded light environments had a strong mating advantage over males from dark environments (Díaz-Fleischer and Arredondo 2011).

In *Conogethes* species moths, red light plays a very important role in the success of mating. Without red light, moths either lay sterile eggs or do not oviposit at all.

## 19.11 Role of Hormones

Several studies on the role of hormones in the insect reproduction have been documented but mostly in cockroaches, mosquitoes, and *Rhodnius* bugs (Raikhel et al. 2005). Over 25 species of Lepidoptera have been examined for the role of hormones in reproduction. The regulation of reproduction in Lepidoptera varies with the species. Usually egg development proceeds between the pupal and pharate adult stages. The physiology of reproduction in the pyralids is similar to that studied in the sphingid (Bombycoidea), *Manduca sexta* Linnaeus. In *M. sexta* vitellogenesis (Vg) begins 3–4 days before adult emergence and proceeds in the absence of the pupal corpora allata (CA); thereafter juvenile hormone (JH) is necessary for oocyte development (Satyanarayana et al. 1994). Vitellogenesis was found in pupal cases of male and female in small amounts and disappeared at pupal ecdysis. Surprisingly again it reappeared in pharate adults. Administration of methoprene induced Vg synthesis and Vg mRNA in prepupae and freshly molted pupae. Injection of ecdysteroids nullified effects of Vg on JH (Satyanarayana et al. 1994). The reproductive physiology of pyralids studied so far follows the above trend. For instance, in *Plodia interpunctella* (Hubner), the diminishing levels of ecdysteroids triggered vitellogenesis in developing pupae and retarded vitellogenin uptake by oocytes (Shirk et al. 1992). Likewise in *Diatraea grandiosella*, choriogenesis is completed with JH in the pharate adult (Shu et al. 1997). It has to be investigated if these patterns of physiological effects are found in *Conogethes* spp. In several species of Lepidoptera, follicle cells produce proteins that in later stages are incorporated into egg yolk, as in *Bombyx mori* (Linn.) (Zhu et al. 1986). Raikhel et al. (2005) have reviewed in detail the reproductive physiology in Lepidoptera, but research and understanding of reproductive physiology in insects, in general, are poor. Studies in future should focus on functional genomics, hormones, and genetic expression. Hitherto the sex determination mechanisms and reproduction have attracted attention mainly from developmental and evolutionary points of view. But currently reproduction has to be understood so that pest species populations in cultivated ecosystems can be effectively reduced. Sterile male technique biological control and genetic control are the areas where knowledge of reproduction can be exploited. The shoot and fruit borer *Conogethes* spp. will be an ideal candidate for such studies.

## 19.12 Hearing

There are several types of sound and vibration reception systems within the moth group. Not all of these hearing systems detect the ultrasonic echolocating cries of bats. Some moth larvae, for example, have sensitive acoustic hairs on the thorax that detect flying wasps (Tautz and Markl 1978), while both larval and adult moths can detect vibrations from a surface on which they are resting (Minnich 1936; Spangler 1987). But in few moths tympanic hearing organs or "ears" are developed in response to pressure from echolocating bats. These ears appear on the metathorax, on the first, second, or seventh abdominal segment (Von and Eggers 1933), or as modified structures on the head as in the hawkmoths (Roeder 1972; Roeder and Treat 1970; Roeder et al. 1968). Evidence for the use of ears in sexual communication has been established for noctuid and pyralid species (Gwynne and Edwards 1986; Spangler 1985; Spangler et al. 1984; Surlykke and Gogala 1986).

Most of the physiological data about lepidopteran hearing organs has been obtained from noctuids and moths of other families within the Noctuoidea. The information on the pyralid ear is more limited, but in pyralids sexual acoustic communication is better established.

## 19.13 The Pyralid Ear

Pyralid moths typically have tympanic hearing organs located on the pleural ventral surface of the first abdominal segment (Agee 1969; Coro 1973; Mullen and Tsao 1971). The organ consists of a thin tympanum of varying transparency tightly stretched over a cavity. The anterior part of the tympanum usually consists of a thicker, less flexible, opaque, papillate membrane referred to as the counter tympanum, which gives the tympanum a somewhat U shape.

The tympanic nerves are smaller than those of noctuid moths and enter the ganglion directly through its dorsal surface. Four sense cells are found in the chordotonal sensillum, and there is existence of a spontaneous discharge from a type B cell. Whereas, in Noctuoidea superfamily has only two acoustical sense cells (AI and A2 cells).

## 19.14 Pyralid Hearing

Studies have been done to determine sound frequencies to which certain pyralids are most sensitive, up to 100 kHz. Two peaks in acoustic sensitivity have been observed in *O. nubilalis*, at 25 and 70 kHz (5). In *A. kuehniella* the peaks occurred at 20 and 60 kHz (84). The required intensity level was near 54 dB at the lower peak and near 47 dB at the upper peak for both species. High sensitivity remained for both moths at the highest test frequencies of about 90 kHz, which suggests that the pyralid ear may be effective at frequencies above 100 kHz.

The physiological basis for this two-peak sensitivity is not obvious. However, pyralids that communicate by ultrasound use high frequencies, which suggests that concurrent pressures have resulted in sensitivity peaks suited to both defense and communication. Pyralid ears tilt (as do noctuid ears) in response to ultrasound. This tilting of the ear is used as indicator of sound reception.

Nakano et al. (2015) reported that audiograms, hearing threshold curves for various sound frequencies, of *C. punctiferalis* and *O. furnacalis* showed similar trends

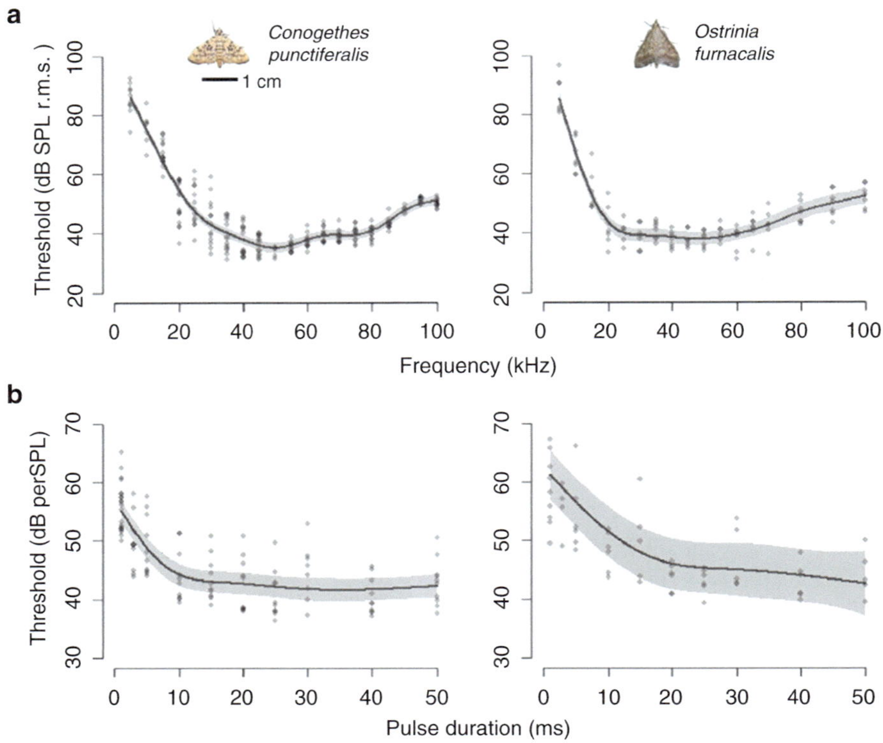


**Fig. 19.9** Hearing threshold curves in two crambid moths. (**a**) Sensitivity for 20 ms duration pulses of 5–100 kHz. Left: *C. punctiferalis* (*N* = 10$ + 7#). Right: *O. furnacalis* (*N* = 6$ + 1#), adapted from Nakano et al. (2008). (**b**) Sensitivity for 50 kHz pulses of 1–50 ms duration. Left: *C. punctiferalis* (*N* = 10$ + 10#). Right: *O. furnacalis* (*N* = 5$ + 4#). Semitransparent gray circles and black lines with semitransparent gray bands denote individual data points and averages with their 95% CI estimated by GAMMs. Top images in (**a**) are adult males of each species

and no significant species difference (LRT in GAMM, v21 = 0.68, $P$ = 0.41) (Fig. 19.9a) and pulse duration significantly affected the hearing thresholds in both crambid species (*C. punctiferalis*, LRT in GAMM, v22 = 49.64, $P < 0.0001$, $N = 20$; *O. furnacalis*, v22 = 16.21, $P$ = 0.00030, $N$ = 10) (Fig. 19.9b).

## 19.15 Sound Production

Insect sound-producing apparatuses are mostly classified into two types: file scraper and tymbal. Structures and locations of these organs are conserved in some phylogenetic groups, e.g., crickets, grasshoppers, and cicadas. However, moths have evolved diversified sound-producing organs, such as wing castanets and proboscis, in addition to the file scraper and tymbal, in each species. In yellow peach moth *C. punctiferalis* mesothoracic tymbal organs are developed never reported so far in insects. Tymbals are male specific and used for generating ultrasonic clicks in mating.

Nakano et al. (2012) found eight to nine striae on the smooth surface of the tymbal membrane, suggesting the production of several clicks by a single buckle of the membrane in association with contraction/relaxation of the mesothoracic muscles. Acoustic data from click sequences support the idea that the series is generated by side-to-side asynchrony with an active/passive half cycle by an inward/outward buckle, and thus in click group (pulse) production, males emit 28 clicks with the right and left tymbals. The click-producing mechanism is similar, but not homologous, to those of other clicking species in five moth families. Thus, moths have acquired tymbal organs through independent and convergent evolution.

## 19.16 Ovipositional Behavior

The ovipositional behavioral process is divided into four behavioral components:

1. Takeoff from the release point
2. Halfway flight
3. Hovering close to the source
4. Landing and egg-laying on substrates

Host plant odors accelerated takeoff, increased orientation to the stimulus source by hovering before landing, and triggered egg-laying. But non-host turnip taproot odors did not accelerate takeoff, even though behavioral components were all induced. Shortly sustained egg-laying under turnip taproot odors indicated involvement of different chemicals from those for other behavioral components. Odors from potato tubers and leaves as an absolute non-host plant never triggered oviposition responses (Honda et al. 1988).

The detailed effects of plant odor stimuli on each behavioral component of oviposition have been investigated by transient odor manipulations in a wind tunnel. Full supply of host plant odors resulted in a lengthier time for egg-laying, but showed no effect on the laying period of single eggs, whereas odor cessation at specific behavioral components (e.g., takeoff) interrupted subsequent responses. Even when plant odors were removed just after they reach halfway flight, female moths showed catenated behavioral components from hovering to egg-laying. Reasonable and nonstop plant olfactory incitements are important to finish the oviposition procedure by upgrading host-discovering productivity regarding both speed and exactness and animating maintained egg-laying after landing as final oviposition site acceptance.

Preoviposition period ranges from 1.22 to 1.65 days with an average of 1.41 ± 0.14 days. Oviposition period ranged from 2.30 to 3.20 days with an average of 2.76 ± 0.30 days.

## 19.17 Fecundity

Fecundity ranges from 80 to 110 eggs/female with a mean of 95.70 ± 10.23 eggs. Viability ranges from 72.80 to 88.55% with an average of 83.55 ± 5.09%. Total developmental period of *C. punctiferalis* occupied 26.29 ± 0.56 d.

## 19.18 Sex Ratio

Mean sex ratio works out to be 1:1.095 (1:1.2) male to female in *C. punctiferalis* (Sithananthan and Subramaniam 1975). Life table studies provide the information on important factors in pest management. Adult longevity studies clearly demonstrated no critical distinction among male and female moths. But laboratory studies with artificial diet indicated contrasts in life span among male and female, i.e., female moths survived 2–3 days more than the male (Sithanantham and Subramaniam 1975; Shanuowr et al. 1993).

## 19.19 Fungi as an Attractant in Oviposition

Attraction and oviposition responses of the yellow peach moth, *C. punctiferalis*, to seven species of fungi were investigated in laboratory and field cages. The *Penicillium* sp. and *Cladosporium* sp. showed the highest attractancy, while the *Aspergillus fumigatus* and *Mucor* species were less attractive. Among four phytopathogenic fungi, *Endothia parasitica* and *Alternaria solani* were also attractive. In a field cage, the *Conogethes* moths oviposited significantly more on the oviposition substrates baited with *Penicillium* sp. than on a fresh codling or unbaited control (Honda et al. 1988).

In *Conogethes* spp. complex reproduction in its entirety is yet to be fully understood. Studies on behavior, ecology, plant relationships, pheromones, etc., to some extent have made reproduction an interesting phenomenon. Since *Conogethes punctiferalis* is a highly polyphagous, widespread species, understanding reproduction process is crucially important for management. It is clear that deeper insights into the reproductive process will contribute to realistic management of this difficult to manage pest.

**Acknowledgment** Authors thank the sources from where the certain figures and data for the chapter were derived, the authorities of the University of Agricultural Sciences, GKVK, Bengaluru, and the MPKV, Rahuri, for encouragement and facilities.

## References

Agee HR (1969) Acoustic sensitivity of the European corn borer, Ostrinia nubialis. Ann Entomol Soc Am 62:1364–1367

Alcock J, Gwynne DT, Dadour IR (1989) Acoustic signalling, territoriality and mating in whistling moths, Hecatesia thyridion (Agaristidae). J Insect Behav 2:27–37
Ando T, Inomata S, Yamamoto M (2004) Lepidopteran sex pheromones. Top Curr Chem 239:51–96
Bailey NW, Zuk M (2008) Acoustic experience shapes female mate choice in field crickets. Proc R Soc Lond B 275:2645–2650
Bailey NW, Gray B, Zuk M (2010) Acoustic experience shapes alternative mating tactics and reproductive investment in field crickets. Curr Biol 20:845–849
Barth M, Hirsch HVB, Heisenberg M (1997a) Rearing in different light regimes affects courtship behaviour in Drosophila melanogaster. Anim Behav 53:25–38
Barth M, Hirsch HVB, Meinertzhagen IA, Heisenberg M (1997b) Experience-dependent developmental plasticity in the optic lobe of *Drosophila melanogaster*. J Neurosci 17:1493–1504
Blest AD, Collett TS, Pye JD (1963) The generation of ultrasonic signals by a New World arctiid moth. Proc R Soc Lond B 158:196–207
Bush GL (1975) Modes of speciation. Annu Rev Ecol Syst 6:339–364
Conner WE (1987) Ultrasound: its role in the courtship of the arctiid moth, Cycnia tenera. Experientia 43:1029–1031
Conner WE (1999) Un chant d'appel amoureaux': acoustic communication in moths. J Exp Biol 202:1711–1723
Coro F (1973) Morfologia del organo timpanico de la polilla del arroz, *Corcyra cephalonica* (Stainton) (Lepidoptera: Galleriidae). Ciene Bioi 45:1–15
Díaz-Fleischer F, Arredondo J (2011) Light conditions affect sexual performance in a lekking tephritid fruit fly. J Exp Biol 214:2595–2602
Endler JA (1993a) Some general comments on the evolution and design of animal communication systems. Philos Trans R Soc Lond B 340:215–225
Endler JA (1993b) The color of light in forests and its implications. Ecol Monogr 63:1–27
Forrest TG (1983) Calling songs and mate choice in mole crickets. In: Gwynne DT, Morris GK (eds) Orthopteran mating systems: sexual competition in a diverse group of insects. Westview Press, Boulder, pp 185–204
Fuller RC (2002) Lighting environment predicts the relative abundance of male colour morphs in bluefin killifish (*Lucania goodei*) populations. Proc R Soc B 269:1457–1465
Gamble S, Lindholm AK, Endler JA, Brooks R (2003) Environmental variation and the maintenance of polymorphism: the effect of ambient light spectrum on mating behaviour and sexual selection in guppies. Ecol Lett 6:463–472
Greenfield MD (2002) Signallers and Receivers: Mechanisms and Evolution of Arthropod Communication. Oxford University Press, Oxford
Greenfield MD, Weber T (2000) Evolution of ultrasonic signalling in wax moths: discrimination of ultrasonic mating calls from bat echolocation signals and the exploitation of an antipredator receiver bias by sexual advertisement. Ethol Ecol Evol 12:259–279
Gwynne DT, Edwards ED (1986) Ultrasound production by genital stridulation in *Syntonarcha iriastis* (Lepidoptera: Pyralidae): long distance signalling by male moths? Zool J Linn Soc 88:363–376
Heller KG, Achmann R (1993) The ultrasonic song of the moth Amyna natalis (Lepidoptera: Noctuidae: Acontiinae). Bioacoustics 5:89–97
Heller KG, Krahe R (1994) Sound production and hearing in the pyralid moth *Symmoracma minoralis*. J Exp Biol 187:101–111
Hirsch HVB, Tompkins L (1994) The flexible fly: experience dependent development of complex behaviors in *Drosophila melanogaster*. J Exp Biol 195:1–18
Hirsch HVB, Barth M, Luo S, Sambaziotis H, Huber M, Possidente D, Ghiradella H, Tompkins L (1995) Early visual experience affects mate choice of *Drosophila melanogaster*. Anim Behav 50:1211–1217
Honda, Mitsuhashi (1989) Morphological and morphometric differences between the fruit and pinaceae-feeding type of yellow peach moth, *Conogethes punctiferalis* (Guenee) (Lepidoptera: Pyralidae). Appl Entomol Zool 24(1):1–10

Honda H, Ishiwatari T, Matsumoto Y (1988) Fungal volatiles as oviposition attractants for the yellow peach moth, *Conogethes punctiferalis* (Guenée) (Lepidoptera: Pyralidae). J Insect Physiol 34:211–209

Hou ML, Sheng CF (2000) Calling behaviour of adult female *Helicoverpa armigera* (Hubner) (Lep., Noctuidae) of overwintering generation and effects of mating. J Appl Entomol 124:71–75

Inoue H, Yamanaka H (2006) Redescription of Conogethes punctiferalis (Guenée) and description of two new closely allied species from Eastern Palaearctic and Oriental regions (Pyralidae, Pyraustinae). Tinea 19:80–91

Jang Y, Greenfield MD (1996) Ultrasonic communication and sexual selection in wax moths: female choice based on energy and asynchrony of male signals. Anim Behav 51:1095–1106

Jia F-Y, Greenfield MD, Collins RD (2001) Ultrasonic signal competition between male wax moths. J Insect Behav 14:19–33

Kaneko J (1978) Abdominal constriction in the copulated female of yellow peach moth, *Dichocrocis punctiferalis* Guenée (Lepidoptera: Pyralidae). Appl Entomol Zool 13:131–133

Kimura T, Honda H (1999) Identification and possible functions of the hairpencil scent of the yellow peach moth, *Conogethes punctiferalis* (Guenée) (Lepidoptera: Pyralidae). Appl Entomol Zool 34:147–153

Kimura T, Honda H (2002) Microstructure and pheromone producing function of male hairpencils in the yellow peach moth, *Conogethes punctiferalis* (Lepidoptera: Pyralidae). Entomol Sci 5:237–247

Kingan TG, Thomas-Laemont PA, Raina AK (1993) Male accessory gland factors elicit change from 'virgin' to 'mated' behavior in the female corn earworm moth *Helicoverpa zea*. J Exp Biol 183:61–76

Konno Y (1986) Time-lag between sex pheromone content and the calling behavior in the yellow peach moth, *Conogethes punctiferalis* (Guenée) (Lepidoptera: Pyralidae). Appl Entomol Zool 21:622–624

Konno Y, Tanaka F (1996) Mating time of the rice-feeding and water-oat-feeding strains of the rice stem borer, *Chilo suppressalis* (Walker) (Lepidoptera: Pyralidae). – Japan. J Appl Entomol Zool 40:245–247

Konno Y, Honda H, Matsumoto Y (1980) Observations on the mating behavior and bioassay for the sex pheromone of the yellow peach moth, *Dichocrocis punctiferalis* Guenée (Lepidoptera: Pyralidae). Appl Entomol Zool 15:321–327

Konno Y, Arai K, Sekiguchi K, Matsumoto Y (1982) (E)-10- Hexadecenal, a sex pheromone component of the yellow peach moth, *Dichocrocis punctiferalis* Guenée (Lepidoptera: Pyralidae). Appl Entomol Zool 17:207–217

Konno Y, Honda N, Matsumoto Y (1981) Mechanism of reproductive isolation between the fruit feeding and pin ace feeding type of yellow peach moth, Dichocrosis punctiferalis. Jpn J Appl Entomol Zool 25:253–258

Krasnoff SB, Yager DD (1988) Acoustic response to a pheromonal cue in the arctiid moth *Pyrrharctia isabella*. Physiol Entomol 13:433–440

Minnich DE (1936) The response of caterpillars to sound. J Exp Zool 72:439–453

Mullen MA, Tsao CH (1971) Tympanic organ of Indian-meal moth, *Plodia interpunctella* (Hubner), almond moth, *Cadra calttella* (Walker) and tobacco moth, *Ephestia elutella* (Hubner) (Lepidoptera: Pyralidae). Int J Insect Morphol Embryol 1:3–10

Nakano R, Ishikawa Y, Tatsuki S, Surlykke A, Skals N, Takanashi T (2006) Ultrasonic courtship song in the Asian corn borer moth, *Ostrinia furnacalis*. Naturwissenschaften 93:292–296

Nakano R, Skals N, Takanashi T (2008) Moths produce extremely quiet ultrasonic courtship songs by rubbing specialized scales. In: Proc Natl Acad Sci U S A, vol 105, pp 11812–11817

Nakano R, Takanashi T, Fujii T (2009a) Moths are not silent, but whisper ultrasonic courtship songs. J Exp Biol 212:4072–4078

Nakano R, Ishikawa Y, Tatsuki S (2009b) Private ultrasonic whispering in moths. Commun Integr Biol 2:123–126

Nakano R, Takanashi T, Skals N, Surlykke A, Ishikawa Y (2010) Ultrasonic courtship songs of male Asian corn borer moths assist copulation attempts by making the females motionless. Physiol Entomol 35:76–81

Nakano R, Takanashi T, Ihara F, Mishiro K, Toyama M, Ishikawa Y (2012) Ultrasonic courtship song of the yellow peach moth, *Conogethes punctiferalis* (Lepidoptera: Crambidae). Appl Entomol Zool 47:87–93. https://doi.org/10.1007/s13355-012-0092-z

Nakano R, Ihara F, Mishiro K, Toyama M, Toda S (2014) Double meaning of courtship song in a moth. Proc R Soc B 281:20140840. https://doi.org/10.1098/rspb.2014.0840

Nakano R, Takanashi T, Surlykke A (2015) Moth hearing and sound communication. J Comp Physiol A 201:111–121. https://doi.org/10.1007/s00359-014-0945-8

Nalnabdov AV (1970) Reproductive physiology. R.J. Taraprorerala and Sons Com, Bombay, p 316

Raikhel AS, Brown MR, Belles X (2005) Hormonal control of reproductive processes. In: Hormones, brain and behavior. Elsevier, Amsterdam, pp 433–491

Rajabaskar D, Regupathy A (2012) Calling behaviour and attractive response of cardamom shoot and capsule borer, *Conogethes punctiferalis* Guenee (Lepidoptera: Pyralidae) to the female crude extract and synthetic blend. J Entomol 2012:1–6

Reinhold K, Greenfield MD, Jang Y, Broce A (1998) Energetic cost of sexual attractiveness: ultrasonic advertisement in wax moths. Anim Behav 55:905–913

Roeder KD (1972) Acoustic and mechanical sensitivity of the distal lobe of the pilifer in choerocampine hawkmoths. J Insect Physiol 18:1249–1264

Roeder KD, Treat AE (1970) An acoustic sense in some hawkmoths (Choerocampinae). J Insect Physiol 16:1069–1086

Roeder KD, Treat AE, Vandeberg JS (1968) Auditory sense in certain sphingid moths. Science 159:1–33

Ryan MJ (1990) Sensory systems, sexual selection, and sensory exploitation. Oxford Surv Evol Biol 7:157–195

Ryan MJ (1998) Receiver biases, sexual selection and the evolution of sex differences. Science 281:1999–2003

Ryan MJ (2007) Sensory ecology: see me, hear me. Curr Biol 17:R1019–R1021

Ryder JJ, Siva-Jothy MT (2000) Male calling song provides a reliable signal of immune function in a cricket. Proc R Soc London Ser Biol Sci 267:1171–1175

Sanderford MV, Conner WE (1995) Acoustic courtship communication in Syntomeida epilais Wlk. (Lepidoptera: Arctiidae, Ctenuchinae). J Insect Behav 8:19–31

Sanderford MV, Coro F, Conner WE (1998) Courtship behavior in Empyreuma affinis Roths. (Lepidoptera, Arctiidae, Ctenuchinae): acoustic signals and tympanic organ response. Naturwissenschaften 85:82–87

Satyanarayana K, Bradfield JY, Bhaskaran G, Dahm KH (1994) Stimulation of Vitellogenin production by methoprene in prepupae and pupae of *Manduca sexta*. Arch Insect Biochem Physiol 25:21–37

Seol KY, Honda H, Matsumoto Y (1986) Mating-behavior and the sexpheromone of the lesser mulberry pyralid, Glyphodes pyloalis Walker (Lepidoptera, Pyralidae). Appl Entomol Zool 21:228–235

Shanuowr TG, Gutierrez AP, Wightman JA (1993) Effect of temperature on development rates, fecundity & longevity of the groundnut leaf miner, *Aproaerema modicella* (Lepidoptera: Gelechiidae), in India. Bull Entomol Res 83(3):413–419

Shashank RP (2012) Biosystematics and pheromone components of *Conogethes punctiferalis* (Guénee) (Lepidoptera: Crambidae) with special reference to populations infesting castor (*Ricinus communis* L.) and cardamom (*Elettaria cardamomum* Maton), Ph.D thesis, UAS, Bangalore, p 66

Shashank RP, Chakravarthy AK, Chandrashekharaiah R, Bhanu KRM (2014) Behavioural studies on Shoot and Fruit Borer, *Conogethes punctiferalis*, Guenée, (Crambidae: Lepidoptera) host-associated populations reveal occurrence of cryptic species. Entomologia Generalis 35(2):103–115

Shirk PD, Zimowska G, Silhacek DL, Shaaya E (1992) Initiation of vitellogenesis in pharate adult females of Indian-meal moth, *Plodia interpuntella*. Arch Insect Biochem Physiol 21:53–63

Shu S, Park YI, Ramaswamy SB, Srinivasan A (1997) Hemolymph juvenile hormone titers in pupal and adult stages of southwestern corn borer (*Diatraea grandiosella* (Pyralidae)) and relationship with egg development. J Insect Physiol 43:719–726

Simmons RB, Conner WE (1996) Ultrasonic signals in the defense and courtship of Euchaetes egle Drury and E. bolteri Stretch (Lepidoptera: Arctiidae). J Insect Behav 9:909–919

Sithanantham S, Subramaniam TR (1975) Sex determination of pupae of sorghum stalk borer, Chilo zonellus (Swinhoe). Madras Agric J 62(2):62–63

Spangler HG (1985) Sound production and communication by the greater wax moth (Lepidoptera: Pyralidae). Ann Entomol Soc Am 78:54–61

Spangler HG (1987) High-frequency sound production by honey bees. J Apic Res 25:213–219

Spangler HG (1988) Ultrasonic communication in *Corcyra cephalonica* (Stainton) (Lepidoptera: Pyralidae). J Stored Prod Res 23:203–211

Spangler HG, Greenfield MD, Takessian A (1984) Ultrasonic mate calling in the lesser wax moth. Physiol Entomol 9:87–95

Surlykke A, Gogala M (1986) Stridulation and hearing in the noctuid moth Thecophora fovea (Tr.). J Comp Physiol A 159:27–73

Takanashi T, Nakano R, Surlykke A, Tatsuta H, Tabata J, Ishikawa Y, Skals N (2010) Variation in courtship ultrasounds of three Ostrinia moths with different sex pheromones. PLoS ONE 5:e13144

Tautz J, Markl H (1978) Caterpillars detect flying wasps by hairs sensitive to airborne vibration. Behav Eco Sociobiol 4:101–110

Von, Eggers P (1933) Die abdominal en Tympanalorgane der Lepidopteren. Zool Jahrb Abt Anat Ontog Tiere 57:1–104

Wang LP, Ji BZ, Liu, Zhao ZP, Cao DD, Xu Z-X (2015) Anatomical study on the morphology of the reproductive systems and the scale brushes of *Dioryctria rubella* Hampson (Pyralidae: Phycitinae). J Lepidopterists Soc 69(1):50–57

Wolfram V, Juusola M (2004) Impact of rearing conditions and short-term light exposure on signaling performance in *Drosophila* photoreceptors. J Neurophysiol 92:1918–1927

Zhu J, Indrasith L, Yamashita O (1986) Characterization of vitellin, egg-specific protein and 30-kDa protein from Bombyx eggs, and their fates during oogenesis and embryogenesis. Biochem Biophys Acta 882:427–436

Zuk M, Kolluru GR (1998) Exploitation of sexual signals by predators and parasitoids. Q Rev Biol 73:415–438

# Host Plant Relationships of the Shoot and Fruit Borer, *Conogethes* spp. (Crambidae: Lepidoptera): Mechanisms and Determinants

20

B. Doddabasappa, K. R. M. Bhanu, and S. Subhash

**Abstract**

Among castor plant types, spineless spike type exhibited resistance to the borer, *Conogethes*. Among cardamom plant types, Malabar clones exhibited borer resistance. Host shift experiments and other laboratory studies revealed that *Conogethes* population feeding on cardamom are genetically different from those feeding on castor. The patterns and relationships between the borer and the above two plants are different and independent. The larval behavioural responses were different on the two plants. A combination of antixenosis and antibiosis together with tolerance mechanism conferred resistance to the plants against the borer, *Conogethes*, attack. *Conogethes* borer-resistant plants play a vital role in pest management as resistant plants offer an ideal means of borer suppression.

**Keywords**

Cardamom · Castor · Antibiosis-antixenosis mechanisms · Host-mediated speciation

B. Doddabasappa (✉)
Department of Entomology, College of Horticulture, Kolar, Karnataka, India

K. R. M. Bhanu
Bio-Control Research Laboratories, A Division of Pest Control (India) Pvt. Ltd., Sriramanahalli, Bangalore, Karnataka, India
e-mail: bhanu.krm@pcil.in

S. Subhash
Department of Agricultural Entomology, University of Agricultural Sciences, Gandhi Krishi Vignan Kendra (GKVK), Bengaluru, Karnataka, India


A. K. Chakravarthy (ed.), *The Black spotted, Yellow Borer, Conogethes punctiferalis Guenée and Allied Species*,
https://doi.org/10.1007/978-981-13-0390-6_20

## 20.1 Introduction

The shoot and capsule borer, *Conogethes punctiferalis* (Guenee) (Lepidoptera: Crambidae), is a polyphagous pest in tropical, subtropical and select temperate countries attacking host plants, namely, fruit, spice crops, plantation crops and forest plants (Thyagaraj 2003; Pena et al. 2002). In high-altitude landscape habitats, where generally biological diversity is high and unique, conservation of biological resources assume importance. In these habitats the endemic species are found which perform multiplicative ecological roles. Gene pools which can be exploited for human welfare are also found in these habitats (Inoue and Yamanaka 2006; Lu et al. 2010). Since *Conogethes* attacks diverse categories of crops, namely, fruit, vegetable, plantation, root and field crops, in diversified habitats, information on host plant relationships is of utmost importance (Sekiguchi 1974) but is glaringly lacking. This group of pests assume significance as larvae directly inflict damage to reproductive plant parts. It is an internal tissue borer, and many of the pesticides do not penetrate into the insect because larvae are concealed under fibrous material and tunnels. These shoot and fruit borers have less predators and parasitoids. Thus, host plant resistance offers an ideal tool for the management of *Conogethes* species complex.

However, an integrative approach is essential to test species delimitations not only based on host plant origin and morphological characters but also on multiple independent molecular markers (Roe and Sperling 2007). Recently, DNA sequence data have been used successfully to elucidate the relationships of different species at the genetic level (Brower 1997; Ren et al. 2002; Murray and Prowell 2005; Wang et al. 2009).

## 20.2 Host Plants and Borer Species Complex

The Japanese and Chinese workers categorized two types of yellow peach moths (*Conogethes* sp.) based on different feeding habits and morphological characters, i.e. the fruit-feeding type (FFT) and Pinaceae-feeding type (PFT) (Koizumi 1963; Chakravarthy et al. 1991). They studied the biological characters between the two types, and host plant specificity tests revealed that FFT female oviposition was reduced when provided with leaves from PFT-preferred hosts (Honda and Matsumoto 1984). Konno et al. (1981) also found that both FFT and PFT types respond to different spectra of host plant constituents and the specificity correlated with the taxonomic grouping of the two species. Furthermore, Honda and Matsumoto (1987) had also studied FFT and PFT larvae preferentially feeding on exudate and carbohydrates derived from their specific host plants. Thus, reproductive isolation between the two suggested that the mechanism underlying sexual isolation must be different (Konno et al. 1981; Honda 1986). The EAG response (electroantennogram response) of FFT and PFT female moths were different towards *n*-alkyl compounds that showed moths of the two types have possibly evolved host plant-specific mechanisms for chemoattraction (Honda et al. 1986). On the other hand, morphological evidence from scanning electron microscope (SEM) showed differences in female reproductive organs, larvae's maxilla, upper lip, pupae's cremaster and labial palpus among FFT and PFT types (Sekiguchi 1974) and agreed with additional characters described by Chai and He (1987). Additionally, developmental variation in eclosion time was observed where 10:00 pm to 8:00 am was preferred for PFT (Kuang et al. 2009), whereas FFT types eclosed between 8:00 pm and 10:00 pm (Wang 1980). Based on these studies the two types were believed to be two different species, i.e. *C. punctiferalis* (fruit-feeding type, FFT) and *C. pinicolalis* (Pinaceae-feeding type, PFT) (Inoue and Yamanaka 2006). In 2018, Shashank et al. (2018) identified the borer species on zingibers in South India as *C. sahyadriensis*.

## 20.3 Host Plant Relationships

### 20.3.1 Germplasm Screening

Jayalaxmi (1996) reported that a total of 232 indigenous entries of castor were screened at 2 locations in Karnataka, South India, against shoot and fruit borer, and based on per cent capsule bored, 48-1 and SHB-556 at Hiriyur Local; JI-130, SKI-129 and SKI-126 at Hiriyur; and SHB-556, SHB-392, JHB-705, JI-130, 48-1 and SKI-126 at Bangalore proved resistant against shoot and capsule borer. Genotypes, namely, SHB-392, Hiriyur local and 48-1, also proved resistant to castor leaf miner and defoliators (Jayalaxmi et al. 2009). Lakshminarayana (2005) reported the morphological characters of castor capsules associated with resistance to major insect pests. Twelve genotypes with different spike characteristics and DCS-9 as control were evaluated against *C. punctiferalis*. Genotypes with compact spikes (86.7% mean capsule damage) and big capsules (85.8%) were highly preferred by the pest. Moderate preference for genotypes with small and non-spiny capsules (mean capsule damage of 42% and 51.7%, respectively) were observed. Genotypes with loose spikes were the least damaged (18.3%–28.3%).

Four hundred and ninety-two accessions of ginger, *Zingiber officinale* Rose, including popular cultivars and high-yielding varieties were screened in the field against the shoot borer *Conogethes* species during 2001–2004 at the experimental farm of Indian Institute of Spices Research (IISR), Peruvannamuzhi, Kerala, India. All the accessions were susceptible to the pest attack. None of the accessions was rated as resistant, whereas 49, 2S1, 130 and 62 accessions were rated as moderately resistant, moderately susceptible, susceptible and highly susceptible, respectively, to the pest (Devasahayam et al. 2010).

The castor genotypes, namely, 48-1, Geeta Kiran and GCH-4, showed less than 10%; GCH-5, Kranti, VI-9, JI-35 and DCH-177 showed 10% to 15%; and Haritha, LRES-17, DCS-5, GAUCH-1, GCH-2, VP-1, SH-72, AKC-1 and DCS-9 showed more than 50% significant borer damage (Directorate of Oilseed Research (DOR), Hyderabad (Hyd), 2004). The castor genotypes with loose (RG 1933, RG 1934) (26.7%), very loose spike (RG 2543, RG 2546) (20%), non-spiny (RG258, RG 250) (51.7%) and small capsules (RG 2635, RG 1636) (42.6%) were found less damaged by the borer, while those with compact spikes (RG1618, RG1627) (86.7%) and big capsules (RG1626, RG 1627) (85.8%) suffered more (DOR, Hyd., 2000) borer damage. Yethapur centre (2004) Andhra Pradesh, South India, reported castor genotypes, namely, RG 1228, 1280, 1266 and 1359, resistant to the borer, while RG 2820 and 2786 showed low borer infestation at DOR, Hyd (2004).

Krishnamurthy et al. (1989) reported that *Conogethes* species completed one life cycle on cardamom (*Elettaria cardamom*) in 25–40 days with five generations a year at Mudigere, South India, at 26 °C and 70% relative humidity (RH). The phenology of the host plant influenced the size, growth and development of borer eggs (Bilapate and Talati 1978; Jacob 1981; Twine 1971). There was a significant difference in per cent egg hatching period from 65.0 ± 0.76 days to 90.5 ± 1.38 days and incubation period from 4.19 ± 0.80 days to 9.35 ± 1.05 days under varied temperature and relative humidity (Thyagaraj 2003; Wang and Cai 1997).

First instar (neonate) larva bore into the pseudostem, rhizome or capsule of the zingiber plant. On the pseudostem, the larva bores at the base of the leaf axil and enters inside the cardamom/ginger/turmeric shoot tissue (Jacob 1981). The excreta plugged at the entry hole on the shoot indicated larval boring (Thyagaraj 2003). Larva that fed on the shoot was light greenish, while those on capsules were dull yellow (Bilapate and Talati 1977; Thyagaraj 2003). The larva remains inside the pseudostem till pupation making it difficult to study the larval instars directly.

The growth and development of different larval instars varied with varying temperature and relative humidity. Each larval instar lasted for 3–4 days. Sloan (1945) reported that larval period of *C. punctiferalis* lasted 3 weeks under normal conditions and 2–3 weeks in winters. Young and Shaw (1962) studied the peach borer biology in China and reported 4–5 generations a year with the larvae overwintering in the flowers, stem, and fallen leaves. The duration of the larval stage varied from 20–23 days in August–September at 21–35 °C to 22–26 days in October–January at 14–28 °C, and larvae were found in the field until February, and also the total larval period extended up to 12–14 days (Bilapate 1977; Jacob 1981; Kondo and Miyahara 1930; Twine 1971; Wang and Cai 1997; Xi et al. 1996). The larval period varied from 12.55 ± 2 days to 19.59 ± 5.50 days, and the per cent survival varied from 49.6 ± 0.18 to 92.8 ± 1.39 and 28 ± 1 °C and 80 ± 5% RH were most favourable for larval development (Thyagaraj 2003). Stanley et al. (2009) reported *C. punctiferalis* to complete life cycle within a shorter period on castor, followed by guava under laboratory conditions. The number of days required for the neonate larva to become adult was 27.76 on castor.

## 20.4 Resistance Mechanisms

### 20.4.1 Antixenosis Mechanism

#### 20.4.1.1 Host Shift Experiments

The null hypothesis while conducting these experiments was that the *Conogethes* infesting castor and cardamom belonged to the same species. The experiments addressed the question whether *Conogethes* infesting on castor and cardamom have

**Table 20.1** Survival of life stages of cardamom borer *C. sahyadriensis* on castor

| Castor plant type | Larval instar | | | | | Pupation (%) |
|---|---|---|---|---|---|---|
| | 1st | 2nd | 3rd | 4th | 5th | |
| Spineless spike | 14.9 ± 1.6 (22.68)[a] | 30.8 ± 1.6 (33.72)[bc] | 15.7 ± 2.4 (23.25)[a] | 41.3 ± 1.6 (39.95)[a] | 48.9 ± 1.9 (44.38)[a] | 13.6 ± 1.45 (21.65)[a] |
| Compact spiny | 19.0 ± 2.9 (25.77)[b] | 32.6 ± 2.3 (34.79)[c] | 31.0 ± 2.6 (33.81)[b] | 51.9 ± 1.4 (46.09)[c] | 54.75 ± 2.9 (47.72)[c] | 23.9 ± 1.7 (29.26)[c] |
| Spiny loose spike | 16.1 ± 2.1 (23.59)[a] | 23.3 ± 2.4 (29.52)[a] | 17.3 ± 3.2 (28.78)[a] | 43.5 ± 0.5 (41.26)[b] | 52.4 ± 1.4 (46.37)[b] | 18.7 ± 2.1 (25.62)[b] |
| S.Em | 1.002 | 0.792 | 1.064 | 0.433 | 0.727 | 0.768 |
| CD@1% | 3.20 | 2.53 | 3.40 | 1.38 | 2.32 | 2.45 |

Figures in parentheses are arcsine-transformed values; $n = 20$; numbers followed by the same letter are not statistically significant ($p < 0.001$) by LSD

the same feeding behaviour and life cycle attributes. If species of *Conogethes* is not the same, then one can expect the differences in feeding behaviour, bioecology and life stages attributes on castor and cardamom. In order to test this hypothesis, *Conogethes* reared on castor were implanted on cardamom and vice versa.

The *Conogethes* species larvae of cardamom were reared on castor using plastic tray (15 × 25 × 35 cm). Newly hatched first instar larvae were transferred to capsules of each castor type, namely, spineless, compact spike, spiny loose spike and reared, in trays. Fresh capsules of each castor type were given as feed once in 4 days, i.e. when dried or eaten by the larvae. The study was conducted under seminatural conditions in laboratory.

*Conogethes* reared on cardamom was implanted on three types of castor genotypes at egg stage. When eggs of the moths reared on cardamom were placed on castor, almost cent per cent hatched on the three types of castor. There were significant differences in larval survival of cardamom *Conogethes* on castor. Larval survival of fourth instar was 41.3 ± 1.6%, 51.9 ± 1.4% and 43.5 ± 0.5% and fifth instar survival was 48.9 ± 1.9, 54.75 ± 2.9, 52.4 ± 1.4% and 13.6 ± 1.45%, 23.9 ± 1.7% and 18.7 ± 2.1% pupation on spineless spike, compact spiny and spiny loose type of castor, respectively (Table 20.1). Though the first instar larvae survived to an extent of 14.9%, they did not complete the life cycle (Fig. 20.1). So first instar larvae suffered 86% mortality. This is crucially important as this is a critical stage in host plant selection for the insect. This suggested that for *Conogethes* larvae reared on cardamom, castor is not the suitable host plant.

There were statistical significant differences in developmental times of each larval instar among the spineless spike, compact spiny and spiny loose type of castor host plants examined. Fourth instar larval period was 2.75 ± 0.3 days, 3.15 ± 0.2 days and 3.06 ± 0.2 days, respectively, and fifth instar larval period was 3.00 ± 0.3 days, 3.27 ± 0.5 days and 3.38 ± 0.3 days, respectively (Table 20.2).

When *Conogethes* reared on cardamom were implanted on castor, larvae suffered almost cent per cent mortality. Besides, larvae completed the period early probably because of physiological stress brought out by host plant shift

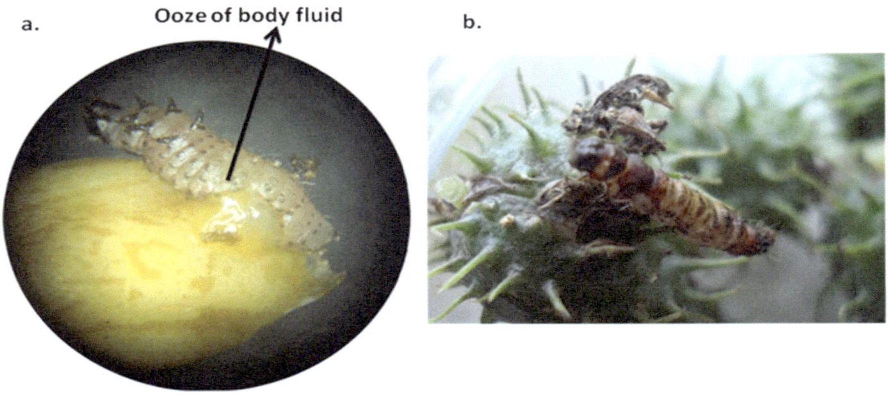


**Fig. 20.1** Microphotograph. (**a**) Mortality of second instar larva and (**b**) malformed pupa (premature pupa) from third instar *Conogethes* larvae reared on castor, implanted on cardamom

**Table 20.2** Developmental time of cardamom borer *C. sahyadriensis*[a] life stages on castor

| | Larval instar (days) | | | | |
|---|---|---|---|---|---|
| Castor plant type | 1st | 2nd | 3rd | 4th | 5th |
| Spineless spike | 1.00 ± 0.4[c] | 2.00 ± 0.3[ba] | 2.50 ± 0.2[c] | 2.75 ± 0.3[a] | 3.00 ± 0.3[a] |
| Compact spiny | 1.35 ± 0.4[a] | 1.94 ± 0.3[a] | 2.51 ± 0.3[a] | 3.15 ± 0.2[a] | 3.27 ± 0.5[a] |
| Spiny loose spike | 1.66 ± 0.4[bc] | 2.14 ± 0.4[a] | 2.97 ± 0.2[bc] | 3.06 ± 0.2[a] | 3.36 ± 0.3[a] |
| S.Em± | 0.102 | 0.086 | 0.066 | 0.063 | 0.098 |
| CD@1% | 0.28 | 0.24 | 0.18 | 0.17 | 0.28 |

Figures in boxes are days ± SD ($n = 20$); numbers followed by the same letter are not statistically significant ($p < 0.001$) by LSD

[a]This means eggs from *Conogethes* moths reared on cardamom were implanted on castor

(11–12 days). When "castor larvae" were implanted on castor, larvae took 14–16 days. This is the normal larval period, and larvae did not suffer from mortality.

There were significant differences in larval survival of castor *Conogethes* on cardamom. Larval survival of first instar was 13.8 ± 2.8%, 19.2 ± 4.4% and 18.1 ± 2.7% on Malabar, Mysore and Vazhuka types of cardamom, respectively. Larval survival of fourth instar was 33.83 ± 2.5%, 46.42.9 ± 1.4% and 43.17 ± 2.4%, and fifth instar survival was 54.58 ± 2.9%, 64.42 ± 2.6% and 63.33 ± 3.3%. Further, 15.08 ± 2.8%, 39.25 ± 0.8% and 17.73 ± 2.7% pupation was observed on Malabar, Mysore and Vazhuka types of cardamom, respectively (Table 20.3). Though the first instar larvae survived to an extent of 13.8%, it did not complete life cycle (Fig. 20.1). This maximum larval mortality was recorded in first instar (85.5%), and this is crucially important for the insect to select a host plant.

There were statistical significant differences in developmental time of each instar among the Malabar, Mysore and Vazhuka types of cardamom host plant examined. Third instar required 3.03 ± 0.6 days, 2.44 ± 0.2 days and 2.16 ± 0.3 days,

**Table 20.3** Survival of life stages of castor borer *C. punctiferalis* on cardamom

| Cardamom plant type | Larval instar 1st | 2nd | 3rd | 4th | 5th | Pupation (%) |
|---|---|---|---|---|---|---|
| Malabar | 13.8 ± 2.8 (21.73)[a] | 23.8 ± 3.8 (29.09)[a] | 15.4 ± 2.7 (23.04)[a] | 33.83 ± 2.5 (35.55)[a] | 54.58 ± 2.9 (47.62)[a] | 15.08 ± 2.8 (22.77)[a] |
| Mysore | 19.2 ± 4.4 (25.83)[b] | 29.3 ± 2.69 (32.71)[c] | 29.1 ± 3.0 (32.60)[b] | 46.42 ± 1.4 (42.94)[c] | 64.42 ± 2.6 (53.62)[c] | 39.25 ± 0.8 (38.79)[b] |
| Vazhuka | 18.1 ± 2.7 (25.09)[ab] | 28.3 ± 3.0 (32.12)[bc] | 15.0 ± 2.8 (22.70)[a] | 43.17 ± 2.4 (41.06)[b] | 63.33 ± 3.3 (52.75)[bc] | 17.73 ± 2.7 (24.47)[a] |
| S.Em± | 1.495 | 1.222 | 1.205 | 0.765 | 1.012 | 1.032 |
| CD@1% | 4.78 | 3.19 | 3.85 | 2.42 | 3.23 | 3.30 |

Figures in parentheses are arcsine-transformed values; $n = 20$; numbers followed by the same letter are not statistically significant ($p < 0.001$) by LSD

**Table 20.4** Developmental time of castor borer *C. punctiferalis*[a] life stages on cardamom

| Cardamom plant type | Larval instar (days) 1st | 2nd | 3rd | 4th | 5th |
|---|---|---|---|---|---|
| Malabar | 1.75 ± 0.6[a] | 2.23 ± 0.3[a] | 3.03 ± 0.6[b] | 3.43 ± 0.7b[a] | 3.88 ± 0.8[a] |
| Mysore | 1.50 ± 0.5[a] | 2.51 ± 0.2b[a] | 2.44 ± 0.2[a] | 2.71 ± 0.3[a] | 5.06 ± 0.7[c] |
| Vazhuka | 1.83 ± 0.53[a] | 2.33 ± 0.2[a] | 2.16 ± 0.3[a] | 2.60 ± 0.4[a] | 4.71 ± 0.7[b] |
| S.Em± | 0.133 | 0.073 | 0.101 | 0.120 | 0.176 |
| CD@1% | 0.37 | 0.20 | 0.28 | 0.34 | 0.49 |

Figures in boxes are days ± SD ($n = 20$); numbers followed by same letter are not statistically significant ($p < 0.001$) by LSD
[a]This means eggs from *Conogethes* moths reared on castor, implanted on cardamom

respectively. Fourth instar required 3.43 ± 0.7 days, 2.71 ± 0.3 days and 2.60 ± 0.4 days, respectively, and fifth instar required 3.88 ± 0.8, 5.06 ± 0.7 and 4.71 ± 0.7 days, respectively (Table 20.4).

The reduced larval survival and failure to pupate may be attributed mainly to antibiotic factors. Antixenosis in combination with antibiosis plays a role in reduced survival of the pest. The antibiotic factors in terms of antimetabolites/antidigestible chemical components and antixenosis in terms of the presence of trichome on castor types may impede larval movement on the host, the spines inhibited consumption of the host material. It is interesting to record here that the mortality of neonate larvae recorded was much higher compared with the later instars larvae.

Data on *Conogethes* larval period when reared on cardamom from castor and cardamom itself is presented in dissertation of Doddabasappa 2012. When *Conogethes* reared on castor were implanted on cardamom, larvae suffered mortality. Larvae completed the period early probably because of physiological stress brought out by host plant shift (13–14 days). When "cardamom larvae" were implanted on cardamom, larvae took 20–22 days. This is the normal larval period, and larvae did not suffer from mortality.

#### 20.4.1.2 Castor *Conogethes* on Three Cardamom Types

In this case *Conogethes* larvae were expected to have rough surface, bluish green to green colour, but *Conogethes* larvae met with relatively smooth surface so as to be able to bore into and penetrate into the shoot.

#### 20.4.1.3 Cardamom *Conogethes* on Three Castor Types

In this case, *Conogethes* larvae acquired parrot green colour but came cross roughened surface which impeded larval movements and larvae feeding on the plant. They exhibited restless behaviour, and because of the spines, larvae were unable to sustain continuous feeding and suffered mortality.

Calcagno et al. (2007) showed that *Zea mays* L. and *Artemisia vulgaris* L. are two genetically different plants and *Ostrinia nubilalis* Hub. females preferentially laid eggs on their native host species. Despite slight differences in the overall larval feeding performance, consistent patterns of local adaptation for survival in the two species of *Conogethes* were noticed. In France, *O. nubilalis* feeding on maize are genetically differentiated from sympatric populations feeding on mugwort (*Artemisia vulgaris* L.) and hop (*Humulus lupulus* L.), but these populations with different hosts are morphologically indistinguishable (Bourguet et al. 2000; Martel et al. 2003; Leniaud et al. 2006).

## 20.5 Biophysical Plant Components

The literature pertaining to the effect of biophysical parameters responsible for resistance against *Conogethes* on castor and cardamom is limited. In India, a number of other cultivars of cardamom are also recognized. In general, they can be considered as ecotypes of varieties Mysore, Malabar or Vazhuka. Most common among them are Munjarabad, Bijapura, Kannielam, Makaraelam, Thara and Nadan (Ravindran and Madhusoodanan 2002) in South India.

Tomlinson (1969) and Mercy et al. (1977) carried out anatomical studies on stem, rhizome, leaf sheath and root of cardamom suckers. The stem is solid with a thin subepidermal layer of sclerenchymatous cells separating the inner vascular bundles from the outer ones. Each vascular bundle is surrounded by a prominent bundle sheath of sclerenchyma cells. There are two to three large metaxylem vessels with a few protoxylem vessels. In the parenchyma cells, rhomboidal crystals of calcium oxalate are occasionally found. The fruit is a capsule developed from an inferior ovary. It is more or less three sided with rounded edges. The shape and size vary. In the variety Malabar, the fruits are short and broadly ovoid, and dried fruits are somewhat longitudinally wrinkled. In variety Mysore, the fruits are ovoid to narrowly ellipsoid or elongate; the surface is more or less smooth (Ravindran and Madhusoodanan 2002). These characters can be related to the borer infestation. The plant structure with thick stem, closely packed vascular bundles and compactly placed tissues protects the plant from borer attack.

Lakshminarayana (2005) reported the plant morphological characters of castor capsules associated with borer resistance. Twelve genotypes with different spike characteristics and DCS-9 as standard check were evaluated against *C. punctiferalis*. Genotypes with compact spikes (86.7% mean capsule damage) and big capsules (85.8%) were highly preferred by ovipositing *C. punctiferalis* moths. Moderate preference for genotypes with small and non-spiny capsules (mean capsule damage of 42% and 51.7%, respectively) were observed. Genotypes with loose spikes were the least damaged (18.3%–28.3%). Jayalaxmi (1996) reported that castor plants with spiny and compact spikes carried higher shoot and capsule borer infestation than those with spineless and loose spikes. Aruna was less susceptible to shoot and capsule borer (on an average 13% capsule damage) than SHB-18, GAUCH-1, VHB-150 and JI-35 (14.2–19.4%) genotypes, because of anatomical differences in plants shoot structure.

Lakshminarayana (2005) reported that the morphological characters imparted antixenosis type of resistance against capsule borer, *C. puncliferalis*. The capsule borer infestation was low in the genotypes with loose spikes (RG-1934 and RG-2543) and small-sized (RG-2635) non-spiny capsules (RG-258).

## 20.6 Anatomical Parameters

The three select types of castor (spineless, compact spike and spiny loose spike) and cardamom (Malabar, Mysore and Vazhuka) tender shoots and capsules (<3 months old) were taken for recording observations on biophysical parameters. The same age (15–18 days old) plant parts were utilized for anatomical parameters. The slides were examined under microscope to study the variation of anatomical characteristics (tissue thickness, pericarp thickness, toughness and pilosity) of selected castor types (spineless, compact spike and spiny loose spike) and erect (Mysore), semi erect (Vazhuka) and prostrate (Malabar) types of cardamom.

Anatomical details of castor capsules were examined under microscope to determine if *Conogethes* infestation is related to capsule anatomy (Table 20.5). Seed coat thickness was inversely related with *Conogethes* infestation. Spineless spike with thick seed coat was recorded with 15% capsule damage, whereas compact spiny with thin seed coat showed 26% by *Conogethes* on castor. Spiny loose spike had moderate seed coat showed moderate level of infestation (14%).

Anatomical details of cardamom shoot were examined in relation to *Conogethes* infestation. Shoot infestation by *Conogethes* was less when the number of primordial layers in the shoot was more (11–16), and when primordial layers were less (7–9), the infestation was high. The relationships between shoot infestation and number of primordial layers in the shoot were inversely related (Table 20.6).

Observations on biophysical parameters revealed that cardamom and castor differed in plant morphology and anatomy. The response of *Conogethes* moths and larvae also varied with plant biophysical parameters, and because of these differences, the *Conogethes* larvae exhibited different patterns of relationships with the two plants. If the *Conogethes* were to be the same species, the patterns of relationship of the insect on plant phenophases would be similar, if not the same. From the foregoing results, the following pattern of relationships can be established for *Conogethes* populations breeding on castor and cardamom.

**Table 20.5** Anatomy of castor capsule in relation to *Conogethes* infestation

| Castor types | Castor capsule characters[a] % capsule infestation | Seed coat | Sclerenchymatous tissue | Parenchymatous tissue | Cotyledon with seed coat |
|---|---|---|---|---|---|
| Spineless spike type | 15.68 ± 2.72 | Thick | Well defined | Thick | Compact |
| Compact spiny type | 26.17 ± 2.45 | Thin | Not so distinct | Thin | Loose |
| Spiny loose spike | 14.13 ± 1.05 | Moderate | Not so well defined | Moderate | Moderate |

[a]In the absence of accurate measurements, the thickness of the seed coat was arbitrarily categorized into thick, thin and moderate after examining several sections ($n$ = 15–20)

**Table 20.6** Anatomy of cardamom shoot in relation to *Conogethes* infestation

| Cardamom types | % shoot infestation | Cardamom shoot characters[a] Number of primordial layers | Lysogenous cavities shape | Lysogenous cavities size | Lysogenous cavities number/layer | Distance between two cavities |
|---|---|---|---|---|---|---|
| Malabar | 11.50 ± 1.0 | 8–16 | Circular | Smaller | 15–20 | Wider |
| Mysore | 19.87 ± 1.9 | 7–9 | Rectangular | Bigger | 22–28 | Narrow |
| Vazhuka | 18.12 ± 1.4 | 10–12 | Circular | Bigger | 20–26 | Narrow |

[a]In the absence of accurate measurements, the thickness of the lysogenous cavity was arbitrarily categorized into smaller and bigger after examining several sections ($n$ = 15–20)

***Conogethes* Breeding on Castor** Trichome number/cm$^2$, trichome length (mm), number of capsules/spike, capsule size (cm) and shoot diameter (mm) were related to oviposition and larval feeding.

***Conogethes* Breeding on Cardamom** The number of capsules/panicle, length of panicle (cm), number of tillers per plant, number of panicles/plant and shoot diameter were related to oviposition and larval feeding. Therefore, the observation revealed that different plant morphological and anatomical characteristics were related to *Conogethes* at egg and larval stages. So, *Conogethes* breeding on castor preferred trichome density compactness, while *Conogethes* on cardamom preferred smooth surface and parrot green colour. Hence, *Conogethes* breeding on castor and cardamom differed in their preference for plant morphological and anatomical factors.

Sharma et al. (2008) revealed that plant height, size and maturity of capsule did not affect castor capsule borer damage. In guava, Bangalore Selection 6/10, Shillong-1 and 6/12 had significantly less capsule damage compared to other guava cultivars/genotypes. Coloured genotypes had the lowest castor capsule borer (*Conogethes*) damage. There was no correlation between borer and capsule breadth, capsule length, palatability rating, pulp, TSS, acidity, vitamin C and total sugars.

## 20.7 Behavioural Responses of the Borer to Host Plants

The literature pertaining to the behavioural responses of *Conogethes* moths and larvae to select plant volatiles from castor and cardamom types is limited. On cardamom, moths laid eggs singly on the top of the leaf axils of young pseudostems; rarely two larvae were found in a pseudostem. This kind of egg-laying habit probably is to avoid larval competition for food within the same pseudostem (Thyagaraj 2003). As soon as the egg hatches, the young caterpillar bores at the base of the pedicle and the base (0.3 m above ground level) of the seedlings in the nursery and at the nodal region of the grown-up suckers. Larvae feed on the central tissue and tunnel the shoot causing a dead heart. The live caterpillar indicated its presence by throwing out faecal matter from the entry hole (Clausen 1927; Thyagaraj 2003). Borer larvae fed on capsules from July to November (Thyagaraj 2003). Second instar larvae bored into the capsule fed on the immature seeds leaving empty capsule and then moved on to another capsule. Fully grown-up larvae moved to the pseudostems for pupation (Ironside and Davis 1969; Ram et al. 1997; Sloan 1945; Smith 1937; Wang and Cai 1997). Occasionally larvae bored into the panicles also and resulted in the panicles drying up. Feeding preference test showed that among the cardamom plant parts tested, tender shoots were preferred the most, followed by young capsules (Thyagaraj 2003). Most insect studies deal with laboratory-reared insects, and most of these studies incorporate insects reared on artificial diets. Details on concerning artificial diet for *Conogethes* sp. have been dealt with in another chapter in this book.

Through behavioural tests, EAG responses and multiple-choice oviposition tests, Xu et al. (2001) revealed the effects of volatiles from Nongda No.1 chestnut (NC) and Heyuan oil chestnut (HC) on the host-selection behaviour of adult *C. punctiferalis*. The studies showed that the moth-eaten rate of NC capsules by adult *C. punctiferalis* was 16.1%–25.3%, while that of HC capsules was less than 5%. For both NC and HC, the EAG responses of female and male moths to capsule volatiles were higher than those to leaf volatiles. The number of eggs laid by female moths was much greater on NC capsules than on NC leaves and on HC capsules and leaves. Honda and Matsumoto (1984) reported that the host plant odours play an important role in ovipositional attraction of the capsule feeding type of the yellow peach moth to host plants. Ovipositional responses were induced not only by odours of the young capsule host but also by leaf odours of non-host plants. Ovipositional responses were also induced by leaf odours of host plants, but those were less effective than the capsule odours.

Huang et al. (2000) studied the contact, fumigant toxicities and antifeedant activity of the essential oil of cardamom, *E. cardamomum*, on two stored product insects, namely, *Sitophilus zeamais* Motschulsky and *Tribolium castaneum* (Herbst). It was found that the oil reduced hatching of *T. castaneum* eggs and the subsequent survival rate of the larvae. Adult emergence was also drastically reduced by cardamom oil. However, it significantly reduced all the nutritional indices of the adults of *S. zeamais*. Wei et al. (2003) studied the insect repellent components from castor leaves. Through GC-MS and UV, IR analysis, four components were identified, i.e.

2-hexanal, cyclohexyl formate, 5-methyl hexanal and 4,4-dimethyl hexanal; among them, 2-hexanal was the main insect repellent.

The most significant component of cardamom is the volatile oil with characteristic aroma, described generally as comphory, sweet, aromatic spicy. The cardamom oil has few mono- or sesquiterpenic hydrocarbons and is predominantly made up of oxygenated compounds. While many of the identified compounds – alcohols, esters and aldehydes – are commonly found in many spice oils, the dominance of the ether, 1, 8-cineole and the esters, α-terpinyl and linalyl acetates in the composition, make the cardamom volatiles a unique combination. The aromatic differences in different sources of cardamom are attributed to the proportion of the esters and 1, 8-cineole (Wijesekar and Jayawardena 1973; Korikanthimath et al. 1997).

Jyothi et al. (1996) reported that the EAGs elicited by aldehydes were greater than the corresponding alcohols in *C. punctiferalis* and *Dysgonia algira* L. The absolute EAG responses of all the volatiles tested were comparatively lower in *C. punctiferalis* than in the noctuid, *Dysgonia algira* L.

Honda et al. (1988) investigated the attraction and ovipositional responses of the gravid yellow peach moths, *C. punctiferalis* on capsules of codling, a host plant. The attractancy was removed from a fresh codling capsule by dipping it in ether, but higher attractancy reappeared subsequently when the capsule was inoculated with fungi. In contrast, a rotten codling or rice cake, infected with bacteria rather than fungi, inhibited oviposition by the moths. Among four phytopathogenic fungi tested, *Endothia parasitica* and *Alternaria solani* were attractive. In a field cage, the moths oviposited significantly more on the oviposition substrates baited with *Penicillium* sp. than on a fresh codling or unbaited control (Honda et al. 1988).

Honda et al. (1985) reported that the EAG responses to 90 n-alkyl compounds with different functional groups and select homologues were compared between the two types of the yellow peach moth, *C. punctiferalis*, fruit-feeding type (FFT) and Pinaceae-feeding type (PFT). Females of both types responded clearly to C5 or C6 compounds in all three series of alkanol, alkanal and alkanoic acid, while males showed prominent bimodal response with C5 or C6 and C9 compounds in aldehyde. The antennae of both sexes in each type more significantly responded to n-hexanal than to the corresponding alcohol and acid. The FFT antennae were twice sensitive to n-hexanol and n-hexanal and 1.5 times more to hexanoic acid than the males the PFT antennae. Females of both types were more sensitive to n-hexanoic acid than the males, particularly in FFT, but no significant difference was observed with n-hexanol and n-hexanal. The EAG responses of both types to ester homologues were unimodal and bimodal in female and male, respectively, similar to those aldehyde homologues. From these differences in EAG responses and those reported earlier, it was concluded that the two types of *Conogethes* populations on peach and the pines are separate species.

Honda (1986) reported that the EAG responses to 21 monoterpene compounds were compared between the 2 types of the yellow peach moth, *C. punctiferalis*. The fruit-feeding (FFT) females responded more favourably to 17 compounds than males, whereas no sexual difference was found in the 2 feeding types of *Conogethes*. Both sexes of PFT responded to alcohols, aldehydes and ketones significantly higher than that of FFT moths. The cluster analysis of EAG responses of each type showed a definite difference in the antennal olfactory spectra between FFT and PFT. From these facts, it was concluded that FFT and PFT are probably taxonomically different species.

Morphology of males and females between these species is slightly different, but some morphometric comparisons are possible in order to segregate both species. Although *C. punctiferalis* larvae were typically polyphagous, their development was delayed on *C. pinicolalis* host, while *C. pinicolalis* lay eggs and feed exclusively on conifer needles including *Pinus parviflora* or *Abies homolepis*. Such host plant preferences may also cause reproductive isolation between the two species. In electroantennogram (EAG), *C. punctiferalis* females responded to 17 more compounds of host plants more than males, whereas there was no sexual difference in *C. pinicolalis*. A cluster analysis of the EAG responses of each species showed a definite difference in the antennal olfactory spectra between *C. punctiferalis* and *C. pinicolalis*. Although the male moths of both species were cross-attracted to calling females and their pheromone gland extracts, a strong homogamic mating preference between both species was confirmed in laboratory test and postmating reproductive isolation by laboratory cross-tests. Female sex pheromone system of *C. punctiferalis* and *C. pinicolalis* are quite similar, which allows cross-attraction by males, and consists of E-10-hexadecenal (E10-16:Ald) and Z-10-hexadecenal (Z10-16:Ald) at 95.4:4.5. The final conspecific sexual recognition in each species is accomplished with a male pheromone. E-2-methyl-2-butenoic acid was identified from hairpencil organs of *C. punctiferalis*, but no pheromonal volatiles from *C. pinicolalis* hairpencils were identified. Recently, two hydrocarbons were found as pheromonal synergists in female pheromone system of *C. punctiferalis*, which functioned at a short distance from pheromone source by calling females. A similar system was also prospected in *C. pinicolalis*, but these new hydrocarbon synergists may have no contribution to reproductive isolation between *C. punctiferalis* and *C. pinicolalis* (Honda 2013).

## 20.8 Larval Behaviour

Separate sets of plastic boxes (0.3 m diameter) were maintained with tender shoots and capsules (<3 months old) of castor and cardamom host plant types. The freshly emerged neonate *Conogethes* larvae were released at the centre of the plastic box (0.3 m) where the fresh plant part pieces (5 cm length) were placed on a spread of moist cotton covered on top with filter paper of same size. The box was placed in a wire mesh insect cage (0.3 m) with zero-watt red bulbs and covered with black paper on all other sides. The larvae were starved for 45 min before release. The movements of the larvae were monitored for 15 min. The larvae were released at different distances from the stimulus to detect as taxis or kinesis. If the larvae made random movements, the behaviour was designated as kinesis.

The behaviour exhibited by *C. punctiferalis* larvae to the host stimulus is summarized in Table 20.7. *Conogethes* larvae reared on castor perceived the host plant

**Table 20.7** Effective distance for perception of stimulus source in castor borer *C. punctiferalis* larvae

| Instar[a] | Distance from the stimulus source (cm) ($n = 10$) | Behavioural responses |
|---|---|---|
| First | 0.2–0.3 | Showed taxis and palpated on the stimulus within 1 s |
| First | 1–1.5 | Larvae can perceive food source |
| Second | 0.5–0.8 | Showed taxis and palpated on the stimulus within 1 s. The larvae perceived the host within 30 s |
| Third | 2.0 | Larvae perceives host from 2 cm distance |
| Fourth | 1.5–2.2 | Showed taxis and palpated on the stimulus within 1 s. The larvae perceived the host within 30 s |
| Fifth | 3.0–3.4 | -do- |

[a]Each test was replicated ten times; larvae starved for 2 h before the test

part at 0.2 cm–2.0 cm from the source. As the age of the larvae increased, the distance also proportionately increased for orientation and perception. This is obviously because the antennal olfactory sense organs in early instar larvae will not be as developed as in the later instar larvae. The effective distance was arrived at after a number of trial and error tests. After the larvae showed taxis, the test was repeated five times to confirm the larval behavioural responses. At effective distance, larvae palpated at the stimulus source and tended to continue feeding without any further movements.

*Conogethes* larvae reared on cardamom showed a wide variation in perception of host plant part. For instance, the effective distance for variation and perfection for first instar varied from 0.3 cm to 0.5 cm. The cardamom *Conogethes* larvae perceived the stimulus from a greater distance than castor *Conogethes* larvae. This may be mostly related to the size of the larvae. The behavioural event associated with the perception of the host was raising the anterior part of the body, legs and head. As the age of larvae increased, the distance from which the larvae could detect the stimulus also increased (Table 20.8).

## 20.9 Orientation and Biting Responses

**Castor *Conogethes* Larvae to Three Castor Types** The percentage of larvae responding to the host stimulus at first instar varied from 10% to 19%. Maximum per cent (19%) of first instar larvae oriented towards compact spiny type castor and minimum per cent (10%) towards spineless spike type. The trend in the second, third, fourth and fifth instars larvae was the same, i.e. maximum larvae orientation towards compact spiny type of castor and the minimum larvae orientation towards the spineless spike type. The three types of castor capsule also sustained biting response where the larvae sustained continued feeding, produced faecal pellets and later instars larvae spun the silken thread (Doddabasappa 2012).

**Table 20.8** Effective distance for perception of stimulus source in cardamom borer *C. sahyadriensis* larvae

| Instar[a] | Distance from the stimulus source (cm) (*n* = 10) | Behavioural responses |
|---|---|---|
| First | 0.3–0.5 | Showed taxis and palpated on the stimulus within 1 s |
| Second | 1.2–1.5 | Exhibited taxis, reached stimulus within 20 s |
| Third | 1.8–2.1 | Exhibited kinesis, intermittently raising legs, anterior part of the body and head |
| Fourth | 3.0–3.2 | Initially larvae made random movements, intermittently raising anterior part of the body, legs and head, sidewards |
| Fifth | 3.5 | Within 25 s the larva reached the stimulus exhibited only directed movements (taxis) |

[a]Larvae were starved for 2 h before the tests; fresh larva was used for every test; each test was replicated ten times

**Castor *Conogethes* Larvae to Three Cardamom Types** The percentage of first instar larvae oriented towards castor was proportionately lower compared with later instar larvae. It is at the first instar that the insect critically discriminate between plants. It may be a shift from one plant to another or from a host to a non-host plant and or vice versa. On Mysore type cardamom a maximum of 47% fifth instar larvae oriented towards castor followed by 30% and 28% towards Malabar and Vazhuka types, respectively, and other instars larvae were in between (Doddabasappa 2012).

**Cardamom *Conogethes* Larvae to Three Cardamom Types** The percentage of larvae responding to the host stimulus at first instar varied from 12% to 18%. Maximum per cent (18%) of first instar larvae oriented towards Mysore type castor and minimum per cent (12%) towards Malabar type. The trend in the second, third, fourth and fifth instars larvae was the same: maximum larvae orienting towards Mysore type of cardamom and the minimum towards the Malabar type. The three types of cardamom shoot also sustained biting response where the larvae sustained continued feeding, produced faecal pellets and later instars larvae spun the silken thread (Doddabasappa 2012).

**Cardamom *Conogethes* Larvae to Three Castor Types** The percentage of first instar larvae orienting towards castor was proportionately lower compared with later instar larvae. It is at the first instar that the insect critically discriminated between plants. It may be a shift from one plant to another or from a host to a non-host plant and or vice versa. On compact spiny type castor maximum of 36% fifth instar larvae oriented towards castor followed by 27% and 15% by spineless spike and spiny loose type, respectively, and other instars larvae were in between.

When non-host plant like *Ocimum* was offered to these *Conogethes* larvae, none of the larvae oriented and none of them showed biting responses. This observation

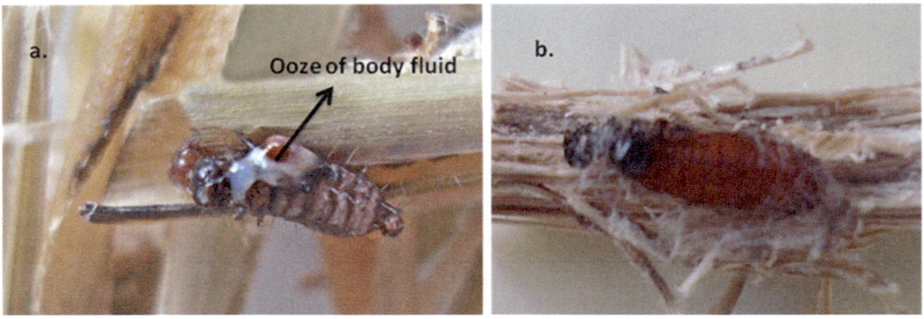


**Fig. 20.2** Cartography. (**a**) Mortality of second instar larva and (**b**) malformed pupa (premature pupa) from third instar *Conogethes* larvae reared on castor, implanted on cardamom

suggested that on the same host plant higher number of larvae oriented towards and moderate number of larvae oriented towards plants within the host range of *Conogethes* species.

The new food plant (cardamom for castor larvae and vice versa) eventually did not sustain continuous feeding and disruption of integument and oozing of body fluid leading to death of larvae was recorded (Fig. 20.2). The larvae unable to feed on secreted silken thread and tried to undergo pupation, though it was unsuccessful. Thus, studies on host shift provided evidence that cardamom is not a suitable plant for castor *Conogethes* and for cardamom *Conogethes* population, castor is not a suitable plant. As reiterated, no such studies on host plant relationships on *Conogethes* have been conducted before. However, studies on host shifts on other species are available. When non-host plant like *Ocimum* was offered to these *Conogethes* larvae, none of the larvae oriented and none of them showed biting responses.

Consideration of the sequential behavioural steps in host selection raises a number of issues that have consequences for host specificity testing. Much of the progress in applying the concepts of insect behaviour to host specificity testing has been made by examining this process (Wapshere 1989; Marohasy 1998; Chakravarthy et al. 2009). Possibly the most important consequences are those that stem from the absence of early steps in the host selection sequence in experimental arenas.

Behavioural changes in larvae (Table 20.9) were also associated with the host shift. *Conogethes* larvae when host shifted did not exhibit normal orientation and feeding behaviour and exhibited random movements, i.e. restless behaviour, secretion of silken threads for induction of pupation, very less feeding, very frequent movements, abnormal growth, oozing, death and drastic time budgeting changes. Under normal conditions the sequential events, viz., orientation and feeding behaviour, i.e. orientation, direct movement (taxis), palpation, nibbling (initial bite), biting, feeding, continuous feeding, defecate and pupation occur (see Flowchart 19.1). These sequential events point to the fact that host plant relationships are specific to each insect and plant and they are different and independent of each other.

**Table 20.9** Behaviour attributes of *Conogethes* larvae on host and on shifted plant

| Attribute | Host plant | Shifted plant |
|---|---|---|
| Movement | Direct (taxis) | Random (kinesis) |
| Time budgeting | Less for movements, more for feeding | Negligible or very less for feeding |
| Energetics | Normal growth and development | Abnormal, highly reduced |
| Establishment on plant | Occurs | Does not occur |
| As food source | Suitable | Unsuitable |
| Feeding behaviour of larvae | Normal | Deviate from the normal |
| Ovipositional behaviour of moths | Normal | Deviate from the normal |

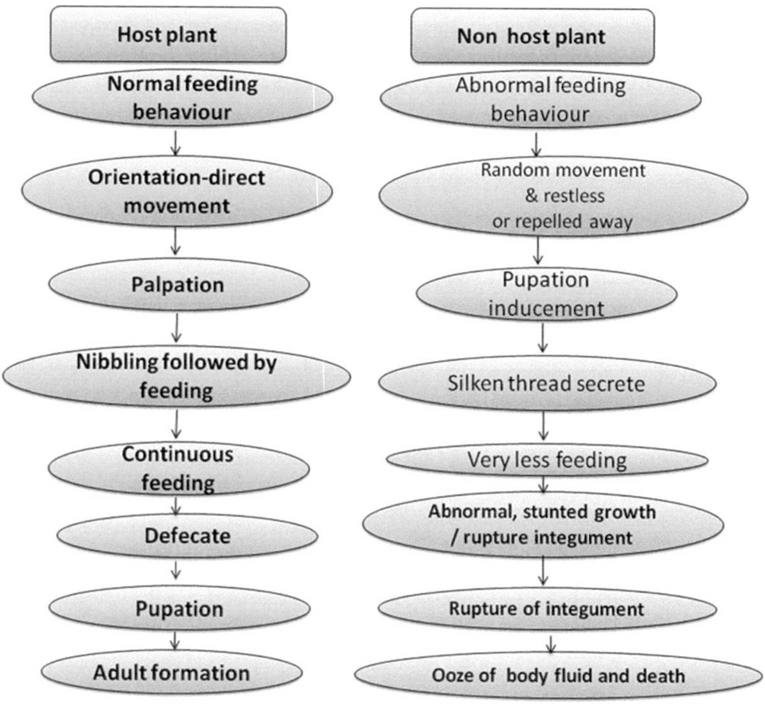


**Flow chart 19.1** Showing the sequence of events in *Conogethes* larval feeding on a host plant and non-host plant

## 20.10 Bioassay: Adults

### 20.10.1 Ovipositional Preference of *Conogethes* Moths on Castor and Cardamom

In order to find out the probable effect of resistance factors on the ovipositional preference of castor and cardamom *C. punctiferalis* moths, two experiments were designed in both "free-choice" and "no-choice" conditions. The objective of this test was to determine if any significant differences existed between two crops and three types of castor and cardamom in terms of attracting or deterring moths of *Conogethes* reared on them at the time of oviposition. Two categories of tests were conducted, i.e. "free choice" and "no choice".

The ovipositional preferences of *Conogethes* moths on their respective plants and on the other plants are not shown in Table 20.10. When castor *Conogethes* moths were subjected to "no-choice" test, it was observed that the moths laid maximum number of eggs (59) on the compact spiny type inflorescence of castor compared to 22 eggs under free-choice conditions in the laboratory. When cardamom *Conogethes* moths were subjected to "no-choice" test, it revealed that the moths laid maximum number of eggs (66) on Mysore type of cardamom compared to 32 eggs under free-choice conditions in laboratory.

When cardamom *Conogethes* moths were subjected to "no-choice" test, it was seen that the moths laid maximum number of eggs (27) on single type castor inflorescence compared to 14 eggs under "free-choice" conditions in laboratory. When castor *Conogethes* moths were subjected to "no-choice" test, it was seen that the moths laid maximum number of eggs (33) on single cardamom type compared to 12 eggs under free-choice conditions in laboratory.

Wind tunnel responses for the natural castor oil and ricinoleic acid showed a complete range of behavioural steps up to source contact in only 3 of the 15 males tested. The solvent, dichloromethane (DCM), which served as control did not elicit any response from the moths. There were statistically significant differences among the treatments at 5% level of significance by chi-square test ($p < 0.01$ at 2 df).

Female *Conogethes* moths showed the complete range of behavioural steps up to source contact in only 3 out of 15 moths compared with the other treatment. There

**Table 20.10** Ovipositional preference of *Conogethes* moths on castor and cardamom

| Test insect | Plant type | £No-choice test | # Free-choice test | *t*-value |
|---|---|---|---|---|
| Castor *Conogethes* | Spineless spike | 22 ± 1.0a | 16 ± 3.0 | 8.49** |
| | Compact spiny | 59 ± 5.0 | 25 ± 6.0 | 7.34** |
| Cardamom *Conogethes* | Malabar | 46 ± 4.0 | 30 ± 2.0 | 9.08** |
| | Mysore | 66 ± 1.0 | 32 ± 1.0 | 6.50** |
| Castor *Conogethes* | Cardamom types | 27 ± 3.0 | 14 ± 3.0 | 5.23** |
| Cardamom *Conogethes* | Castor types | 33 ± 3.0 | 12 ± 4.0 | 6.81* |

df, 1; **, significant at 1%; *, significant at 5% ; NS, non-significant; £, means only single of plant type; #, means combining both the plant types; a, mean of four replications were statistically significant differences among the treatments at 5% level of significance by $\chi^2$ test ($p < 0.01$ at 2 df).

In this bioassay, the cardamom oil and α-terpinyl acetate showed the complete range of behavioural steps up to source contact in only 3 of the 15 male moths tested. The solvent (DCM) which served as control did not elicit response from the moths. There were statistically significant differences among the treatments at 5% level of significance by $\chi^2$ test ($p < 0.01$ at 2 df).

Female *Conogethes* moths showed a complete range of behavioural steps up to source contact in only 4 of the 15 female moths compared to the other treatments. There were statistically significant differences among the treatments at 5% level of significance by $\chi^2$ test ($p < 0.01$ at 2 df). Cardamom *Conogethes* moths reared on castor to select plant volatiles and castor *Conogethes* moths reared on cardamom to select plant volatile showed complete range of behavioural responses up to source contact to natural castor oil and ricinoleic acid for both male and female *Conogethes* moths.

### 20.10.2 Antibiosis Mechanism

The literature pertaining to the antibiosis mechanism in the *Conogethes* populations infesting castor and cardamom types is limited. In order to identify the possible role of plant's nutritional elements in conferring resistance, estimation and comparison of polyphenols, soluble sugars, proteins and silica in two crops was conducted at seedling stage of the crop. Host plant shoot and capsule of the same age free of *Conogethes* larvae and which were heavily infested with *Conogethes* were analysed for their biochemical profiles. The damaging stage of the shoots, capsules and seeds of three types of cardamom (1 year old) and castor (<100 days old) were collected. These samples were put in an oven at 40 °C for 4 days to get dry matter and then were removed from the oven for conducting the studies.

The results revealed that in spineless spiny type castor, the phenol, silica, soluble sugars and proteins showed significantly negative and positive correlation at $p < 0.01$ with the plant infestation showing *r*-values of −0.884, −0.866, 0.769 and 0.800, respectively (Table 20.11). The results revealed that in compact spiny type castor, the phenol, silica, soluble sugars and protein showed significantly negative and positive correlation at $p < 0.01$ with the plant infestation showing *r*-values of −0.964, −0.878, 0.852 and 0.855, respectively. The results revealed that in spiny loose type castor, the phenol, silica, soluble sugars and proteins showed significantly negative and positive correlation at $p < 0.01$ with the plant infestation showing r-values of −0.851, −0.869, 0.860 and 0.848, respectively.

The results revealed that in Malabar type cardamom the phenol, silica, soluble sugars and protein showed significantly negative and positive correlation at $p < 0.01$ with the plant infestation showing *r*-values of −0.894, −0.893, 0.951 and 0.895, respectively (Table 20.11). The results revealed that in Mysore type cardamom phenol, silica, soluble sugars and protein showed significantly negative and correlation

**Table 20.11** Correlation coefficient between capsule infestations by *Conogethes* in castor and cardamom and biochemical constituents

| | Correlation coefficient (*r*-value) | | | | | |
|---|---|---|---|---|---|---|
| Biochemical constituents | Spineless spike type | Compact spiny type | Spiny loose spiny | Malabar | Mysore | Vazhuka |
| % infestation | 15.68 ± 68 | 26.17 ± 2.45 | 14.13 ± 1.05 | 8.32 ± 1.2 | 18.78 ± 1.0 | 11.20 ± 2.5 |
| Phenol | −0.884** | −0.964** | −0.851** | −0.894** | −0.926** | −0.599* |
| Soluble sugars | 0.769** | 0.852** | 0.860** | 0.951** | 0.879** | 0.548* |
| Protein | 0.800** | 0.855** | 0.848** | 0.895** | 0.871** | 0.596* |
| Silica | −0.866** | −0.878** | −0.869** | −0.893** | −0.868** | −0.646* |

*, significant at 5%; **, significant at 1%

at $p < 0.01$ with the plant infestation showing $r$-values of −0.926, −0.868, 0.879 and 0.871, respectively. The results revealed that in Vazhuka type cardamom, phenol, silica, soluble sugars and proteins showed significantly negative and positive correlation at $p < 0.05$ with the plant infestation showing $r$-values of −0.599, −0.646, 0.548 and 0.596, respectively.

The soluble sugars and proteins estimated were positively correlated with *Conogethes* larval infestation on capsule. That is to say that sugars and proteins helped the borer in utilizing the plant better and rendered it susceptible. Phenol and silica were negatively associated with borer infestation. Therefore, phenol and silica could enhance or induce resistance or tolerance to the plant against borer infestation. The results were consistent across three types of castor, namely, spineless spiny, compact spiny and spiny loose spike, and cardamom types, namely, Malabar, Mysore and Vazhuka types.

Antibiosis is the resistance mechanism that operates after the insects have colonized and started utilizing the plant. When an insect feeds on an antibiotic plant, its growth, development, reproduction and survival are affected. The antibiotic effects may result in decline in insect size or weight, reduction in metabolic processes, increase in restlessness and greater larval or pre-adult mortality (Panda and Kush 1995). In certain cases, it is difficult to distinguish between antixenotic and antibiotic mechanisms of resistance, and they often overlap between the morphological and biochemical bases of resistance.

Sooraj (2005) reported 32 different cardamom germplasm accessions biochemical parameters like moisture, oil content, protein, amino acid, carbohydrates, starch, reducing sugar and phenolics. Cardamom accessions APG-174, APG-81 and APG-173 were considered superior because of their high volatile oil distribution and increase in other biochemical constituents.

Cardamom capsules are valued commercially for their volatile oil. The major constituents of oil are 1, 8-cineole and alpha-terpinyl acetate. Based on high oil recovery and the desirable trait of oil terpinyl acetate, APG 273, 277 and 257 were identified as the best accessions. Total carbohydrate and starch content indicated that high-quality cardamom accessions contained relatively high starch, carbohydrate and GLC separated components (Helna 2005).

Keezheveettil et al. (2010) reported the relative levels of antioxidant activity, total flavonoid content, total phenolic content and reducing power of different organic and aqueous extracts with hexane sequentially extracted, dichloromethane, ethyl acetate, methanol and water in four different varieties of cardamom, namely, Mysore, Malabar, Vazhuka and Guatemala. Ethyl acetate extract of all varieties showed greater activity. Based on the results Malabar variety was identified as the best source of antioxidant compounds. Chemical compositions of the essential oil of seed powder of these varieties were studied by GC and GC-MS. The main constituents identified were terpinyl acetate ranging between 61.65% and 68.19% followed by cineol (7.23–11.76%).

### 20.10.3 Host Plant Suitability

Field-collected *C. punctiferalis* larvae/pupae were reared on castor, *R. communis*, using plastic tray (15 × 25 × 35 cm). Ten pairs of moths were sexed and released in ovipositional cages (76 × 36 × 36 cm). Cotton pads soaked in 10% sugar solution served as adult food. Castor inflorescence (panicle) with flowers and immature capsules were kept as ovipositional substrate with the cut ends dipped in water in conical flask (250 ml) and shoots of each cardamom type.

Newly hatched first instar larvae were transferred to the capsules of each castor type, namely, spineless, compact spike, spiny loose spike and shoots of each cardamom type, namely, Malabar, Mysore and Vazhuka as food in trays (15 × 25 × 35 cm) (Fig. 20.1). Fresh capsules were given as food once in 4 days. Spikes of each castor type were replicated three times. Observations were recorded on the weight of food consumed by the larval stage, larval and pupal weights and moth emergence.

For the detailed data on host plant suitability of the third instar *Conogethes* larvae were reared on three select types of castor in laboratory by Doddabasappa (2012). On spineless type the weight of plant material consumed was the least (72 mg) compared to 95 mg and 125 mg on spiny loose spike and compact spiny types, respectively. The same trend was obtained with respect to larval and pupal weights. The weights of plant material consumed reflect its nutritional suitability for the insect. Therefore, the compact spiny type was the most preferred castor type and spiny loose spike was the least preferred and spineless spike type was in between.

On Malabar type the weight of plant material consumed was the least (275 mg) compared to 340 mg and 450 mg on Vazhuka and Mysore types, respectively. The same trend was obtained with respect to larval and pupal weights. The weights of plant material consumed reflect its nutritional suitability for the insect. Thus, the Mysore type of cardamom was the most preferred and Malabar type was the least preferred and Vazhuka type was in between.

Host plant suitability for host shift studies of the third instar cardamom *Conogethes* larvae on three select castor types in laboratory is presented in Table 20.12. On the spiny loose type, the weight of plant material consumed was the least (38 mg) compared with 40 mg and 75 mg on spineless spike and compact spiny type, respectively. The same trend was obtained with respect to larval and pupal weights (malformed). The weights of plant material consumed reflect the nutritional suitability for the insect. So, the compact spike type was the most preferred castor type and spiny loose type was the least preferred and spineless spike type was in between.

Host plant suitability for host shift studies on the third instar castor *Conogethes* larvae on three select types of cardamom in laboratory is presented in Table 20.12. In contrast to castor types, *Conogethes* equally preferred the cardamom types as the amount of plant material consumed was more or less the same. On Malabar, the weight of plant material consumed was the least (68 mg) compared to 80 mg and 81 mg on Vazhuka and Mysore types, respectively. The same trend was obtained with respect to larval and pupal weights (malformed). The weights of plant material

**Table 20.12** Host plant suitability of third instar cardamom *Conogethes* larvae reared on castor plant types and vice versa in laboratory*

| Plant type | Weight of plant material consumed (mg) | Weight (mg) | | Moth emergence | Remarks |
|---|---|---|---|---|---|
| | | Larvae | Pupa | | |
| Spineless spike type | 40[b] | 30[c] | 35[b] | – | Malformed pupae |
| Compact spiny type | 75[a] | 58[a] | 41[a] | – | Malformed pupae |
| Spiny loose spike | 38[b] | 42[b] | 38[ab] | – | Malformed pupae |
| Malabar | 68[b] | 55[b] | 45[a] | – | Malformed pupae |
| Mysore | 81[a] | 71[a] | 48[a] | – | Malformed pupae |
| Vazhuka | 80[a] | 42[c] | 30[b] | – | Malformed pupae |

*Observation recorded 48 h after release of larvae; plant material offered ad libitum

consumed reflect its nutritional suitability for the insect. Thus, the Mysore type was the most preferred cardamom type and Malabar was the least preferred and Vazhuka type was in between.

When the larvae originating from cardamom were fed with castor, there was substantial reduction in the amount of plant material consumed and vice versa. This suggested that plant material, like castor and cardamom, were unsuitable as food material for *Conogethes* reared on either of the two host plants. Consequent to the lower food plant material consumed, there were substantial reduction in larval and pupal weights. This could lead to the lower moth emergence and the emerged moths would be less fecund or infecund.

### 20.10.4 Tolerance Mechanism

Egg incubation period (in days) was the least (2.67 days) on spineless spike castor as also the larval period (26.68 days) compared to compact spiny and spiny loose castor types. The trend in the egg incubation, larval and pupal periods was consistent across three castor types. *Conogethes* larvae preferred compact spiny castor type for feeding over other two types.

*Conogethes* larvae and pupae gained more weight on compact spiny castor type than on the spineless and spiny loose castor types. This is obvious because compact spiny type was more suitable to *Conogethes* than the other two castor types. There were statistically significant differences between the spineless spike and spiny loose spike types and the compact spiny types at 1% ($p < 0.001$) level of significance. Female pupae weighed more than the male. Egg incubation period (in days) was the least (2.68 days) on Vazhuka, as also the larval period (29.40 days) compared to Malabar and Mysore cardamom types. The trend in the egg incubation, larval and pupal periods was consistent across three cardamom types. The observations recorded showed that on Malabar type, the insect completed the life cycle earlier than on the Mysore and Vazhuka types. Studies on the biology of *Conogethes* on castor and cardamom revealed that the insect completed the life cycle on the three types of the castor and the cardamom.

The insect required longer period (in days) of time to complete the life cycle on cardamom than on castor, i.e. on an average 34.5 days on castor compared to 39 days on cardamom. Weight of the life stages of *Conogethes* revealed the susceptibility and the amount of the plant material consumed by the pest through different life stages. *Conogethes* larvae and pupae gained more weight on compact spiny castor type than on the spineless and spiny loose castor types. This is obvious because compact spiny type was more suitable to *Conogethes* than the other two castor types. There were statistically significant differences between the spineless spike and spiny loose spike types and the compact spiny type at 1% ($p < 0.05$) level of significance.

## References

Bilapate GG (1977) Studies on the life tables of *Dichocrocis punctiferalis* Guenee on castor. J Maharashtra Agril Univ 2:246–248

Bilapate GG, Talati GM (1977) Field observations on the incidence of *Dichocrocis punctiferalis* Gn. on castor. Madras Agril J 64:407–408

Bilapate GG, Talati GM (1978) Some studies on bionomics of castor shoot and capsule borer *Dichocrocis punctiferalis* (Guenee). J Maharashtra Agril Univ 3:47–49

Bourguet D, Bethenod MT, Trouve C, Viard F (2000) Host-plant diversity of the European corn borer *Ostrinia nubilalis*: what value for sustainable transgenic insecticidal Bt maize? Proc R Soc Lond B 267:1177–1184

Brower AVZ (1997) The evolution of ecologically important characters in Helicnius butterflies (Lepidoptera: Nymphalidae): a cladistic reiew. Zool J Linnean Soc 119:457–472

Calcagno V, Thomas Y, Bourguet D (2007) Sympatric host races of the European corn borer: adaptation to host plants and hybrid performance. J Evol Biol 20:1720–1729

Chai XM, He WH (1987) The *Dichorcis punctiferalis* Guenée in masson pine. Entomol Knowl 24:99–100

Chakravarthy AK, Honda H, Thyagaraj NE (1991) Comparison of containers for larval rearing in stalk and fruit feeding type of *Conogethes punctiferalis* (Guenee) (Lepidoptera: Pyralidae). Placrosym IX:127–131

Chakravarthy AK, Basavaraju BS, Gangappa E (2009) Orientational responses of shoot and fruit borer larvae, *Conogethes punctiferalis* Guenee (Pyralidae: Lepidoptera) towards selected plants in laboratory. In: Shakunthala S, Nagachaitanya B, Chakravarthy AK, Prabhakara Shetty TK (eds) Recent trends in animal behavior. New India publishing Agency, New Delhi, pp 191–200

Clausen CP (1927) The citrus insects in Japan. Technol Bull USDA 15:15

Devasahayam S, Jacob TK, Abdulla Koya KM, Sasikumar B (2010) Screening of ginger *(Zingiber officinale)* germplasm for resistance to shoot borer (*Conogethes punctiferalis*). J Med Aromat Plant Sci 32:137–138

Doddabasappa B (2012) Factors associated with resistance mechanism with shoot and fruit borer Conogethes punctiferalis Guenee (Lepidoptera, Crambidae) population infesting caster and cardamom. Ph.D dissertation, UAS, GKVK, Bangalore

Helna K (2005) Biochemical characterization of cardamom (*Elettaria cardamomum*) accessions and database of chemical compounds and metabolic pathways in cardamom volatile oil. M.Sc. thesis, Bharathiar University, Coimbatore, Tamil Nadu

Honda H (1986) Post-mating reproductive isolation between fruit- and Pinaceae-feeding types of the yellow peach moth, *Conogethes punctiferalis* (Guenée) (Lepidoptera: Pyralidae). Appl Entomol Zool 21:489–491

Honda H (2013) Bio systematics of Japanese *Conogethes* spp. With special reference to host plant preference and reproductive isolation. In: Chakravarthy AK, Ashok Kumar CT, Varghese A, Thyagaraj NE (eds) New horizons in insect science. Abstracts ICIS-2013. Springer, Bangalore, p 136

Honda H, Matsumoto Y (1984) Oviposition responses of the fruit-feeding type of yellow peach moth, *Conogethes punctiferalis* Guenée (Lepidoptera: Pyralidae) to host-plant odors. Jpn J Appl Entomol Zool 28:82–86

Honda H, Matsumoto Y (1987) Larval feeding responses of the fruit- and pinaceae- feeding type of the yellow peach moth, *Conogethes punctiferalis* (Guenée) (Lepidoptera: Pyralidae) to the host-plant extracts and their sugar components. Jpn J Appl Entomol Zool 31:28–35

Honda H, Matsumoto Y, Matsumoto Y (1985) Comparison in EAG response to n-Alkyl compounds between the fruit and pinaceae feeding type of yellow peach moth, *Conogethes punctiferalis* (Guenee) (Lepidoptera: Pyralidae). Appl Ent Zool 21(1):126–133

Honda H, Maruyama Y, Matsumoto Y (1986) Comparisons in EAG response to n-alkyl compounds between the fruit- and Pinaceae-feeding type of yellow peach moth, *Conogethes punctiferalis* (Guenée) (Lepidoptera: Pyralidae). Appl Entomol Zool 21:126–133

Honda H, Ishiwatari T, Matsumoto Y (1988) Fungal volatiles as oviposition attractants for the yellow peach moth, *Conogethes punctiferalis* (Guenée) (Lepidoptera: Pyralidae). J Insect Physiol 34:205–209

Huang Y, Lam SL, Ho SH (2000) Bioactivities of essential oil from *Elletaria cardamomum* (L.) Maton. to *Sitophilus zeamais* Motschulsky and *Tribolium castaneum* (Herbst). J Stored Produ Res 36:107–117

Inoue H, Yamanaka H (2006) Redescription of *Conogethes punctiferalis* (Guenée) and descriptions of two new closely allied species from Eastern Palaearctic and Oriental Regions (Pyralidae, Pyraustinae). Tinea 19:80–91

Ironside DA, Davis JJ (1969) Major Macadamia pests. Queensland Agric J 95:468–471

Jacob SA (1981) Biology of *Dichocrocis punctiferalis* (Guenee.) on turmeric. J Plantn Crops 9:119–123

Jayalaxmi NH (1996) Screening of Castor (*Ricinus communis* L.) germplasm to major insect pests, M.Sc. thesis, University of Agricultural Sciences, Bangalore, India, p 167

Jayalaxmi NH, Chakravarthy AK, Gangiger PC (2009) Screening of Castor germplasm against leafminer, Semilooper and shoot and capsule borer in South Karnataka. Curr Biotica 3(3):386–396

Jyothi KN, Prasuna AL, Sighamony S, Prasad AR, Yadav JS (1996) Comparative electroantennogram study of *Parallelia algira* (Lepidoptera: Noctuidae) and *Conogethes punctiferalis* (Lepidoptera: Pyralidae) to general plant volatiles. J Entomol Res 20:189–195

Keezheveettil PAPA, Maheswari PR, Indu S, Nisha VNP (2010) Chemical composition, flavonoid – phenolic contents and radical scavenging activity of four major varieties of cardamom. Int J Biol Med Res 1:20–24

Koizumi K (1963) Yellow peach moth, *Dichocrocis punctiferalis* (Guenée). Kobe Plant Protection News 323:58

Kondo T, Miyahara T (1930) Fruit borer and a summary of their life histories in Knangtung. China J Plant Prot 17:85–94

Konno Y, Honda H, Matsumoto Y (1981) Mechanisms of reproductive isolation between the fruit-feeding and the Pinaceae-feeding types of the yellow peach moth, *Dichocrocis punctiferalis* Guenée (Lepidoptera: Pyralidae). Jpn J Appl Entomol Zool 25:253–258

Korikanthimath VS, Ravindra M, Hosmai MM (1997) Preliminary evaluation of elite clones of cardamom (*Elettaria cardamomum* Maton) for yield parameters in clonal nursery. J Spices Aromat Crops 6:34–41

Krishnamurthy KK, Khan MM, Avadhani KK, Venkatesh J, Siddaramaiah AL, Chakravarthy AK, Gurumurthy BR (1989) Three decades of cardamom research, Regional Research Station, Mudigere (1958–1988). Station Tech Bull, pp 44–48

Kuang MH, Liu SW, Ji BZ, Gao JY, Gao YG, Wang GX (2009) Investigation on the overwintering status and bionomics of needle feeding type of *Conogethes punctiferalis*. Chinese Bull Entomol 46:569–573

Lakshminarayana M (2005) Studies on antixenosis in castor, *Ricinus communis* L., against major insect pests. Indian J Plant Protec 33:216–219

Leniaud L, Audiot P, Bourgue D, Frerot B, Genestier G, Lee SF, Malausa T, Lepallec AH, Souqual MC, Ponsard S (2006) Genetic structure of European and Mediterranean maize borers on several wild and cultivated host plants. Entomol Exp Appl 120:51–62

Lu JQ, Wang ZY, He KL, Liu Y (2010) Research history, progresses and prospect on the yellow peach moth, *Conogethes punctiferalis* (Guenée) (Lepidoptera: Crambidae). Plant Protect 36:31–38

Marohasy J (1998) The design and interpretation of host specificity tests for weed biological control with particular reference to insect behaviour. Bio-control News Inform 19:13–20

Martel C, Rejasse A, Rousset F, Bethenod MT, Bourguet D (2003) Host-plant-associated genetic differentiation in Northern French populations of the European corn borer. Heredity 90:141–149

Mercy ST, Babu LC, George MK (1977) Anatomical studies of the aerial stem, rhizome, leaf sheath and root of *Elettaria cardamomum* Maton. Agric Res J Kerala 15:10–12

Murray D, Prowell DP (2005) Molecular phylogenetic and evolutionary history of the neotropical satyrine subtribe Euptychiina (Nymphalidae: Satyrinae). Mol Phylogenet Evol 34:67–80

Panda N, Khush GS (1995) Host plant resistance to insects. CAB International, Wallingford, pp 1–431

Pena JE, Nadel H, Pereira MB, Smith D (2002) Pollinators and pests of annona species. In: Pena J, Sharp J, Wysoki M (eds) Tropical fruit pests and pollinators: biology, economic importance, natural enemies and control. CABI Publication, London, p 208

Ram KT, Hugar PS, Somasekhar PBV (1997) Record of *Conogethes punctiferalis* (Guenee.) on grapes. Pest Manag Hortic Ecosyst 3:37–39

Ravindran PN, Madhusoodanan KJ (2002) Cardamom -The genus *Elettaria*. Taylor & Francis Inc, New York, p 368

Ren ZM, Ma EB, Guo YP (2002) The studies of the phylogeny of acridoidea based on mtDNA sequence. Acta Genet Sin 29:314–321

Roe AD, Sperling FAH (2007) Population structure and species boundary delimitation of cryptic Dioryctria moths: an integrative approach. Mol Ecol 16:3617–3633

Sekiguchi K (1974) Morphology, biology and control of the yellow peach moth, *Dichocrocis punctiferalis* Guenée (Lepidoptera: Pyraustidae). Bulletin of the Ibaraki Horticultural Experiment Station, Special Issue, p 89

Sharma DR, Bal JS, Chahill BS (2008) Assessment of reaction of guava germplasm against fruit fly and castor capsule borer. Indian J Hort 65:145–151

Shashank PR, Kammar V, Mally R, Chakravarthy AK (2018) A new Indian species of shoot and capsule borer of the genus *Conogethes* (Lepidoptera: Crambidae), feeding on cardamom. Zootaxa 4374(2):215–234

Sloan WJS (1945) Insect pests of grain sorghum. Queensland Agric J 61:221–239

Smith JH (1937) Insects and allied pests of the Papua. Queensland Agric J 48:553–557

Sooraj PK (2005) Characterization of small cardamom (*Elettaria cardamomum*) germplasm for its biochemical constituents. Indian Institute of Spices Research, Calicut, p 97

Stanley J, Chandrasekaran S, Preetha G (2009) *Conogethes punctiferalis* (Lepidoptera: Pyralidae) its biology and field parasitization. Indian J Agric Sci 79:906–909

Thyagaraj NE (2003) Integrated Management of some important cardamom pests of hill region of Karnataka, South India. Ph.D thesis, Dr. B.R Ambedkar University, Uttara Pradesh, India, p 213

Tomlinson PB (1969) In: Metcalf CR (ed) Anatomy of the monocotyledons, Commelinales and Zingiberales, vol 3. Oxford University Press, Oxford, pp 343–358

Twine PH (1971) Insect pests of maize. Queensland Agric J 97:523–531
Wang PY (1980) Record of China Economy Insect, Lepidoptera, Pyralidae. Science Press, Beijing, pp 146–148
Wang YX, Cai RT (1997) The emergence and control of *Dichocrocis punctiferalis* Guenee from Younai plum variety. South China Fruits 26:45
Wang JY, Guo JF, Zhao DY, Wang C, Zhang Y, Wang HZ, Wu Y (2009) The genetic diversity and phylogenetic relationship among pig breeds of Shandong Province based on mtDNA CytB gene. Sci Agric Sin 42:1761–1767
Wapshere AJ (1989) A testing sequence for reducing rejection of potential biological control agents for weeds. Ann Appl Biol 114:515–526
Wei TX, Jian Y, Qing XP (2003) Preliminary Report on *Dichocrocis punctifera* resistance as shown by different cultivars of Chinese Chestnut. Econ Forest Res 11:45–59
Wijesekar ROB, Jayawardena AL (1973) Recent developments in production of spices and their essential oils in Ceylon. In: Proceedings of the conference on spices. Tropical Product Institute, London, pp 159–167
Xi K, Xu FY, Dai YS (1996) Studies on the Chinese chestnut fruit life table. J Jiangsu For Sci Technol 23:6–10
Xu CB, Zhi DY, Wen LG, Heng L (2001) Effects of volatiles from chestnut on host preference of adult *Conogethes punctiferalis* (Lepidoptera: Pyralidae). Chinese J Appl Ecol 14:24–29
Young HY, Shaw KY (1962) Preliminary studies on the peach borer *Dichocrocis punctiferalis*. Entomol Sin 1(2):109–134

# Extraction and Identification of Pheromones of the Borer, *Conogethes punctiferalis* (Crambidae: Lepidoptera)

21

J. Stanley, A. R. N. S. Subbanna, and G. Preetha

**Abstract**

The borer, *Conogethes punctiferalis*, is an internal tissue borer, and management of such borers using pheromones offers a scope for its population suppression. Though solvent extraction by abdominal tip excision was used earlier to extract pheromones, air entrainment/trapping volatiles using adsorbents paves a way for extraction of pheromone without contaminants. The chemical components of pheromone of *C. punctiferalis* have been identified. Few selected blends have even proved promising under experimental conditions in China, Japan, Korea and India. However, varied results in attraction obtained in these studies speculate differences in insect populations. Elucidation and identification of pheromone components like non-polar fraction of pheromone glands are necessary for effective mass trapping of *Conogethes* males under field conditions. Further research on basic and applied aspects of *Conogethes* pheromone technology applicable in diversified cultivated ecosystems is urgently required.

**Keywords**

*Conogethes* · Air entrainment · Pheromone · Extraction

J. Stanley (✉) · A. R. N. S. Subbanna
Vivekananda Institute of Hill Agriculture, Indian Council of Agricultural Research, Almora, Uttarakhand, India

G. Preetha
Department of Entomology, Tamil Nadu Agricultural University, Coimbatore, Tamil Nadu, India


A. K. Chakravarthy (ed.), *The Black spotted, Yellow Borer, Conogethes punctiferalis Guenée and Allied Species*,
https://doi.org/10.1007/978-981-13-0390-6_21

## 21.1 Introduction

### 21.1.1 *Conogethes punctiferalis*: A Polyphagous Pest

*Conogethes punctiferalis* Guenee is a polyphagous pest reportedly infesting 36 crop plants belonging to 23 families (Thyagaraj et al. 2003). Jacob (1981) and Ravichandran (1988) listed the host plants of *C. punctiferalis*.

*Conogethes* was found all through the year on castor with high population from August to March (Mishra and Teotia 1965). During the off season, it occurred on sorghum (April–June) feeding on the grains (David et al. 1964), and fully grown larvae entered into diapause by winter (Srivastava and Aswathi 1961). They occur heavily in cardamom fields during December–January, March–April and July–October. Biology of CSCB was studied in cardamom (Ravichandran 1988; Thyagaraj 2002; Thyagaraj et al. 2003), turmeric (Jacob 1981) and castor (David et al. 1964; Bilapate and Talati 1978). Kodo and Miyahara (1930) reported two generations per year on plum in China. On contrary, Yang and Shaw (1962) reported 4.5 generations per year on peach in China. Kodoi and Kaneda (1990) reported that *C. punctiferalis* complete its life cycle in 33 days on apple and 16 days on maize in Japan. Wang and Cai (1997) reported five generations per year on youni plum and five generations in cardamom in China. For *C. punctiferalis* castor is the preferred host (CPCRI 1985; Stanley et al. 2009).

### 21.1.2 Management Options

Though the borer management is normally attained by chemical pesticides (Kumaresan et al. 1978; Koya et al. 1991; Renuka et al. 2002; Rajabaskar et al. 2003; Regupathy et al. 2003; Stanley et al. 2010), it requires heavy dose and repeated applications that make the insecticidal management expensive and less effective (Chakravarthy et al. 2012). *C. punctiferalis* is difficult to manage because of its internal feeding behaviour and concealed nature inside the plant tissue. Pest management using pheromones for mass trapping or mating disruption offers a good scope for such internal tissue borers (Breth and Tee 2007). Moreover, the use of pheromone is more effective for detection of infestation and determination of crop loss (Shorey and McKelvey 1977). Pheromones are semiochemicals that elicit strong behavioural reactions even at small amounts and they are species specific.

### 21.1.3 Reproductive Behaviour

The virgin females of different ages were studied for pheromone release behaviour on many pests. In *Helicoverpa armigera* Hub., no calling was observed on the first scotophase following the emergence and initiated calling from 2 to 5 days (Kou and Chow 1987). In *Earias insulana* Fab., the female initiated calling on a day during the eighth hour of scotophase with peak calling activity being noticed during the last hour of scotophase and the peak calling was observed on 5-day-old females (Tamhankar 1995). Renuka et al. (2000) reported that in *Leucinodes orbonalis* Guenee, 95% of the females called 1 day after emergence.

Information on calling behaviour of insects is the basis for predicting sex, age and time of release. The calling behaviour is said to be synchronous with pheromone production that is common for many moth species, viz. *Helicoverpa zea* (Raina et al. 1986), *Heliothis subflexa* (Heath et al. 1991), *Sesamia nonagrioides* (Babilis and Mazomenos 1992), *Heliothis assulta* (Kamimura and Tatsuki 1993b), *Cydia pomonella* (Backman et al. 1997) and *Palpita unionalis* (Mazomenos et al. 2002). In *P. unionalis* maximal calling and pheromone production were observed on the fourth day (Mazomenos et al. 2002). In *C. punctiferalis* peak calling was observed 3 h prior to the end of scotophase (Rajabaskar 2003). The maximum pheromone (E-10-hexadecenal) production of about 13 ng per 3-day-old female was obtained 5 h after light-off, while no pheromone was detected in light phase. The appearance of pheromone in the pheromone gland was approximately 5 h earlier than the occurrence of calling (Konno 1986).

### 21.1.4 Female Calling

Female calling in *C. punctiferalis* is noticed only after 24 h of emergence and up to 10-day-old females. A definite periodicity in calling under 15 L-9D photoperiod cycles is also reported. One-to-two-day-old moths start and end calling slightly later than their older counterparts (Konno et al. 1980). Adult moths usually perch in cages during daytime without any movements or flights if undisturbed. At 19:30 h, virulent flights by males and females can be noticed, and sometimes they land on the sides of the cages for a few seconds moving the antenna vigorously. During calling, the females usually rest upside down under the top-screen surface or on the walls of the cage (Fig. 21.1). The receptive females exhibited a typical calling posture by bending the abdomen dorsally with extrusion of the apical segments, and this process continues for 2 h (Konno et al. 1980). The moths slowly extend and retract the terminal abdominal segments at regular intervals, during calling (Rajabaskar 2003). This type of calling is also noticed in other lepidopteran moths like *H. armigera* (Kamimura and Tatsuki 1993a; Hou and Shong 2000), *Synanthedon tipuliformis* (Buda and Karalias 1985), *Earias insulana* (Tamhankar 1995) and *L. orbonalis* (Renuka et al. 2000). Kimura and Honda (1999) reported a less mating success in females which are antennectomised, revealing the importance of a male pheromone chemical also. Antennectomised females were reportedly refusing final contact or copulation with males.

### 21.1.5 Male Calling, Hair Pencils and Its Component

Many species of nocturnal moths produce and release odorous compounds from specialised scales such as hair pencils, hair brushes and androconias (Kimura and

**Fig. 21.1** Calling and copulation postures of male and female *Conogethes* sp. moths

Honda 1999) not only from abdominal tips but also from the abdomen, thorax, legs and wings that are involved in short-range mating behaviours (Baker et al. 1981; Dussourd et al. 1991; Heath et al. 1992; Jacquin et al. 1991; Nishida et al. 1982; Teal and Tumlinson 1989). Pronounced and clear calling by male moths of *C. punctiferalis* are also noticed 1 day after emergence by curving the abdomen dorsally and extrusion of hair pencils at the abdominal tip (Stanley 2008). At around 21:00 h, the moths rest on the substrate with their legs firmly fixed and raised with the whole body up and down extruding the abdominal tips, and moths become sexually active. The component identified from hair pencils is 2-methyl 2-butenoic acid (tiglic acid), an unsaturated fatty acid. Males with hair pencils washed with hexane or removed hair pencils showed significantly lower mating success (21% and 28%) than the control ones with 71% mating success despite normal calling by females in both the cases (Kimura and Honda 1999). Thus the hair pencil or the compounds make the male accepted by females, and this is also evident in *Pseudaletia unipuncta* (Fitzpatrick and McNeil 1988), *Heliothis virescens* (Hillier et al. 2006; Hillier and Vickers 2007), *Utetheisa ornatrix* (Conner et al. 1981) and other insects. Tiglic acid is reported as indispensable for establishing normal mating behaviours and thus defined as male sent/sex pheromone. Female moths may discriminate conspecific males from other species because tiglic acid is not found in the sibling species, *Conogethes* sp., which have almost the same female pheromone. Such discrimination followed by repulsion was also reported in *H. virescens* for H. *subflexa* (Hillier et al. 2007; Hillier and Vickers 2004, 2011). Another possible function of tiglic acid is the selection of new virgin males from the male population (Kimura and Honda 1999).

The male scent of the lepidopteran insects is said to possess various kinds of biological functions. Male scents (pheromones) of some lepidopterans function as female attractants (Schneider et al. 1998) and aphrodisiac to the females (Pliske and Eisner 1969); suppress pheromone release in females (Hendricks and Shaver 1975), inhibitors of approach and copulatory behaviours in conspecific competitors (Hirai et al. 1978); and elicit female sexual behaviours, attracting sex pheromone releasing females to a short distance (Baker et al. 1981; Nishida et al. 1982) and exhibiting other courtship behaviours. Display of hair pencils by multiple rhythmic extrusions and retractions accompanied by "puffs" of wind generated by the vibrating wings and directed towards the female (45 and 90 cm/s.) by oriental fruit moth, *Grapholita molesta*, allows females choice for selection of mate (Baker and Carde 1979).

### 21.1.6 Copulatory Behaviour

Copulation occurs in the latter half of the dark phase (Kaneko 1978), and the peak period of copulation is at 6 h after light-off. Males start flying randomly in the presence of a calling female. Then males approach the females from posterior with extruded hair pencil, hover and have occasional contacts with the female body with the antenna (Konno et al. 1980). Precopulatory behaviours of females include extension of proboscis, upward curvature of the abdomen, extension of telescopic abdominal tip and rapid fluttering of wings which indicate readiness for pairing (Stanley 2008). During copulation (Fig. 21.1), the male lands beside the female and ceases fluttering of wings and extrudes the abdomen laterally to reach the genital portion of female and copulate. The male turns 180° and assumes posterior to posterior copulation (Konno et al. 1980). Copulation lasts for 30 min to 1 h, followed by a significant abdominal constriction of the female and unpairing.

## 21.2 Extraction of Pheromones

Though many pheromone extraction methods are available, solvent extraction of gland/abdominal tip and trapping of pheromones by adsorbents (air entrainment/head space sampling) are successfully used by many scientists for extracting the pheromones of lepidopteran insects.

### 21.2.1 Solvent Extraction of Gland/Abdominal Tip

Pheromones of lepidopterans can be extracted by clipping the abdominal tips of virgin adult moths in a solvent, like the hexane. Usually the last two segments of the abdominal tips (succuli laterals) are cut and immersed in solvents (Kuwahara and Casida 1973). In *Conogethes*, male and female moths to be used for solvent extraction of pheromone were kept separately after emergence to avoid mating. The insect can be sexed at pupal stage also. The moths were maintained separately at 23 ± 1 °C, 70–80% RH and 15 L–9D photoperiod regime (Konno et al. 1980). The moths were put back in the dark for about 6 h before extraction to enhance the pheromone yield. Pheromone was extracted by clipping the abdominal tips of 3-day-old unmated male and female moths, separately in 1 ml of dichloromethane and left to soak for an hour at 25 °C. The abdomen is squeezed with a pair of fine forceps gently to extrude the ovipositor, and the terminal two to three segments were excised with dissection scissors directly into dichloromethane (Stanley et al. 2007). Four-day-old females were used by Rajabaskar and Regupathy (2012). Ovipositors of 4-day-old females were extruded and excised just anterior to the pheromone gland at 4 h after light-off to get maximum pheromone yield (Konno et al. 1982). The unmacerated ovipositors were allowed to steep in redistilled dichloromethane for 15 min., filtered and condensed for further studies. Extraction of pheromones from the excised abdominal tips is done by (i) soaking in a solvent, (ii) homogenizing in a tissue grinder and then centrifugation and (iii) sonicating using a sonicator. Studies showed that all the three methods yield similar amount of pheromones (Daley et al. 1978) and thus soaking or immersing in a suitable solvent is being followed widely.

In case of male moths, hair pencils are extruded by slight finger pressure on the abdomen and removed using ophthalmic scissors at 4 h after light-off and put in hexane for 5 min to extract the male scent (Kimura and Honda 1999). The corematal scales of *Utetheisa ornatrix* were extracted by pressing the abdomen of males and excised using a micro-scissor and macerated in carbon disulphide to extract the male scent (Conner et al. 1981). Hair pencils of oriental fruit moth, *Grapholita molesta*, are located in the seventh and eighth abdominal segments and associated with the claspers, which are extracted by freezing the male and excising the terminal three abdominal segments including hair pencils and extracting in redistilled dichloromethane (Carde et al. 1975). Weatherston and Percy (1976) everted the hair pencils of male bertha armyworm, *Mamestra configurata*, by injecting air into their abdomen for extraction of male pheromone. Extraction of pheromones by excising the abodominal tips is given below (Table 21.1).

## 21.2.2 Pheromone Trapping Using Adsorbents

The techniques in chemical ecology for identification and quantitation of pheromone are deployed after the pre-concentration technique of head space analysis. This method involves trapping of the volatile components produced by the insect over a period of time either in a cold trap or on a solid adsorbent. In dynamic head space analysis, the sample is confined in an entrainment chamber, and a carrier gas is passed over the sample. The volatile chemicals released by the sample are trapped usually by porous organic polymer such as Porapak Q, Tenax TA, HayeSep Q or activated charcoal, and the analytes are desorbed using a solvent (solvent desorption) or rapid heat treatment (thermal desorption) (Agelopoulos and Pickett 1998). Solid-phase microextraction (SPME) is a new isolation method used to extract and concentrate pheromones (Malosse et al. 1995).

Dynamic headspace analysis with solvent desorption is better for pheromone extraction because it can be stored and used whenever needed (Agelopoulos and Pickett 1998). Porapak Q and Tenax TA are good and equally effective adsorbents and can be used for volatiles in air entrainment studies (Table 21.2). Pheromone of *Tyta luctuosa* was extracted by both the methods, viz. whole gland extract and air entrainment technique by Cao et al. (2003) and stated that the ratio of pheromone components are markedly different. Head space samples contain fewer nonvolatile contaminants since it collects air surrounding the sample and so the better method for pheromone extraction. Moreover, the head space is the most reliable method to investigate actual pheromones, but it is difficult to collect large amounts of pheromone needed for the identification (Tatsuki and Sugie 1992).

As stated above in the table, Tenax and HayeSep are used mostly for kairomone extractions and Porapak for sex pheromones but also for the aggregation pheromone (Jumean et al. 2004) herbivore-induced plant volatiles (Miresmailli et al. 2010), plant volatiles (Sidney et al. 2006; Hossaert-McKey et al. 2010) and so on.

**Collection of *Conogethes* Pheromones**

The pheromone is collected by passing a stream of purified air over the volatile emitting (calling) insect and the emitted volatile compounds/pheromones trapped in an adsorbent, i.e. Porapak Q (ethylvinylbenzene-divinylbenzene copolymer). The air entrainment apparatus used in this study consisted of two glass cylindrical tubes (entrainment chamber) of 38 cm length and 5 cm ID with ground joints (can be dismantled) on both sides and one tube for blank with 28 cm length and 5 cm ID

**Table 21.1** Abdominal tip excision for extraction of pheromones in different lepidopteran insects

| Insect | Family | Time of excision | Solvent | References |
|---|---|---|---|---|
| *Conogethes punctiferalis* | Crambidae | 4 days old | Dichloromethane | Konno et al. (1980, 1982) |
| *C. punctiferalis* | Crambidae | 4 days old | Hexane | Chakravarthy and Thyagaraj (1998) |
| *C. punctiferalis* | Crambidae | 3 days old; 3 h scotophase | Hexane | Rajabaskar (2003) |
| *C. punctiferalis* | Crambidae | 3 days old; 4 h scotophase | Hexane | Rajabaskar and Regupathy (2012) |
| *Spodoptera litura* | Noctuidae | – | Dichloromethane | Tamaki et al. (1973) |
| *Cadra cautella, Plodia interpunctella, Anagasta kuehniella, Ephestia elutella* | Pyralidae | – | Benzene | Kuwahara and Casida (1973) |
| *Archips argyrospilus* and *A. mortuanus* | Tortricidae | 2–3 days old | Dichloromethane | Carde et al. (1977) |
| *Bombyx mori* | Bombycidae | Newly emerged | Benzene | Kasang et al. 1978 |
| *Chilo partellus* | Crambidae | 1 days old; 1 h scotophase | Dichloromethane | Nesbitt et al. (1979) |
| *Ostrinia furnacalis* | Pyralidae | 2 days old | Dichloromethane | Ando et al. (1980) |
| *Earias insulana* | | 1–2 days old; 3 h scotophase | Ether | Hall et al. (1980) |
| *Euproctis similis* | Lymantriidae | 3 days old; 3–4 h scotophase | Hexane | Yasuda et al. (1994) |
| *Euproctis taiwana* | Lymantriidae | 2 days old; 1 h photophase | Hexane | Yasuda et al. 1995 |
| *Diatraea considerata* | Crambidae | 1–2 days old; 1–2 h scotophase | Hexane | Gries et al. (1998) |
| *Choristoneura parallela* | Tortricidae | 1–3 days old; 4–6 h scotophase | Dichloro methane | Sridhar and Lonergon (1998) |
| *Neoleucinodes elegantalis* | Crambidae | 2–3 days old; 5–9 h scotophase | Hexane | Eiras (2000) |
| *Argyrotaenia pomililiana* | Tortricidae | 2–3 days old; 3 h scotophase | Hexane | Cichon et al. (2004) |

(continued)

**Table 21.1** (continued)

| Insect | Family | Time of excision | Solvent | References |
|---|---|---|---|---|
| *Eucosma notanthes* | Tortricidae | 1–4 h photophase | Hexane | Chu et al. (2005) |
| *Eupithecia assimilata* | Geometridae | 3 days old; 4 h scotophase | Hexane | Campbell et al. (2007) |
| *Deanolis sublimbalis* | Crambidae | 3 days old; 2–3 h scotophase | Hexane | Andrew et al. (2007) |
| *Glyphodes perspectalis* | Pyralidae | 1–3 days old | Hexane | Kim and Park (2013) |
| *Proeulia auraria* | Tortricidae | 1–3 days old; 2 h scotophase | Hexane | Reyes-Garcia et al. (2014) |
| *Holocacista capensis* | Heliozelidae | 2 days old; 2–4 h scotophase | Hexane | Wang et al. (2015) |
| *Archips goyerana* | Tortricidae | 2–3 days old; 2–3 h scotophase | Hexane | Sullivan et al. (2015) |
| *Hypsipyla robusta* | Pyralidae | 2–3 days old | Hexane | Ma et al. (2015) |

**Table 21.2** Some of the adsorbents used for pheromone collection from lepidopteran insects

| Adsorbent | Insect | References |
|---|---|---|
| Porapak Q | *Agrotis ipsilon* | Hill et al. (1979) |
| | *Pseudaletia unipuncta* and *P. separata* males | Hirai (1980) |
| | *Choristoneura occidentalis* | Silk et al. (1982) |
| | *Helicoverpa armigera* | Kehat and Dunkelblum (1990) |
| | *Achroia innotata* males | Nemoto et al. (1990) |
| Tenax | Used mostly for kairomone extraction and tested on insect pests | Hartlieb and Rembold (1996), Huang and Mack (2002), and Natale et al. (2003) |
| HayeSep | Used mostly for kairomone extraction and tested on insect pests | Megido et al. (2014), Strapasson et al. (2014), and Sarkar et al. (2015) |
| Charcoal | *Agrotis segetum* | Lofstedt et al. (1982) |
| | *Chilo partellus* | Nesbitt et al. (1979) |
| | *Heliothis subflexa* and *H. virescens* | Heath et al. (1991) |

(Fig. 21.2). On one side of the cylindrical glass tube, a small glass tube (1 cm length and 0.05 cm ID), packed with 2 g of adsorbent Porapak Q (60–80 mesh size), was fixed. Porapak Q acts as a filter on which volatiles were adsorbed and were held in place by silanised glass wool. In addition to these three glass tubes, two small glass tubes one of which was filled with glass beads with glass wool on both sides act as molecular sieve and the other filled with activated charcoal served to purify

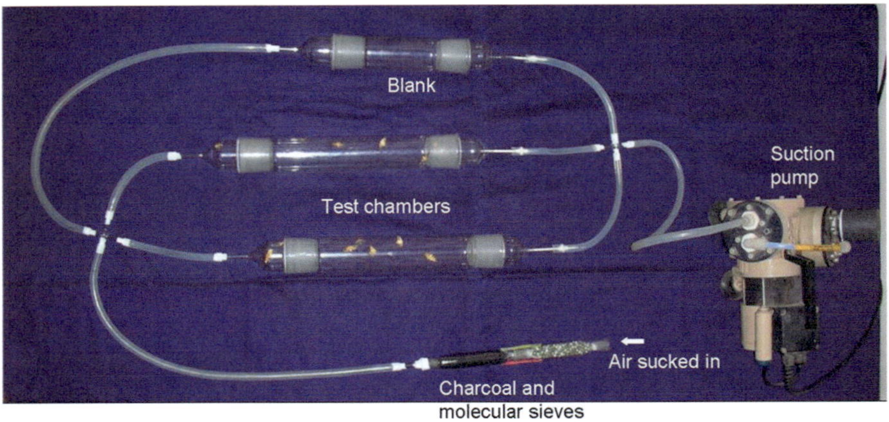


**Fig. 21.2** Air entrainment used for collecting *Conogethes* pheromones

air. All the tubes were connected to suction pump with silicon tube. When suction pump was operated, the atmospheric air entered through the molecular sieve then to activated charcoal and passed over the materials kept in the air entrainment chamber and finally exchanged through the Porapak Q, where the volatile compounds were adsorbed (Stanley 2008).

Adult male and female insects (2 days after emergence) each ten in number were placed separately in two air entrainment chambers and kept for 3 days. A blank was also maintained for comparison. The experiment was carried out for 72 h, for the collection of pheromone released by the insect. The entrained volatiles were eluted from Porapak Q filters by means of HPLC grade dichloromethane. The tube containing Porapak Q was clamped vertically, and solvent was pipetted in at the top and collected under gravity in a sample tube placed at the base. The sample tubes were immediately capped after flushing with $N_2$ using screw caps and stored at −4 °C till analysis. Glasswares used were kept clean by washing in teepol solution, followed by water then rinsed with acetone and dried in oven at 250 °C for 2 h before and after the experiments (Stanley 2008).

### 21.2.3 Solid-Phase Microextraction

Insects especially lepidopterans produce minute quantities of pheromones, and extracting them from a single calling female insect is extremely difficult as it gets adsorbed in the scales or glass apparatus. In such cases, solid-phase microextraction (SPME) is a novel way to trap the pheromones by gentle rubbing of the gland with the fibre (Frerot et al. 1997). SPME is successfully used in the absorption of volatile compounds from insects in a static head space (Malosse et al. 1995). SPME adsorbs organic compounds directly from the sample onto a fused-silica fibre, coated with an appropriate inert, polymeric adsorbent. The fibre is then desorbed into the injector of a gas chromatograph for identification (Nemer et al. 2014). In a typical solid-phase microextraction of pheromone of apple leaf roller, *Bonagota cranaodes* (Meyrick), the tip of the syringe with SPME fibre was kept a few millimetres away from the protruding abdominal gland of a virgin calling female in a small glass vial for 2–3 h for the pheromones to get adsorbed in the SPME fibre (Eiras et al. 1999).

Polydimethylsiloxane (PDMS)-coated fibres are best adapted to sample low to medium polarity VOCs (Pawliszyn 1997) and successfully used in the extraction of many lepidopteran pheromones (Karlson-Borg and Mozuraitis 1996; Frerot et al. 1997; Nemer et al. 2014) apart from other insects (Malosse et al. 1995; Rochat et al. 2000). Both the solid-phase microextraction (SPME) and the gland excision in solvent were performed for pheromone collection in *Eucosma notanthes*, and SPME was reported as much better than solvent extraction (Chu et al. 2005). SPME extraction yields good amount of pheromones as it is reported that the amount of volatiles released by 1 calling female during 3 h and collected on a polydimethylsiloxane fibre (SPME) was as large as the amount extracted from the glands of 20 females of *Phyllonorycter sylvella* Haworth, moths. Further, this technique gives an opportunity of continuously monitoring the release of behaviour-mediated signals from weak scented living organisms (Borg-Karlsona and Mozuraiti 1996).

## 21.3 Determination of Sex Pheromones

Sex pheromones are usually extracted using standard procedures and determined using GC (mostly ECD) and GC-MS. Long before, Kuwahara and Casido (1973) have given four important aspects for pheromone determination which are still somewhat relevant, though head space sampling instead of abdominal tip excision is reported more efficient and results in less contaminants and GC-MS instead of GC for identification of pheromone components which are used nowadays.

(a) Abdominal tips can be the source material for extraction because the pheromone is localised or occurs only in this body region.
(b) No chlorinated solvents are used to avoid interferences in analysis.
(c) The pheromone can be hydrolysed to the alcohol and re-esterified again achieving purification.
(d) The requirements for analytical sensitivity and specificity are met by trichloroacetylation of the alcohol and determination of the trichloroacetate derivative by electron-capture gas chromatography.

### 21.3.1 Estimation of Biological Activity of Pheromone

Olfactory receptor potentials can be recorded from an isolated antenna, positioned between microelectrodes connected to an amplifier and a recording instrument. EAG records the potential difference between the recording (distal end of antenna) and the indifferent electrodes (Roelofs 1984). The principle behind EAG is that when biologically active volatile chemicals pass through the microtubules of insect antennae, it binds with the receptor proteins, initiating reactions in the dendrites of the olfactory cells as the receptor membrane is depolarised. EAG measures the olfactory receptor potentials recorded by an electrode, connected to the sensory epithelium of the insect.

#### 21.3.1.1 Antennal Stimulation Studies

Before starting on any attempt to extract and study the pheromone-induced response in an insect antenna, it is necessary to detect whether the insect is responding to cue compounds (House et al. 1998). For this, insect response to different alcohols and aldehydes are to be measured by placing the antenna of both male and female insect's antenna in an electroantennogram. The chemicals which normally elicit antennal response of lepidopteran insect's, viz. heptanol, hexanol, pentanol, octanol, nonanol, linalool, dodecane, tridecane, propyl acetate, isoamyl acetate, etc., and solvents like dichloromethane and ethyl acetate were used to know the response of *Conogethes* sp. moths antenna. The test chemicals (5 μl) were adsorbed on to a filter paper presented in EAG to study the antennal deflection. Male antenna response was more pronounced than that of female (Stanley et al. 2007). This is a common phenomenon in insects, where male antennae are more responsive to odorous substances (Suckling et al. 1996; Deng et al. 2004).

EAG responses of 90 n-alkyl compounds, with different functional groups, were compared with fruit feeding and pinaceae feeding *C. punctiferalis* moths. Females of both the types (fruit and pinaceae feeding) reportedly responded to $C_5$ or $C_6$ compounds of alkanol, alkanal and alkanoic acids. Both the sexes respond significantly to n-hexanal than the corresponding alcohol and acid, and the females were more sensitive to n-hexanoic acid than males (Honda et al. 1986). The response of 1-day-old male *Conogethes* was high to heptanal, nonanal, heptanol, hexanol, propyl acetate and octanal and the values being −7.660, −7.278, −6.354, −6.354, −5.950 and − 5.748 mV, respectively. Female response was high to propyl acetate (−4.454 mV), pentanol (−4.454 mV), hexanol (−4.296 mV) and heptanol (−4.297 mV) (Stanley et al. 2007).

#### 21.3.1.2 Antennal Stimulation to Pheromone Compounds

The pheromones extracted using standard procedures such as gland excision and solvent extraction or extraction by air entrainment or by solid-phase microextraction are tested with electroantennogram for antennal response/deflection to the extracts. The antenna is placed so that the tip of the lamella touches one of the electrodes and the scape is fixed to the other (Reinecke et al. 2005) using a conductive gel. Fresh and charcoal-filtered air is continuously flushed over the antenna. Along with the airstream, the solvent and the pheromone compounds are puffed on to the antenna. The antennal response to pheromone compounds are usually tested and recorded from ten adults with three replications per antenna (Stanley et al. 2007). The extracts made through different extraction techniques can be tested to find the efficient extraction method. The EAG response of actual pheromone compound is obtained after corrected for solvent and other background effects by subtracting the averaged EAG responses of the solvent responses recorded before and after each sample as described by Visser et al. (1979). EAG response of head space extract was reported as better than that of gland or abdominal tip excision in many studies. The EAG response was more than two times to the head space sample extract than that of abdominal tip extract of *C. punctiferalis* (Stanley et al. 2007).

### 21.3.2 Estimation/Determination of Pheromone Components

#### 21.3.2.1 GC-EAD and ECD/FID

Gas chromatography followed by electroantennographic detector (EAD) is more convenient to detect pheromone components. Direct coupling of GC and electroantennogram (EAG) has added a new dimension to electrophysiological studies in insect pheromones (Arn et al. 1975; Cork et al. 1992). Here, the column effluent from the capillary GC is divided by a splitter mostly at 1:1 ratio into two portions. One portion can be made to pass continuously over the antenna and the EAG response recorded. The second portion is introduced to the FID/ECD, and the signal is simultaneously recorded. Thus the GC peaks which stimulated the antenna to respond are clearly identified. Since mass spectrometry is not used here, the compounds are identified based on the retention times in different columns in comparison with authentic standards. Pheromone gland extracts of *A. pomililiana* was analysed using GC-EAD and FID with HP-INNOWAX column (Cichon et al. 2004). Pheromone extracts of *D. considerata* were subjected to GC analysis with both FID and EAD using DB-5/ DB-210/ DB-2C columns for assessing the antennal response, isolation and identification of the components (Gries et al. 1998). Here, the EAD detects and reveals the behavioural active component, whereas the FID identifies the component of the pheromone. The advantage over wind tunnel bioassay is that it requires few insects, i.e. less than 1% of that required for wind tunnel assay, and conditioning of insect is not necessary.

#### 21.3.2.2 GC-EAD and MS Studies

This technique of electrophysiological assessment by EAD and identification of pheromone components and confirmation through GC-MS is used by many scientists. In 1971 itself, Roelofs and his co-workers assessed the antennal response of *Cydia pomonella* L. to monounsaturated compounds and found trans-8, trans-10-Dodecadien-1-ol as the attractant compound. Sex effluents of virgin *Callosobruchus analis* (F.) were analysed by Cork et al. (1991) in GC-EAG, and the EAG active compound was analysed in GC-MS and identified the attractant as 3-methyl-substituted heptenoic acid. Even the crude extracts of *Chilo suppressalis* (Walker) when analysed in GC, the mass spectrum gives the chemical structure of the pheromone components (Tatsuki and Sugie 1992). To find out the absolute configuration of sex pheromone of *Euproctis pseudoconspersa*, Ichikawa et al. (1995) used GC-MS technique with DB-Wax column. Similarly Polavarapu and Lonergan (1998) identified the sex pheromone components of *C. parallela* using GC-EAD and GC-MS with SupelcoWax 10 columns. Abdominal tip extracts of *Euproctis taiwana* (Shiraki) was analysed in GC-EAD using DB-Wax column and GC-MS using DB-1 and HP-5 columns, whereas DB-23 was used for *Euproctis similis* (Fuessly) sex pheromone identification (Yasuda et al. 1994, 1995). SPME-collection of calling females of *Bonagota cranaodes* (Meyrick) gave only one peak in GC-MS whereas two peaks with strong and weak antennal response in GC-EAD (Eiras et al. 1999).

Crude extracts of 1000 virgin females of *C. punctiferalis* were extracted and purified in Florisil column chromatography. The behavioural active substance is isolated and analysed in GC and EAD and then with MS. The mass spectrum of the biologically active compound corresponds to E-10 hexadecenal (Konno et al. 1982).

## 21.4 Sex Pheromone of *C. punctiferalis*

Sex pheromones which attract *C. punctiferalis* males are used in pest monitoring and management long before (Cai and Mu 1993; Liu et al. 1994a, b; Du et al. 2014). The usage of female pheromone of *C. punctiferalis* in attracting males was first identified by Konno et al. (1980) using a preliminary model of Y-tube olfactometer and wind tunnel device. Initial laboratory male trapping tests with synthetic (E)-10-hexadecenal proved it is the sole component of sex pheromone of *C. punctiferalis* (Konno et al. 1982). However, further evaluation of the compound under field conditions showed maximum activity only when associated with (Z)-10-hexadecenal. Liu et al. (1994a, b) identified (E)-10-hexadecenal and (Z)-10-hexadecenal as the major pheromone compounds existing in *C. punctiferalis* in China. The attraction efficiency of this blend was four times greater than the (E)-10-hexadecenal alone (Konno et al. 1982). Earlier studies also reported the existence of a blend of geometric isomers as sex pheromones of European corn borer, *Ostrinia nubilalis* (Hubner), and red-banded leaf roller moth, *Argyrotaenia velutinana* (Walker) (Klun et al. 1973). Three compounds, (E)-10-hexadecenal (E10-16: Ald) and (Z)-10-hexadecenal (Z10-16: Ald) and hexadecanal (16: Ald), were identified in the female gland extract of *C. punctiferalis*. Among the various combinations E10-16:Ald and Z10-16:Ald at 70:30–80:20 were the most attractive to males in wind tunnel experiments and field trapping experiments in orchards (Jung et al. 2000).

Majority of the lepidopteran sex pheromone components are classified into two types. The Type I components are the major ones that include $C_{10}$-$C_{18}$ alcohols, acetates, and aldehydes. The Type II components are not readily classified and consist of $C_{17}$-$C_{23}$ polyunsaturated hydrocarbons or their epoxide derivatives and a number of miscellaneous components (Millar 2000; Ando et al. 2004). Recent studies in pyralids (Millar et al. 2005; Strong et al. 2008) and crambids (Cabrera et al. 2001; Gibb et al. 2007) proposed more complex systems of sex pheromones or sex attractants involving a blend of both Type I and Type II compounds. An earliest study by Konno et al. (1982) identified (E)-10-hexadecenal and (Z)-10-hexadecenal, at 9:1, as an effective female sex pheromone of *C. punctiferalis.* Identification of this pheromone chemistry leads to widespread usage of synthetic lure in traps for monitoring, mass trapping and eventual damage reduction by this pest in China and

Korea (Cai and Mu 1993; Liu et al. 1994a, b). However, the same blend of these binary aldehydes did not perform well on some Japanese populations (Kondo et al. 2008), indicating involvement of some more essential components with low volatility. In Korea, best attraction of males to various synthetic sex pheromone blends was obtained at 80:20 ratio of E10-hexadecenal and Z10-hexadecenal (Jung et al. 2000). However, the field trials in Karnataka, India, showed male moths from cardamom plantations were highly attractive at 9:1 ratio (Chakravarthy and Thyagaraj 1998). Additionally, Shashank et al. (2015) reported a completely different trend in moth catches of *C. punctiferalis* in castor and cardamom plantations using the same blend of sex pheromones. They observed no moth catches in cardamom fields suggesting possible variability in sex pheromone composition of cultivar or crop-adopted populations of *C. punctiferalis*. Boo (1998) found that *C. punctiferalis* consists of two populations, with a group found in Japan and China responding to a 100:8–100:11 blend and a group in Korea and China to a ratio of 100:43. Keeping in view the specificity of sex pheromones, the variability observed in these studies suggests the use of different ratios of the same blend by different geographical and or host-adopted populations leading to speciation in near future. This type of chemosensory based speciation was reviewed by Smadja and Butlin (2009). The geographical variation in pheromone composition and attraction efficiency were common in other lepidopteran pests like *Hemileuca eglanterina* (Boisduval) with two distinct pheromone types (McElfresh and Millar 2001), *Ostrinia nubilalis* (Hubner) with two different isomeric ratios (Bethenod et al. 2005), *Ostrinia scapulalis* (Walker) with pheromonal blends of E and Z type (Takanashi et al. 2005) and *Diatraea saccharalis* (Fabricius) with differences in composition of major components (Cortes et al. 2010).

A deep insight into the attraction chemistry of *C. punctiferalis* also showed existence of synergistic non-polar fractions of pheromone gland extracts and body wax extracts of females. They include monounsaturated hydrocarbon, (Z)-9-heptacosene (Z9-27:HC, Xiao et al. 2011) and polyunsaturated hydrocarbons, (Z3, Z6, Z9)-tricosatriene (Z3, Z6, Z9-23:HC) and (Z3, Z6, Z9)-heptacosatriene (Z3, Z6, Z9-27:HC, Xiao et al. 2012) which were found to enhance male responses close to pheromone sources (Xiao and Honda 2010). Such combinations are widespread motif in lepidopteran pheromone chemistry (Millar et al. 2005). Additionally, Kimura and Honda (1999) reported an unsaturated fatty acid, a tiglic acid from male hair pencils as male sex pheromone of *C. punctiferalis*. The tiglic acid also helps in selection of new virgin males from a given male population.

Cuticular waxes are also known to be chemical communication agents in insects and serve as species and gender recognition signals, nest mate recognition signals, task-specific cues and other signals (Howard 1993; Howard and Blomquist 2005). The non-polar fraction of female pheromone extracts of *C. punctiferalis* alone induced no male responses other than starting flight, but this fraction showed marked enhancement in response when mixed with the polar fraction or the synthetic blend. These results clearly indicated that unknown non-polar components are essential for stimulating the final approach of males to calling females in sexual communication of the yellow peach moth. Various hydrocarbons have been identified as cofactors or minor components of sex pheromone gland extracts, but exclusive production and location of these components in the sex pheromone glands remain unclear.

The primary olfactory sensor of insects is antennae, and the surface of antennae is covered with several sensilla. The dendrites of olfactory receptor neurons of sensilla extend into antennal haemolymph where peripheral olfactory signal transduction occurs (Jia et al. 2016). Previous studies reported diverse olfactory proteins, including odorant-binding proteins (OBPs), odorant receptors (ORs), chemosensory proteins (CSPs), sensory neuron membrane proteins (SNMPs), ionotropic receptors (IRs) and odorant-degrading enzymes (ODEs) involved in different odour perception steps of signal transduction pathway (Bruyne and Baker 2008; Korsching 2002; Leal 2013). However, based on the different expression profiles of ORs in male and female antennae of *C. punctiferalis* (Xiao et al. 2016), Jia et al. (2016) suggested that male antennae expressed ORs are involved in sex pheromone detection and female antennae expressed ORs play important roles in locating suitable host plants and oviposition sites.

### 21.4.1 Attraction Studies on *C. punctiferalis*

Insects respond to different olfactory cues like volatiles from plants (e.g. phytophagous insects), host (e.g. parasitoids and predators) and pheromones for mate searching and aggregation. Y-tube olfactometer and wind tunnels are often used to monitor the responses of insects to these odour cues (Ranjith 2007). Most importantly, the Y-tube olfactometer was chiefly used to study the attraction at qualitative levels, whereas the wind tunnel experiment provides quantitative estimates of attraction efficiency (Noldus and van Lenteren 1985). In both the experiments, ventilation plays a major role in estimation of attraction statistics as the ad lib flight of males before ventilation was changed to oriented movement towards pheromonal source associated with hovering and intermittent antennae tapping (Konno et al. 1980). A detailed description of methodology and usage of these techniques in pheromonal studies of *C. punctiferalis* are described below.

#### 21.4.1.1 Y-tube Olfactometer

A Y-tube olfactometer is a simple and ready-to-use methodology that is extensively used by various chemical ecologists (Xiao et al. 2011, 2012; Du et al. 2016) to test dual-choice responses of *C. punctiferalis* to single volatiles and/or an array of blends. The olfactometer consists of two similar glass chambers (2.5 cm diameter, 10 cm long) that converge into a 13-cm-long common arm (2.5 cm diameter) giving the apparatus a "Y"-shaped appearance. The upper 10 cm long two arms are the source emitters that contain volatiles (either singly or blends) at the closing end. The lower 13 cm long arm was used to place the test insect. If the test volatiles were in crude concentrate form, they can be serially diluted to require concentrations using laboratory-grade mineral oil (Du et al. 2016) as no attraction was reported in *C. punctiferalis*. Thus prepared dilutions of the volatiles can be applied on to either a piece of filter paper (approx. 1 cm × 5 cm) or fresh cotton rolls of 0.03 m long (Chakravarthy and Thyagaraj 1998) and placed into one of the test chambers. Another test chamber with mineral oil coated filter paper or cotton roll served as control. Charcoal-filtered air pumping system to both the test chambers (at a rate of 500 ml/min controlled by flow metres) carries volatiles to test insect to initiate a response. Flushing fresh air and rinsing with 75% ethanol clean the tubes for further use.

Particularly with *C. punctiferalis*, attraction studies of sex pheromones are carried out between 19.00 and 23.00 h under a 25 W red light lamp. For virgin females and males, 4 days old is found the best age for sex pheromonal studies (Konno et al. 1980). The test for each moth lasts for 3 min, and the behavioural response was classified as attraction, when the test moth passes over 1/3 the length of the Y-tube arm associated with bait/stimuli and stayed there for more than 1 min. Conversely, no choice was assigned if the test moth remained in the common arm for 3 min (Du et al. 2016).

#### 21.4.1.2 Wind Tunnel Assay

An ideal wind tunnel apparatus for *C. punctiferalis* (Sundararaju et al. 1994) consist of two plastic boxes (bait box used to keep bait/volatiles and insect box used to keep the test insect) connected by a transparent tube (100 cm length and 7 cm dia) made up of Mylar film sheet. Each box was provided with a lid at the top and two circular holes (6.5 cm in dia) on opposite sides. The outer hole of both the boxes are covered with muslin cloth to confine the insects, and other ends were connected via the transparent tube to allow free movement of test insect towards bait box and make observations. Before conducting the test, the adult moths should be preadopted to confined cages, and during the testing 10% sugar solution using cotton pad should be provided. A portable personal mini-fan with an electronic regulator was kept in front of the bait box to carry regulated flow of air to the test box. All wind tunnel tests were conducted at 25 ± 1 °C and 40–60% RH under a red incandescent light at ca. 2 lux. Male moths were allowed to leave the cage just after setting a stimulus source at the upwind end of the tunnel and followed by four behaviours recorded for 3 min: (1) flight initiation, (2) orientation towards the plume, (3) remained close to source and (4) source contact. The time spent close to the source and the numbers of source contacts can also be recorded (Xiao and Honda 2010; Xiao et al. 2012).

Tiglic acid identification as sex pheromone with functions like conspecific male recognition and as aphrodisiac pheromone was done using attraction behaviours of antennectomised females and hair pencil extracts in wing tunnel apparatus (Kimura and Honda 1999). Xiao and Honda (2010) identified inert non-polar fractions from crude pheromone extract and female body wax extract which synergise the polar fractions. The study also revealed that these fractions were involved in males attraction closes to the source, thereby, functioning as a cue for the final recognition of females. They also proposed a full set of the sex pheromone system of yellow peach moth as E10-16:Ald, Z10-16:Ald for a long-range attraction and Z9-27:CH, Z3Z6Z9-23:CH for the final recognition of females by males in near-pheromone source. These short- and long-range observations were possible only in large-sized apparatus where authors used Plexiglas wind tunnel with 2 m length, 0.3 m diameter. In addition, studies also reported wind tunnels are more useful in identification of calling, premating and courtship behaviours (Rajabaskar and Ragupathy 2012; Nakano et al. 2012).

## 21.4.2 Field Evaluation of Sex Pheromones

### 21.4.2.1 Funnel Traps with Plastic Plates

Chakravarthy and Thyagaraj (1998) used this trap to study the attraction efficiency of synthetic components of (E)-10-hexadecenal and (Z)-10-hexadecenol at 9:1 in cardamom plantations of Arehalli and Goudahalli of Karnataka, India. The trap consisted of a white plastic funnel (0.21 m dia) that is hinged to a yellow plastic plate with a moth collecting polybag (0.28 × 0.45 m) below. The pheromone blend was suspended at the centre below the funnel.

### 21.4.2.2 Delta Traps

The trap consists of a triangle-shaped “house” made up of water-resistant material and a sticky insert for trapping the respondents. The lure is placed in the middle of the ridge of the roof using a hanger. Male adults of *C. punctiferalis* are attracted by the pheromone lure and get stuck on the sticky insert. Many studies reported the use of these delta traps in monitoring (Kondo et al. 2008), mass trapping and mating disruption (Kim et al. 2013) studies of *C. punctiferalis*. Kim et al. (2013) also found dose and regional dependent variability in field attraction of sex pheromones of *C. punctiferalis* using delta traps.

## References

Agelopoulos GN, Pickett JA (1998) Head space analysis in chemical ecology: effects of different sampling methods on ratio of volatile compounds present in head space samples. J Chem Ecol 24:1161–1172

Ando T, Inomata S, Yamamoto M (2004) Lepidopteran sex pheromones. Top Curr Chem 239:51–96

Ando T, Saito O, Arai K, Takahashi N (1980) (Z) and (E)-12-tetradecenyl acetates: sex pheromone components of oriental corn borer (Lepidoptera: Pyralidae). Agric Biol Chem 44(11):2643–2649

Andrew G, Bruno P, David T, Annastasia K, Barry B, Padmaja R, Suckling D (2007) (Z)-11-hexadecenal and (3Z,6Z,9Z)-tricosatriene: sex pheromone components of the red banded mango Caterpillar *Deanolis sublimbalis*. J Chem Ecol 33:579–589

Arn H, Stadler E, Rauscher S (1975) The electroantennographic detector – a selective tool in the gas chromatographic analysis of insect pheromones. Z Naturf 30:722–725

Babilis NA, Mazomenos BE (1992) Pheromone production in *Sesamia nonagrioides*: Diel periodicity and effect of age and mating. J Insect Physiol 38:561–564

Backman AC, Bengtsson M, Witzgall P (1997) Pheromone release by individual females of codling moth, *Cydia pomonella*. J Chem Ecol 23:807–815

Baker TC, Carde RT (1979) Endogenous and exogenous factors affecting periodicities of female calling and male sex pheromone response in *Grapholitha molesta* (Busck). J Insect Physiol 25(12):943–950

Baker TC, Nishida R, Roelofs WL (1981) Close-range attraction of female oriental fruit moths to herbal scent of male hairpencils. Science 214:1359–1361

Bethenod MT, Thomas Y, Rousset F, Frerot B, Pelozuelo L, Genestier G, Bourguet D (2005) Genetic isolation between two sympatric host plant races of the European corn borer, Ostrinia nubilalis Hübner. II: assortative mating and host-plant preferences for oviposition. Heredity 94(2):264–270

Bilapate GG, Talati GM (1978) Some studies on bionomics of castor shoot and capsule borer, *Dichocrocis punctiferalis* (Guenee). J Maharashtra Agric Univ 3:47–49

Boo KS (1998) Variation in sex pheromone composition of a few selected lepidopteran species. J Asia Pac Entomol 1(1):17–23

Borg-Karlsona A, Mozuraiti R (1996) Solid phase micro extraction technique used for collecting semiochemicals. Identification of volatiles released by individual signalling *Phyllonorycter sylvella* moths. Z Naturforsch C 51(7–8):599–602

Breth DI, Tee E (2007) Progress in managing internal lepidopteran pests in tree fruit. NY Fruit Q 15(1):19–22

Bruyne M, Baker TC (2008) Odor detection in insects: volatile codes. J Chem Ecol 34:882–897

Buda V, Karalias V (1985) Calling behaviour of females of current clearwing moth *Synanthedon tipuliformis* (Clerk) (Lepidoptera: Sesiidae). Z Ang Ent 100:297–302

Cabrera A, Eiras AE, Gries G, Gries R, Urdaneta N, Mirás B, Badji C, Jaffe K (2001) Sex pheromone of tomato fruit borer, Neoleucinodes elegantalis. J Chem Ecol 27:2097–2107

Cai RX, Mu ZL (1993) Trap trails for *Dichocrocis punctiferalis* Guenee with sex pheromone in citrus orchards. China Citrus 22:33

Campbell CAM, Tregidga EL, Hall DR, Ando T, Yamamoto M (2007) Components of sex pheromone of the currant pug moth, *Eupithecia assimilata* a re-emergent hop pest in UK. Entomol Exp Appl 122:265–269

Cao W, Charlton RE, Nechols JR, Horak MJ (2003) Sex pheromone of the noctuid moth, *Tyta luctuosa* (Lepidoptera: Noctuidae), a candidate biological control agent of field binweed. Environ Entomol 32:17–22

Carde RT, Baker TC, Roelofs WL (1975) Ethological function of components of a sex attractant system for oriental fruit moth males, *Grapholitha molesta* (Lepidoptera: Tortricidae). J Chem Ecol 1(4):475–491

Carde RT, Carde AM, Hill AS, Roelofs WL (1977) Sex pheromone specificity as a reproductive isolating mechanism among the sibling species *Archips argyrospilus* and *A. mortuanus* and other sympatric tortricine moths (Lepidoptera: Tortricidae). J Chem Ecol 3(1):71–84

Chakravarthy AK, Thyagaraj NE (1998) Evaluation of selected synthetic sex pheromones of the cardamom shoot and capsule borer, *Conogethes punctiferalis* Guenee (Lepidoptera: Pyralidae) in Karnataka. Pest Manage Hortic Ecosyst 4:78–82

Chakravarthy AK, Shashank PR, Doddabasappa B, Thyagaraj NE (2012) Status of shoot and fruit borer, *Conogethes* spp (Lepidoptera: Crambidae) in the orient: biosystematics, biology and management. J Insect Sci 25(2):107–117

Chu TY, Hung CC, Hsu CY (2005) Solid-phase microextraction for the investigation of sex pheromone of *Eucosma notanthes* Meyrick. Talanta 65(3):743–749

Cichon LI, Trematerra P, Coracini MDA, Fernandez D, Bengtsson M, Witzgall P (2004) Sex pheromone of *Argyrotaenia pomililiana* (Lepidoptera: Tortricidae), a leafroller pest of apples in Argentina. J Econ Entomol 97(3):946–949

Conner WE, Eisner T, Vander-Meer RK, Guerrero A, Meinwald J (1981) Precopulatory sexual interaction in an arctiid moth (*Utetheisa ornatrix*): role of a pheromone derived from dietary alkaloids. Behav Ecol Sociobiol 9:227–235

Cork A, Hall DR, Blaney WM, Simmonds MSJ (1991) Identification of a component of the female sex pheromone of *Callosobruchus analis* (Coleoptera: Bruchidae). Tetrahedron Lett 32:129–132

Cork A, Boo KS, Dunklelblum E, Hall DR, Rajunga KJ, Kehat M, Jie EK, Park KC, Tepgidagarn P, Liu X (1992) Female sex pheromones of oriental tobacco budworm, *Helicoverpa assulta* (Guenee) (Lepidoptera: Noctuidae); identification and field testing. J Chem Ecol 18:403–418

Cortes AMP, Zarbin PH, Takiya DM, Bento JMS, Guidolin AS, Consoli FL (2010) Geographic variation of sex pheromone and mitochondrial DNA in Diatraea saccharalis (Fab., 1794) (Lepidoptera: Crambidae). J Insect Physiol 56(11):1624–1630

CPCRI (1985) Annual report for 1985. Central Plantation Crops Research Institute, Kasaragod, p 249

Daley RC, Kislow CJ, Brady UE (1978) Comparison of the levels of sex pheromone extracted from female *Cadra cautella* by three methods [Homogenization, sonication, leaching]. J Georgia Entomol Sci. Assessed through http://agris.fao.org/agris-search/search.do?recordID=US7925884

David BV, Narayanasami PS, Murugesan M (1964) Bionomics and control of the castor shoot and capsule borer, *Dichocrosis punctiferalis* Guenee in Madras state. Indian Oil seeds J 8:146–158

Deng JY, Wei HY, Huang YP, Du JW (2004) Enhancement of attraction to sex pheromones of *Spodoptera exigua* by volatile compounds produced by host plants. J Chem Ecol 30:2037–2045

Du YL, Zhang MZ, Ma YQ, Sun SL, Wang JB, Liu JL (2014) Formulation screening of sex pheromones and field trapping tests for the yellow peach moth, *Conogethes punctiferalis* (Guenée) (Lepidoptera: Pyralidae). Acta Phytophy Sin 41:187–191

Du Y, Zhang J, Yan Z, Ma Y, Yang M, Zhang M et al (2016) Host preference and performance of the yellow peach moth (*Conogethes punctiferalis*) on chestnut cultivars. PLoS One 11(6):e0157609

Dussourd DE, Harvis CA, Meinwald J, Eisner T (1991) Pheromonal advertisement of a nuptial gift by a male moth (*Utetheisa ornatrix*). Proc Natl Acad Sci USA 88:9224–9227

Eiras EA (2000) Calling behaviour and evaluation of sex pheromone glands extract of *Neoleucinodes elegantalis* Guenee (Lepidoptera: Crambidae) in wind tunnel. An Soc Entomol Bras 29:157

Eiras AE, Kovaleski A Vilela EF, Chambon JP, Unelius CR, Borg-Karlson A, Liblikas I, Mozuraitis R, Bengtsson M, Witzgall P (1999) Sex pheromone of the Brazilian apple Leafroller, *Bonagota cranaodes* Meyrick (Lepidoptera, Tortricidae). Z Naturforsch 54:595–601

Fitzpatrick SM, McNeil JN (1988) Male scent in lepidopteran communication: the role of male pheromone in mating behaviour of *Pseudaletia unipuncta* (Lepidoptera: Noctuidae). Mem Entomol Soc Can 120(146):131–151

Frerot B, Malosse C, Cain A (1997) Solid phase microextraction (SPME): a new tool in pheromone identification in Lepidoptera. J High Resolut Chromatogr 20:340–342

Gries R, Dunkelblum E, Gries G, Baldilla F Hernandez C, Alvarez F, Perez A, Velasco J, Oehlschlager AC (1998) Sex pheromone components of *Diatraea considerata* (Heinrich) (Lepidoptera: Pyralidae). J Appl Entomol 122:265–268

Hall DR, Beevor PS, Lester R, Nesbitt BF (1980) (E,E)-10,12-Hexadecadienal: a component of the female sex pheromone of the spiny bollworm, *Earias insulana* (Boisd.) (Lepidoptera: Noctuidae). Experientia 36(2):152–154

Hartlieb K, Rembold H (1996) Behavioral response of female *Helicoverpa armigera* (Lepidoptera: Noctuidae) moths to synthetic pigeonpea (*Cajanus cajan* L.) kairomone. J Chem Ecol 22(4):821–837

Heath RR, Mclaughlin JR, Proshold F, Teal PEA (1991) Periodicity of female sex pheromone titer and release in *Heliothis subflexa* and *H. virescens* (Lepidoptera: Noctuidae). Ann Entomol Soc Am 84(2):182–189

Heath RR, Landolt PJ, Dueben BD, Murphy RE, Schneider RE (1992) Identification of male cabbage looper sex-pheromone attractive to females. J Chem Ecol 18:441–453

Hendricks DE, Shaver TN (1975) Tobacco budworm: male pheromone suppressed emission of sex pheromone by the female. Environ Entomol 4(4):555–558

Hill AS, Rings RW, Swier SR, Roelofs WL (1979) Sex pheromone of the black cutworm moth, *Agrotis ipsilon*. J Chem Ecol 5(3):439–457

Hillier NK, Vickers NJ (2004) The role of heliothine hairpencil compounds in female Heliothis virescens (Lepidoptera: Noctuidae) behavior and mate acceptance. Chem Senses 29:499–511

Hillier NK, Vickers NJ (2007) Physiology and antennal lobe projections of olfactory receptor neurons from sexually isomorphic sensilla on male *Heliothis virescens*. J Comp Physiol A Neuroethol Sens Neural Behav Physiol 193:649–663

Hillier NK, Vickers NJ (2011) Hairpencil volatiles influence interspecific courtship and mating between two related moth species. J Chem Ecol 37:1127–1136

Hillier NK, Kleineidam C, Vickers NJ (2006) Physiology and glomerular projections of olfactory receptor neurons on the antenna of female *Heliothis virescens* (Lepidoptera: Noctuidae) responsive to behaviorally relevant odors. J Comp Physiol A Neuroethol Sens Neural Behav Physiol 192:199–219

Hillier NK, Kelly D, Vickers NJ (2007) A specific male olfactory sensillum detects behaviorally antagonistic hairpencil odorants. J Insect Sci 7:4

Hirai K (1980) Male scent emitted by armyworms, *Pseudaletia unipuncta* and *P. separata* (Lepidoptera: Noctuidae). Appl Entomol Zool 15(3):310–315

Hirai K, shorey HH, gaston LK (1978) Competition among courting male moths: male-to-male inhibitory pheromone. Science 202(4368):644–645

Honda H, Maruyama Y, Matsumoto Y (1986) Comparisons in EAG response to n-aIkyl compounds between the fruit- and pinaceae-feeding type of yellow peach moth, conogethes punctiferalis (Guenee) (Lepidoptera: Pyralidae). Appl Entomol Zool 21(1):126–133

Hossaert-McKey M, Soler C, Schatz B, Proffit M (2010) Floral scents: their roles in nursery pollination mutualisms. Chemoecology 20(2):75–88

Hou ML, Shong CF (2000) Calling behaviour of adult female *Helicoverpa armigera* (Hubner) (Lepidoptera: Noctuidae) of overwintering generation and effects of mating. J Appl Entomol 124:71–75

House PE, Stevens IDR, Jones OT (1998) Insect pheromone and their use in pest management. Chapman and Hall, London, p 183

Howard RW (1993) Cuticular hydrocarbons and chemical communication. In: Stanley-Samuelson DW, Nelson DR (eds) Insect lipids—chemistry. biochemistry & biology. University Nebraska Press, Lincoln/London, pp 179–226

Howard RW, Blomquist GJ (2005) Ecological, behavioral, and biochemical aspects of insect hydrocarbons. Annu Rev Entomol 50:371–393

Huang X, Mack TP (2002) Collection and determination of lesser cornstalk borer (Lepidoptera: Pyralidae) larval attractant from peanut plants. Environ Entomol 31(1):15–21

Ichikawa A, Yasuda T, Wakamura S (1995) Absolute configuration of sex pheromone for tea tussock moth, *Euproctis pseudoconspersa* (Strand) via synthesis of (R) and (S) 10,14-dimethyl-1 pentadecyl isobutyrates. J Chem Ecol 21:627–634

Jacob SA (1981) Biology of *Dichocrosis punctiferalis* Guenee on turmeric. J Plant Crop 9:119–123

Jacquin E, Nagnan P, Frerot B (1991) Identification of hairpencil secretion from male *Mamestra brassicae* (L) (Lepidoptera, Noctuidae) and electroantennogram studies. J Chem Ecol 17:239–246

Jia XJ, Wang HX, Yan ZG, Zhang MZ, Wei CH, Qin XC, Ji WR, Falabella P, Du YL (2016) Antennal transcriptome and differential expression of olfactory genes in the yellow peach moth, *Conogethes punctiferalis* (Lepidoptera: Crambidae). Sci Rep 6:29067

Jumean Z, Rowland E, Judd GJR, Gries G (2004) Male and female *Cydia pomonella* (Lepidoptera: Olethreutidae) larvae produce and respond to aggregation pheromone. Can Entomol 136(6):871–873

Jung JK, Han KS, Choi KS, Boo KS (2000) Sex pheromone composition for field-trapping of *Dichocrocis punctiferalis* (Lepidoptera: Pyralidae) males. Korean J Appl Entomol 39:105–110

Kamimura M, Tatsuki S (1993a) Effects of photoperiodic changes on calling behavior and pheromone production in the oriental tobacco budworm moth, *Helicoverpa assulta* (Lepidoptera: Noctuidae). J Insect Physiol 40:731–734

Kamimura M, Tatsuki S (1993b) Diel rhythms of calling behavior and pheromone production of oriental tobacco budworm moth *Helicoverpa assulta* (Lepidoptera: Noctuidae). J Chem Ecol 19:2953–2963

Kaneko J (1978) Abdominal constriction in the copulated female of yellow peach moth, *Dichocrocis punctiferalis* Guenee (Lepidoptera: Pyralidae). Appl Entomol Zool 13:131–133

Karlson-Borg AK, Mozuraitis R (1996) Solid phase microextraction technique used for collecting semiochemicals: identification of volatiles released by individual signalling *Phyllonorycter sylvella* moths. Z Naturforsch C 51:599–602

Kasang G, Kaissling KE, Vostrowsky O, Bestmann HJ (1978) Bombykal, a second pheromone component of the silkworm moth *Bombyx mori* L. Angw Chem 17(1):60

Kehat M, Dunkelblum E (1990) Behavioral responses of male Heliothis armigera (Lepidoptera: Noctuidae) moths in a flight tunnel to combinations of components identified from female sex pheromone glands. J Insect Behav 3(1):75–83

Kim J, Park I (2013) Female sex pheromone components of the box tree pyralid, *Glyphodes perspectalis*, in Korea: field test and development of film-type lure. J Asia Pac Entomol 16:473–477

Kim YJ, Kim HK, Kang KN, Kim YM, Moon SJ, Kim GH (2013) Attractive effect using pheromone trap of various conditions against the peach pyralid moth, *Dichocrocis punctiferalis*. Korean J Pestic Sci 17(3):200–205

Kimura T, Honda H (1999) Identification and possible functions of the hairpencil scent of the yellow peach moth, *Conogethes punctiferalis* (Guenee) (Lepidoptera: Pyralidae). Appl Entomol Zool 34:147–153

Klun JA, Chapman OL, Mattes KC, Wojtkowski PW, Beroza M, Sonnet PE (1973) Insects sex pheromones: minor amount of opposite geometrical isomer critical to attraction. Science 181:661–663

Kodo T, Miyahara T (1930) J Plant Protect 17:85–94

Kodoi M, Kaneda M (1990) Res bull plant protec service. Japan 26:61–63

Kondo A, Nagata K, Mochizuki F (2008) Geographical differences in pheromone trap performance in the yellow peach moth, *Conogethes punctiferalis* (Guenée) (Lepidoptera: Pyralidae) occurring in Japanese peach orchards. Jpn J Appl Entomol Zool 50:35–38

Konno Y (1986) Time lag between sex pheromone content and the calling behaviour in the yellow peach moth, *Conogethes punctiferalis* (Guenee), Lepidoptera: Pyralidae. Appl Entomol Zool 21(4):622–624

Konno Y, Honda H, Matsumoto Y (1980) Observations on the mating behavior and bioassay for the sex pheromone of the yellow peach moth, *Dichocrocis punctiferalis* Guenée (Lepidoptera: Pyralidae). Appl Entomol Zool 15(3):321–327

Konno Y, Arai K, Sekiguchi K, Matsumoto Y (1982) (E)-10-hexadecenal, a sex pheromone component of the yellow peach moth, *Dichocrocis punctiferalis* Guenee (Lepidoptera: Pyralidae). Appl Entomol Zool 17(2):207–217

Korsching S (2002) Olfactory maps and odor images. Curr Opin Neurobiol 12:387–392

Kou R, Chow YS (1987) Calling behavior of the cotton bollworm, *Heliothis armigera* (Lepidoptera: Noctuidae). Ann Entomol Soc Am 80:490–493

Koya AKM, Devasahayam S, Kumar PT (1991) Insect pests of ginger (*Zingiber officinale* Rosc.) and turmeric (*Curcuma longa* Lin.) in India. J Plant Crop 19:1–13

Kumaresan D, Regupathy A, George KV (1978) Control of the cardamom stem borer *Dichocrocis punctiferalis* (G.). J Plant Crop 6:85–86

Kuwahara Y, Casida JE (1973) Quantitative analysis of the sex pheromone of several phycitid moths by electron-capture gas chromatography. Agric Biol Chem 37(3):681–684

Leal WS (2013) Odorant reception in insects: roles of receptors, binding proteins, and degrading enzymes. Annu Rev Entomol 58:373–379

Liu M, Tian Y, Li Y (1994a) Identification of minor components of the sex pheromone of yellow peach moth *Dichocrocis punctiferalis* Guenée and field trials. Insect Sci 1(2):150–155

Liu MY, Tian Y, Li YX (1994b) Identification of minor components of sex pheromone of yellow peach moth, *Dichocrosis punctiferalis* Guenee (Lepidoptera: Pyralidae) and field trials. Entomol Sin 1:15–155

Lofstedt C, Van-Der-Pers JNC, Lofqvist J, Lanne BS, Appelgren M, Bergstrom G, Thelin B (1982) Sex pheromone components of the turnip moth, *Agrotis segetum* chemical identification, electrophysiological evaluation and behavioral activity. J Chem Ecol 8(10):1305–1321

Ma T, Liu Z, Lu J, Sun Z, Li Y, Wen X, Cui Y (2015) A key compound: (Z)-9-tetradecen-1-ol as sex pheromone active component of *Hypsipyla robusta* (Lepidoptera: Pyralidae). Chemoecology 25:325–330

Malosse C, Ramirez-Lucas P, Rochat D, Morin JP (1995) Solid-phase microextraction: an alternative method for the study of airborne insect pheromones (*Metamasius hemipterus*, Coleoptera, Curculionidae). J High Resolut Chromatogr 18:669

Mazomenos BE, Konestantopouious M, Stefaneu D, Skareas S, Tzeiranakis LC (2002) Female calling behaviour and male response to the synthetic sex pheromone components of *Palpita unionalis* (Lepidoptera: Pyralidae). IOBC WPRS Bull 25:1–10

McElfresh JS, Millar JG (2001) Geographic variation in the pheromone system of the saturniid moth, *Hemileuca eglanterina*. Ecology 82(12):3505–3518

Megido RC, De-Backer L, Ettaib R, Brostaux Y, Fauconnier M, Delaplace P, Lognay G, Belkadhi MS, Haubruge E, Francis F, Verheggen FJ (2014) Role of larval host plant experience and solanaceous plant volatile emissions in *Tuta absoluta* (Lepidoptera: Gelechiidae) host finding behavior. Arthropod Plant Interact 8(4):293–304

Miresmailli S, Gries R, Gries G, Zamar RH, Isman MB (2010) Herbivore-induced plant volatiles allow detection of *Trichoplusia ni* (Lepidoptera: Noctuidae) infestation on greenhouse tomato plants. Pest Manag Sci 66(8):916–924

Mishra SC, Teotia TPS (1965) Studies on the biology of castor shoot and capsule borer, *Dichocrosis punctiferalis* Guenee. Indian Oil seed J 9:184–187

Millar JG (2000) Polyene hydrocarbons and epoxides: second major class of lepidopteran sex attractant pheromones. Annu Rev Entomol 45:575–604

Millar JG, Grant GG, Mcelfresh JS, Strong W, Rudolph C, Stein JD, Moreira JA (2005) (3Z,6Z,9Z,12Z,15Z)- Pentacosapentaene, a key pheromone component of the fir coneworm moth, Dioryctria abietivorella. J Chem Ecol 31:1229–1234

Nakano R, Takanashi T, Ihara F, Mishiro K, Toyama M, Ishikawa Y (2012) Ultrasonic courtship song of the yellow peach moth, *Conogethes punctiferalis* (Lepidoptera: Crambidae). Appl Entomol Zool 47(2):87–93

Natale D, Mattiacci L, Hern A, Pasqualini E, Dorn S (2003) Response of female Cydia molesta (Lepidoptera: Tortricidae) to plant derived volatiles. Bull Entomol Res 93(4):335–342

Nemer N, Bal J, Bechara E, Frerot B (2014) Pheromone identification of the cedar shoot moth *Dichelia cedricola* Diakonoff (Lepidoptera: Tortricidae). Int J Entomol 50(3–4):367–371

Nemoto T, Kuwahara Y, Suzuki T (1990) Isolation and identification of male sex pheromone present in the wing gland of *Achroia innotata obscurevittella* (Lepidoptera : Pyralidae). Appl Entomol Zool 25(2):261–264

Nesbitt BF, Beevor PS, Hall DR, Lester R, Davies JC, Seshureddy KV (1979) Components of the female sex pheromone of the spotted stalk-borer, Chilo partellus (Swinhoe) (Lepidoptera: Pyralidae) identification and preliminary field trials. J Chem Ecol 5:153–156

Nishida R, Baker TC, Roelofs WL (1982) Hairpencil pheromone components of male oriental fruit moths, *Grapholitha molesta* (Lepidoptera, Tortricidae). J Chem Ecol 8:947–959

Noldus LPJJ, Van Lenteren JC (1985) Kairomones for the egg parasite Trichogramma evanescens Westwood. 1. Effect of volatile substances released by two of its hosts, Pieris brassicae L. and Mamestra brassicae L. J Chem Ecol 11:781–791

Pawliszyn J (1997) SMPE, solid phase microextraction, theory and practice. Wiley–VCH, New York

Pliske TE, Eisner T (1969) Sex pheromone of the queen butterfly: biology. Science 164(3884):1170–1172

Polavarapu S, Lonergan G (1998) Sex pheromone of *Choristoneura parallela* (Lepidoptera: Tortricidae) components and development of a pheromone lure for population monitoring. Environ Entomol 27:1242–1249

Raina AK, Klun JA, Stadelbacher EA (1986) Diel periodicity and effect of age and mating on female sex pheromone titer in *Heliothis zea* (Lepidoptera: Noctuidae). Ann Entomol Soc Am 79:128–131

Rajabaskar D (2003) Studies on the evaluation of IPM modules against Conogethes punctiferalis Guenee and Sciothrips cardamomi Ramk. on *cardamom*. Unpublished Ph.D thesis, The Tamil Nadu Agricultural University, Coimbatore, India, p 198

Rajabaskar D, Regupathy A (2012) Calling behavior and attractive response of cardamom shoot and capsule borer *Conogethes punctiferalis* Guenee (Lepidoptera: Pyralidae) to the female crude extract and synthetic blend. J Entomol 9:296–301

Rajabaskar D, Sureshkumar RS, Regupathy A, Chozhan K (2003) Evaluation of a new insecticide lambda cyhalothrin 5EC for the management of shoot and capsule borer *Conogethes punctiferalis* (Guen.) in cardamom. In: National symposium on sustainable insect pest management- 1, Feburary 6–7, Chennai. Abstract, p 54

Ranjith AM (2007) An inexpensive olfactometer and wind tunnel for Trichogramma chilonis Ishii (Trichogrammatidae: Hymenoptera). J Trop Agric 45:63–65

Ravichandran R (1988) Bio ecology and chemical control of cardamom shoot and capsule borer, *Conogethes punctiferalis* (Guenee) (Pyralidae: Lepidoptera). Unpublished M.Sc (Ag.) thesis, The Tamil Nadu Agricultural University, Coimbatore, India, p 113

Regupathy A, Palanisamy S, Chandramohan N, Gunathilagaraj K (2003) A guide on crop pests. Sheeba Printers, Coimbatore, p 276

Reinecke A, Ruthar J, Hilker M (2005) Electrophysiological and behavioural responses of *Melolontha melolontha* to saturated and unsaturated aliphatic alcohols. Entomol Exp Appl 115:33–40

Renuka S, Regupathy A, Chozhan K (2000) Studies on the sex pheromone release behaviour of *Leucinodes orbonalis*. In: Entomocongress 2000, November 5–7, Trivandram, Kerala

Renuka S, Regupathy A, Rajabaskar D (2002) Bio-efficacy of profenofos (Curacron 50EC) against cardamom thrips *Sciothrips cardomomi* (Ramk). In: Rethinam P, Khan HH, Reddy VM, Mandal PK, Suresh K (eds) Proceeding of PLACROSYM XIV,12–15 December 2000, Hyderabad, India, pp 491–492

Reyes-Garcia L, Cuevas Y, Ballesteros C, Curkovic T, Lofstedt C, Bergmann J (2014) A 4-component sex pheromone of the Chilean fruit leaf roller *Proeulia auraria* (Lepidoptera: Tortricidae). Cienc Inv Agr 41(2):187–196

Rochat D, Ramirez-Lucas P, Malosse C, Aldana R, Kakul T, Morina J (2000) Role of solid-phase microextraction in the identification of highly volatile pheromones of two rhinoceros beetles *Scapanes australis* and *Strategus aloeus* (Coleoptera, Scarabaeidae, Dynastinae). J Chromatogr A 885(1–2):433–444

Roelofs WL (1984) Electroantennogram assays: rapid and convenient screening procedures for pheromones. In: Techniques in pheromone research. Springer, New York, pp 131–159

Sarkar N, Mukherjee A, Barik A (2015) Attraction of *Epilachna dodecastigma* (Coleoptera: Coccinellidae) to *Momordica charantia* (Cucurbitaceae) leaf volatiles. Can Entomol 147(2):169–180

Schneider D, Schulz S, Priesner E, Ziesmann J, Francke W (1998) Autodetection and chemistry of female and male pheromone in both sexes of the tiger moth *Panaxia quadripunctaria*. J Comp Physiol A 182(2):153–161

Shorey HH, Mckelvey JJ (1977) Chemical control of insect behaviour: theory and application. Wiley Inter Science, New York

Sidney M, Gries R, Danci A, Judd GJR, Gries G (2006) Almond volatiles attract neonate larvae of *Anarsia lineatella* (Zeller) (Lepidoptera: Gelechiidae). J Entomol Soc B C 103. Available at: http://journal.entsocbc.ca/index.php/journal/article/view/108

Silk PJ, Wiesner CJ, Tan SH, Ross RJ, Grant GG (1982) Sex pheromone chemistry of the western spruce budworm *Choristoneura occidentalis* free. J Chem Ecol 8(2):351–362

Smadja C, Butlin RK (2009) On the scent of speciation: the chemosensory system and its role in premating isolation. Heredity 102(1):77–97

Sridhar P, Lonergan G (1998) Sex pheromone of *Choristoneura parallela* (Lepidoptera: Torticidae) components and development of a pheromone lure for population monitoring. Environ Entomol 27:1242–1249

Srivastava AS, Aswathi GP (1961) The preferred temperature of castor capsule borer larvae. Proc Natl Acad Sci India 13:30–31

Stanley J (2008) Chemical and behavioural approaches for pest management in cardamom. Unpublished PhD thesis, Tamil Nadu Agricultural University, Coimbatore, p 210

Stanley J, Chandrasekaran S, Preetha G (2009) Conogethes punctiferalis (Lepidoptera: Pyralidae) its biology and field parasitization. Indian J Agric Sci 79(11):906–909

Stanley J, Chandrasekaran S, Preetha G, Kuttalam S (2010) Physical and biological compatibility of diafenthiuron with micro/macro nutrients fungicides and biocontrol agents used in cardamom. Arch Phytopathol Plant Protect 43(14):1396–1406

Stanley J, Chandrasekaran S, Subaharan K (2007) EAG and air entrainment studies on cardamom shoot and capsule borer, Conogethes punctiferalis Guenee. Abstracts and Souvenir of National conf on Applied zoology and sustainable development, 13–17 July 2007, Indian Institute of Chemical Technology, Hyderabad, p 41

Strapasson P, Pinto-Zevallos DM, Paudel S, Rajotte EG, Felton GW, Zarbin PHG (2014) Enhancing plant resistance at the seed stage: low concentrations of methyl jasmonate reduce the performance of the leaf miner *Tuta absoluta* but do not alter the behavior of its predator *Chrysoperla externa*. J Chem Ecol 40(10):1090–1098

Strong WB, Millar JG, Grant GG, Moreira JA, Chong JM, Rudolph C (2008) Optimization of pheromone lure and trap design for monitoring the fir coneworm, Dioryctria abietivorella. Entomol Exp Appl 126:67–77

Suckling DM, Karg G, Gibb AR, Bradley SJ (1996) Electroantennogram and oviposition responses of *Epiphyas postvittana* Lepidoptera: tortricidae to plant volatiles. NZ J Crop Hortic Sci 24:323–333

Sullivan BT, Allison JD, Goyer RA, Shepherd WP (2015) Sex pheromone of the baldcypress leafroller (Lepidoptera: Tortricidae). J Econ Entomol 108(1):166–172

Sundararaju D, Bakthavatsalam N, John NJ, Vidyasagar PSPV (1994) Presence of female sex pheromone in tea mosquito bug, Helopeltis antonii Sign. (Heteroptera: Miridae). Entomon 19:47–51

Takanashi T, Huang Y, Takahasi KRYO, Hoshizaki S, Tatsuki S, Ishikawa Y (2005) Genetic analysis and population survey of sex pheromone variation in the adzuki bean borer moth, *Ostrinia scapulalis*. Biol J Linn Soc 84(1):143–160

Tamaki Y, Noguchi H, Yushima U (1973) Sex pheromones of *Spodoptera litura* (F.) (Lepidoptera: Noctuidae): isolation, identification and synthesis. Appl Entomol Zool 8(3):200–203

Tamhankar AJ (1995) Sex pheromone gland and calling behaviour of female spiny bollworm *Earias insulana* (Boisduval). Entomon 20:1–4

Tatsuki S, Sugie H (1992) Rice insect pheromone identification and utilization in monitoring. In: Heinrichs EA, Miller TA (eds) Rice insects: management strategies. Springer, New York, p 347

Teal PE, Tumlinson JH (1989) Isolation, identification, and biosynthesis of compounds produced by male hairpencil glands of *Heliothis virescens* (F.) (Lepidoptera: Noctuidae). J Chem Ecol 15:413–427

Thyagaraj NE (2002) Integrated management of some important cardamom pests in hill region of Karnataka, South India. Unpublished PhD thesis, Dr. B.R. Ambedkar University, Agra, p 213

Thyagaraj NE, Singh PK, Chakravarthy AK (2003) Bioecology of cardamom shoot and fruit borer, *Conogethes punctiferalis* Guenee. Curr Res 32:3–4

Visser JH, Straaten SV, Maarse H (1979) Isolation and identification of volatiles in the foliage of potato, *Solanum tuberosum,* a host plant of Colorodo potato beetle, *Leptinotarsa decemlineata*. J Chem Ecol 5:13–25

Wang Y, Cai RT (1997) The emergence and control of Dichocrocis punctiferalis Guenee for younai plum variety. S China Fruits 26:3–45

Wang H, Geertsema H, Van-Nieukerken EJ, Lofstedt C (2015) Identification of the female-produced sex pheromone of the leafminer, *Holocacista capensis* infesting grapevine in South Africa. J Chem Ecol 41(8):724–731

Weatherston J, Percy JE (1976) The biosynthesis of phenethyl alcohol in the male bertha armyworm *Mamestra configurata*. Insect Biochem 6:413–417

Xiao W, Honda H (2010) Non-polar body waxes enhance sex pheromone activity in the yellow peach moth, Conogethes punctiferalis (Guenée) (Lepidoptera: Crambidae). Appl Entomol Zool 45:449–456

Xiao W, Honda H, Matsuyama S (2011) Monoenyl hydrocarbons in female body wax of the yellow peach moth as synergists of aldehyde pheromone components. Appl Entomol Zool 46(2):239–246

Xiao W, Matsuyama S, Ando T, Millar JG, Honda H (2012) Unsaturated cuticular hydrocarbons synergize responses to sex attractant pheromone in the yellow peach moth, Conogethes punctiferalis. J Chem Ecol 38(9):1143–1150

Yang HV, Shaw KY (1962) Acto Qccon Ent Sino 12:109–134

Yasuda T, Yoshii S, Wakamura S (1994) Identification of sex attractant pheromone of the brown tail moth, *Euproctis similis* (Fuessly). Appl Entomol Zool 29:21–30

Yasuda T, Wakamura S, Arakaki N (1995) Identification of sex attractant pheromone components of the tussock moth, *Euproctis taiwana* (Shiraki) (Lepidoptera: Lymantriidae). J Chem Ecol 21:1813–1822

# Novel Tools for the Management of *Conogethes punctiferalis* Guenée (Crambidae: Lepidoptera)

# 22

N. E. Thyagaraj, K. S. Jagadish, and Naveen Kumar

**Abstract**

The borer, *Conogethes punctiferalis* species infests a wide variety of crops and plant parts, is difficult to manage with insecticides. However, newer compounds, spinetoram and cyazypyr, have shown some promise which can be selectively applied at appropriate time and dose. Timely harvests, clean cultivation, encouragement of pollinators and natural enemies, fruit thinning and bagging and balanced nutrition can greatly help in managing the *Conogethes* populations. Mass trapping of adult moths by pheromone traps and botanical formulations can lead to realistic management of the pest resulting in sustainable crop yields. Such a set of management tools can be practical, cheap and environmentally sound.

**Keywords**

*Conogethes* · Beneficial conservation · Integrated pest management · Pheromone traps

N. E. Thyagaraj (✉)
Department of Entomology, Collage of Agriculture, Hassan, Karnataka, India

K. S. Jagadish
Department of Agricultural Entomology, University of Agricultural Sciences, Gandhi Krishi Vignan Kendra (GKVK), Bengaluru, Karnataka, India

N. Kumar
Denthottu House, Ballamanja, Manchina, Belthangady, Mangaluru, Karnataka, India





A. K. Chakravarthy (ed.), *The Black spotted, Yellow Borer, Conogethes punctiferalis Guenée and Allied Species*,
https://doi.org/10.1007/978-981-13-0390-6_22

## 22.1 Introduction

*Conogethes sahyadriensis* has been identified on zingibers from India (Shashank et al. 2018). *Conogethes* (= *Dichocrocis*) *punctiferalis* (Guenee) is presently considered as one of the most important and widely spread lepidopterous pests causing economical losses to cultivated crops as well as the wild plant species. Larvae of this pest are polyphagous, feeding more than 120 plant species belonging to 16 families (Sekiguchi 1974). *C. punctiferalis* has been a species complex (Solis 1999) and difficult to identify at the species level. There has been no taxonomic revision or change to separate the species within genus (Robinson et al. 1994). Armstrong (2010) described *Conogethes* spp. and mentioned as species complex. This pest is commonly called as yellow peach moth, but there are several other common names. This pest has been considered a potential hazard (CPC 2007). Feeding behaviour of this pest is different from host to host, and there is varied life table (Chakravarthy et al. 2012; Thyagaraj 2003).This pest breeds throughout the year in India and parts of Australia (USDA 1957). Most adults are active at night, and they remain on the back of host leaves and bark (CPC 2007) during the day. Both male and female moths feed on the nectar of the host and surrounding plants (CPC 2007). Till date, 15 species have been identified by DNA bar codes and deposited in the Barcode of Life Data Systems (BOLD).

The advantage of IPM tools for the efficient management of C. *punctiferalis* has been tried by several workers throughout the world on crops this pest attacks. An attempt is being made to present in this chapter an overview on management of the pest. Other chapters in this book also cover management aspect, but briefly. *C. punctiferalis* is clearly potential and hazardous pest throughout its geographical range of occurrence. The methods, in practice, are far from satisfactory for the pest.

## 22.2 Cultural Practices

If simple, practicable management tools are taken at appropriate time, then substantial yield losses can be prevented. Cultural practices followed in different crops play an important role in reducing pest damage. In cardamom plantations, some of the practices reduced the pest incidence effectively. Following clean cultivation, i.e. removal of alternate host plants, maintaining proper shade, proper weeding and removal of affected plant parts significantly *C. punctiferalis*, currently identified as *C. sahyadriensis* on zingiber crop plants from South India (Shashank et al. 2018) damage to panicle, shoot and capsules. The use of recommended N fertilizer reduces the incidence of pest attack wherein increased dose increases the pest damage. Plant population also contribute to the pest incidence, i.e. the higher the population, the higher the pest damage has been recorded. Therefore, plant density played a role in increasing or decreasing the borer infestation (Thyagaraj 2003) (Tables 22.1and 22.2). Sloan (1945) suggested that harvesting of sorghum grains as soon as possible and drying and threshing immediately aid in reducing infestation.

**Table 22.1** Effect of cultural management practices on the thrips and shoot and fruit borer on cardamom

| | Thrips and shoot and fruit borer damage (%) | | | | | |
|---|---|---|---|---|---|---|
| | 2000–2001 | | | 2001–2002 | | |
| Treatment details | (1)[a] | (2)[a] | 3)[b] | (1)[a] | (2)[a] | 3)[b] |
| T1 – clean cultivation (removal of alternate host plants + shade regulation + weeding + removal of affected plants) | 18.6 | 4.89 | 5.19 | 19.99 | 3.19 | 6.16 |
| T2 – increased N (100 kg/ha) dose over recommended (75 kg/ha) dose | 31.39 | 7.62 | 16.39 | 29.32 | 7.36 | 18.49 |
| T3 – decreased N (50 kg/ha) over recommended dose (75 kg/ha) | 24.59 | 6.18 | 11.33 | 23.15 | 8.05 | 09.99 |
| T4 – recommended N (75 kg/ha) with normal package | 21.66 | 6.66 | 09.19 | 21.33 | 6.00 | 09.33 |
| T5 – control (no treatment) | 26.38 | 9.79 | 11.79 | 21.49 | 08.19 | 10.29 |
| CD @ 5% | 2.874 | 1.369 | 1.985 | 3.189 | 1.765 | 2.050 |

*Source*: Thyagaraj (2003)
[a]Mean of thousand capsules of 5 harvests; 3 replications, [b]Calculated from 1000 randomly selected clumps, *1* = Kajji (Itch); *2* = capsule damage; *3* = shoot damage

**Table 22.2** Effect of plant population on thrips and shoot and fruit borer infestation on cardamom

| Spacing (m) | Thrips damage[c] (%) | Shoot and fruit borer damage (%) | |
|---|---|---|---|
| | | Shoot damage[d] | Capsule[d] |
| 1.8 × 0.6[a] | 33.45 | 16.39 | 15.73 |
| 1.8 × 0.9 | 28.16 | 14.66 | 10.52 |
| 1.8 × 1.8[b] | 21.33 | 05.99 | 03.29 |
| 2.7 × 2.7 | 15.69 | 03.19 | 02.59 |
| CD @ 5% | 5.82 | 1.35 | 1.21 |

*Source*: Thyagaraj (2003)
[a]High-density planting, [b]Recommended spacing, [c]Mean of 1000 capsules of 7 harvests; 4 replications, [d]Calculated from 100 randomly selected plants

Appropriate cultural practices coupled with clean management of farm and orchards prevent pest incidence to some extent (Chakravarthy and Khan 1987). Chakravarthy and Thyagaraj (1999) reported that increased levels of nitrogen fertilizers significantly increased pest damage on cardamom. Sridharan et al. (1990) reported the higher pest damage on cardamom in closer planting plantations. Phosphorus fertilization has been known to reduce the incidence of pest, and low potassium supply frequently favours pest damage. Application of biofertilizers makes the plants hardier to attack. Excessive use of nitrogenous fertilizers attracts pests, and cultivation of trap crops is a basic strategy in which major population of pest infesting main crop can be reduced by trapping. Crop rotation is an effective cultural tool for reducing pest population (Singh et al. 2009), as also in *Conogethes*. Crop rotation is difficult to accomplish in perennial crops like plantation and spice crops. Here, as in cardamom plantations, mixed cropping can be adopted.

## 22.3 Host Plant Resistance and Transgenic

Plant resistance is a natural and useful strategy that can be exploited in the field of pest management. It constitutes a cheap and practically feasible input in the IPM system. Sharma et al. (1995) evaluated select cultivars of castor against yield potential and resistant once to *C. punctiferalis* and found positive results. The delta toxin of the Bt is expressed in many genetically modified crops for control of major pests in several countries (Singh et al. 2009). Devasahayam et al. (2010) tested some of the resistant varieties in ginger germplasm against *Conogethes* sp. and reported that none of them showed resistance. Wang Jiang et al. (2013) studied the molecular bar coding for differentiating species of *Conogethes* in order to induce effective management practices. Hence, the same tool can be adopted in the management of *C. punctiferalis* elsewhere. There is a scope to produce GM crops in different plant species which are attacked by *C. punctiferalis*. Li De Yu et al. (2015) showed the effect of different host plants on the development and reproduction of yellow peach moth, *C. punctiferalis*, and suggested this could be well exploited in the pest management programmes. Oviposition was induced by plant olfactory stimuli, but also that many non-host plants potentially stimulate oviposition. Even in host plants, biophysical plant features can restrict oviposition, suggesting that the real host range for oviposition may be narrower than that expected from laboratory assessments based on olfactory responses of female moths (Luo and Honda 2015). Li De Yu et al. (2015) reported the effects of different host plants on the development and the reproduction of *C. punctiferalis*; this would help in understanding the population dynamics on different host plants. Screening of new isolates of Bt for cry1 genes and testing of toxicity against *C. punctiferalis* proved effective. New isolates of Bt, T27, showed 100% larval mortality on the fifth day after treatment (Manikandan et al. 2016). Joseph Rajkumar et al. (2002) conducted a study on luminal protease in *C. punctiferalis* and identified potential targets for proteinaceous biopesticides, such as protease inhibitors. They have explained the possible incorporation of protease inhibitors into transgenic plants. Shashank et al. (2014) conducted behavioural studies on *C. punctiferalis* and revealed that there are host-associated populations with occurrence of cryptic species. Shetty et al. (2015) studied the correlation coefficient and occurrence of pest and diseases in turmeric crop in hill zone of Karnataka and found that varietal characters significantly influenced the pest damage. This study would certainly be useful in developing resistant varieties through genetic engineering against the borer damage.

## 22.4 Mechanical Methods

Trapping adult moths using components of pheromone is an effective tool for the management. Walsh (1867) reported that insect speciation could be helped by shifting and adapting to new host plants. Larvae of *C. punctiferalis* can feed upon more than hundred host plants such as fruit trees, annual cultivated crops, spices crops, vegetables and wild plant species as well (Lu et al. 1995). Thyagaraj (2003) showed

**Table 22.3** Mean[a] number of male moths (*Conogethes* sp.) trapped in (E)-10- and (Z)-10-hexadecenal at 9:1 trap

| | #Male moths trapped | |
|---|---|---|
| Location | Cup trap | Borer infestation[b] (%) |
| RRS Mudigere | | |
| 2000–2001 | 10 | 11.00 |
| 2001–2002 | 13 | 19.50 |
| | Daradahalli | |
| 2000–2001 | 16 | 30.00 |
| 2001–2002 | 24 | 55.00 |
| | Gowdahalli | |
| 2000–2001 | 19 | 45.00 |
| 2001–2002 | 14 | 26.50 |

*Source*: Thyagaraj (2003)

[a]Mean of 2 years trapping, [b]Calculated from 1000 randomly selected cardamom clumps during September to October each year

that the use of pheromone substances to trap *C. punctiferalis* adult moths in the field conditions is an effective tool (Table 22.3). Boo (1998) showed variations in sex pheromone composition of a few lepidopteran species as it is species specific. Konno et al. (1980) extracted pheromone compounds from female moths of *C. punctiferalis* and field evaluated and good response in trapping adult moths. Konno et al. (1981) studied the mating behaviour in the laboratory and found that artificial pheromone compounds can be successfully used to trap adult moths. Kondo et al. (2008) evaluated different sex pheromones and reported that there are differences in species based on location. Accordingly, the traps are to be designed. Mori et al. (1990) synthesized two biological pheromone of *C. punctiferalis* to use as mechanical methods of trapping adult moths and found promising. Cai and Mu (1993) designed cup traps for attraction of moths as one of the IPM tools. Liu et al. (1994) suggested sexual attraction as effective mechanical methods for pest management. Tomomatsu et al. (1995) suggested the mechanical eradication of *C. punctiferalis* larvae and eggs using methyl bromide fumigation and reported 100% motility. High degree of response uniformity between the species and the sexes was observed (Jyothi et al. 1996). Chakravarthy and Thyagaraj (1997 and 1998) showed synthetic pheromone compounds are effective in trapping adult moths in cardamom plantations. Kimura and Honda (1999) trapped *C. punctiferalis* moths in the peach plantation using pheromone traps.

Luo and Honda (2015) suggested segmentation method for touching pest image could improve the segmentation performance and had a remarkable significance for the future extraction, identification and management. From these reports it could be that mechanical methods of collection of *C. punctiferalis* adult moths is wonderful tool successfully employed in the pest management. Wang NianFeng et al. (2016) showed the efficacy of trap crops in peach orchards as a potential tool for the management of *C. punctiferalis*. Gaur (2014) and Akashe et al. (2015) showed that meteorological parameters affect the growth and development; this key factor can well be exploited in the management of *C. punctiferalis*. Disinfecting chestnuts by validation of radio frequency treatments as an alternative non-chemical method proved effective against *C. punctiferalis* (Hou LiXia et al. 2015). Hou LiXia et al. (2015) showed that thermal death kinetics of *C. punctiferalis* at different life stages, heating rate and temperature are essential for developing postharvest treatments to control pests in chestnuts. This information is useful and effective in developing protocols for postharvest pest management. Jia XiaoJian et al. (2015) showed that the cDNA cloning, expression profiling and binding affinity assay of the pheromone-binding protein Cpun-PBP1 in *C. punctiferalis* help in developing artificial traps for the *Conogethes* management.

## 22.5 Use of Botanicals/Biorationals

Several plant species possess insecticidal properties, but only few are exploited for pest management. None of the tested botanicals are effective against *C. punctiferalis* (Thyagaraj et al. 2002). Particularly, the use of neem- and tobacco-based insecticides is less effective. Thyagaraj (2003) tested few botanicals to suppress *C. punctiferalis* in cardamom plantations and proved ineffective (Table 22.4). Eapan (1994) reported that tested botanicals failed to suppress the *Conogethes* population. Gopakumar et al. reported that neem insecticides failed to reduce pests on cardamom. Varadarasan (2001) evaluated botanicals to control cardamom pests and reduced the cost of cultivation in Kerala, South India. Joseph Rajkumar et al. (2002) tested biorationals against cardamom pests and found that none of them were effective except fish oil insecticidal soaps (FOIS) 2.5 % + tobacco extract 2.5 %, but they found that lowest number of honey bees visited; this again affected the capsule

**Table 22.4** Effect of botanicals against cardamom thrips and borers

| | Thrips and shoot and fruit borer damage (%) | | | | | |
|---|---|---|---|---|---|---|
| | 2000–2001 | | | 2001–2002 | | |
| Treatment details[b] | (1)[a] | (2)[a] | (3)[b] | (1)[a] | (2)[a] | (3)[b] |
| Bioneem 0.1% | 23.55 | 5.30 | 8.50 | 18.55 | 5.50 | 10.00 |
| Neem gold 0.1% | 21.45 | 6.50 | 9.20 | 19.35 | 4.35 | 7.50 |
| Nimbecidine 0.1% | 21.00 | 7.50 | 9.10 | 20.50 | 6.01 | 11.55 |
| Neemark 0.1% | 21.55 | 5.20 | 10.00 | 19.58 | 7.91 | 10.11 |
| Tobacco decoction 0.1% | 19.58 | 5.10 | 10.65 | 17.14 | 6.50 | 13.65 |
| Nicotine sulphate 40% | 11.68 | 5.00 | 8.35 | 10.05 | 4.95 | 12.65 |
| Monocrotophos 36% SL 0.05 %+phosalone 35% EC @ 20 ml/l of water | 06.55 | 2.50 | 2.55 | 7.19 | 2.00 | 7.05 |
| Control (no spray) | 29.55 | 7.50 | 10.00 | 26.39 | 9.50 | 10.19 |
| CD @ 5% | 2.984 | 1.210 | 2.050 | 3.198 | 1.750 | 2.150 |

*Source*: Thyagaraj (2003)

[a]Mean of 1000 capsules of 5 harvests; 3 replications, [b]Each treatment was repeated 3 times, [c]Calculated from 1000 randomly selected clumps, *1* = Kajji (Itch); 2 = Capsule damage; 3 = Shoot damage

yield. Bhat (2000) also reported adverse effects of neem on areca nut production with phytotoxicity on inflorescence and button shedding as well. Rajabaskar and Regupathy (2013) found that none of the botanicals were effective against pests with chemical molecules. They also found that the use of any individual components of IPM was less effective suppressing in controlling *C. punctiferalis* population than harmonious blend of components.

## 22.6 Biological Agents

Numbers of biopesticides are also available commercially for better management of *C. punctiferalis*. Devasahayam (2000) evaluated biopesticides against cardamom pests and reported some of the promising ones that can be used in the pest management. Biopesticides meet the needs of the export rules where the importation laws are straight and stringent. But they have certain limitations also as their efficacy under field conditions always includes depending on several factors like inactivation by sunlight, low persistence and slow in action. Feeding behaviour, stage of the crops, plant species and ecological factors govern the efficacy of biopesticides and bio-agents. Certain fungal spores when sprayed on buds and young developing fruits attract ovipositing *Conogethes* moths that deposit eggs on fruits/buds, but later succumb to the action of the fungus. Ali et al. (2014) reported for the first time larval parasitoids on *C. punctiferalis* from Tamil Nadu Agricultural University (TNAU), Coimbatore, India. They have not mentioned any impact assessment on the extent of pest control. Rashid Pervez et al. (2014) isolated and identified positive strains of entomopathogenic nematodes (EPNs) and reported as they are of great potential for biological control of *C. punctiferalis*. Ingle et al. (2016) tested host range assay of entomopathogenic fungi *Nomuraea rileyi* and *Metarhizium anisopliae* on major insect pest of common occurrence in Vidarbha (Maharashtra) in select crop ecosystems. They proved that effective on *C. punctiferalis*. Golden-backed woodpecker *Dinopium benghalensis* L. was found feeding on larvae of *C. sahyadriensis* in cardamom plantations at Mudigere and surrounding areas as reported by Chakravarthy (1988) from South India. Patel and Gangrade (1971) found *Microbracon hebtor* parasitizes larvae of *C. punctiferalis*. Jacob (1981) reported that *Myosoma* spp., *Xanthopimpla australis* and a nematode parasitized larval stages of *C. punctiferalis*. Details on biological control are also found in a separate chapter in this book.

## 22.7 Chemical Methods

Since *C. punctiferalis* exhibits highly diverse feeding behaviour, several workers evaluated different molecules and reported their efficacy. Thyagaraj (2003) evaluated few chemical molecules and reported their efficacy and importance in cardamom ecosystems, but their ill-effects on useful insects have not been assessed (Table 22.5). Wilson et al. (1978) evaluated few chemical molecules and reported their efficacy on the cardamom pests. Mandal et al. (1978), Nari et al. (1979),

**Table 22.5** Effect of different chemical insecticides against cardamom thrips and borers

| | Thrips and shoot and fruit borer damage (%) | | | | | |
|---|---|---|---|---|---|---|
| | 2000–2001 | | | 2001–2002 | | |
| Treatment details[c] | (1)[a] | (2)[a] | (3)[b] | (1)[a] | (2)[a] | (3)[b] |
| Thimet 10%G @ 205 kg/ha | 18.55 | 04.00 | 16.90 | 17.39 | 05.69 | 11.39 |
| Ekalux 25% EC @ 0.03% | 21.55 | 05.58 | 17.69 | 20.69 | 06.30 | 13.48 |
| Demecron 100% EC @ 0.01% | 16.33 | 08.19 | 09.98 | 19.38 | 05.55 | 15.59 |
| Methyl parathion 50% EC @ 0.02% | 22.59 | 06.18 | 10.13 | 21.75 | 04.49 | 14.66 |
| Phosalone 32% EC @ 0.05% | 12.35 | 04.39 | 09.33 | 16.48 | 03.99 | 09.19 |
| Monocrotophos 36% SL @ 0.05% | 06.28 | 01.55 | 04.60 | 06.80 | 02.00 | 03.95 |
| Dimethoate 30% EC @ 0.05% | 09.66 | 03.19 | 12.33 | 10.99 | 03.39 | 11.98 |
| Control (no spray) | 27.85 | 06.85 | 21.55 | 25.65 | 06.95 | 17.99 |
| CD @ 5% | 2.390 | 1.210 | 2.050 | 2.740 | 1.750 | 2.150 |

*Source*: Thyagaraj (2003)
[a]Mean of 1000 capsules of 5 harvests; 3 replications, [b]Calculated from 1000 randomly selected clumps, [c]Each treatment was repeated 3 times, *1* = Kajji (Itch); *2* = Capsule damage; 3 = Shoot damage

Kumaresan (1982, 1983), Krishnamurthy et al. (1989), He (1997), Chakravarthy and Thyagaraj (1998) and Sharma et al. (1992) tested different chemical molecules in different crops against *C. punctiferalis* and reported the most effective one. Chethan et al. (2016) evaluated some of the chemical molecules against *C. sahyadriensis* and found effective, but no information on pollinators was provided. This is because pollinators play an important role in cardamom pollination as it is 100 % cross-pollinated crop. Pollination is affected particularly by a large number of hymenopterans. From the reports made available in the publications, it is clear that select chemical molecules might have given effective control, but their effect on the ecological systems and useful entities have not been mentioned anywhere. This aspect should be considered as many of the crops solely depend on the activity of pollinators for the pollination. The acute and selective toxicity of profenofos against the *Conogethes* borer on small cardamom was much higher compared to endosulfan which is now banned in India (Renuka and Regupathy 2008).

Spinetoram is new class of chemicals in the spinosyn group. It is a semi-synthetic insecticide recovered from fermenting substances of *Saccharopolyspora spinosa*. Spinetoram is a broad-spectrum insecticide with effectiveness against a broad range of pests (Table 22.1). The formulation (01ANA) has good insecticidal properties with rapid action and short preharvest interval (Report, Sumitomo Chemical Co. Ltd. 2012). This formulation has proved effective against whiteflies, aphids, thrips, leaf miners and lepidopteran pests. In field trials at GKVK farm and IIHR farm, Bengaluru, during 2016–2017, spinetoram @50 mg a.i./litre solution or 47 ppm has proved effective against *C. punctiferalis* on castor (Chakravarthy AK, 2015, Pers observ). Spinetoram interferes with nicotinic acetylcholine receptors and gamma-aminobutyric acid (GABA) receptors. One of the important properties of spinetoram is that it kills pests through direct contacts but also those that feed on treated plant parts. Spinetoram is a multicomponent compound but comprising mainly of major and minor components (Fig. 22.1).

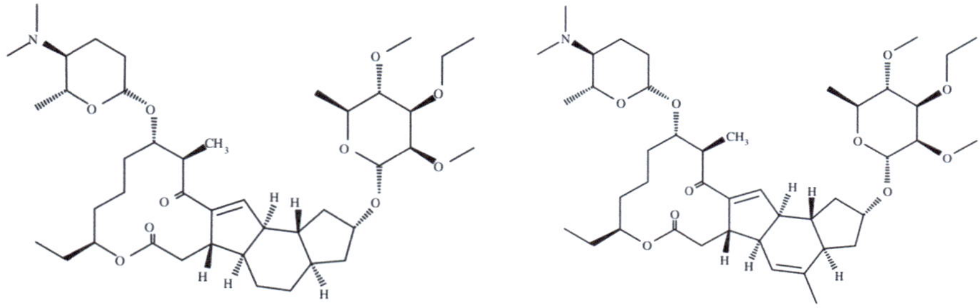

XDE-175-J (Major component) XDE-175-L (Minor component)

**Fig. 22.1** Major and minor components of spinetoram. (*Source*: Sumitomo Chemical Co. Ltd. 2012)

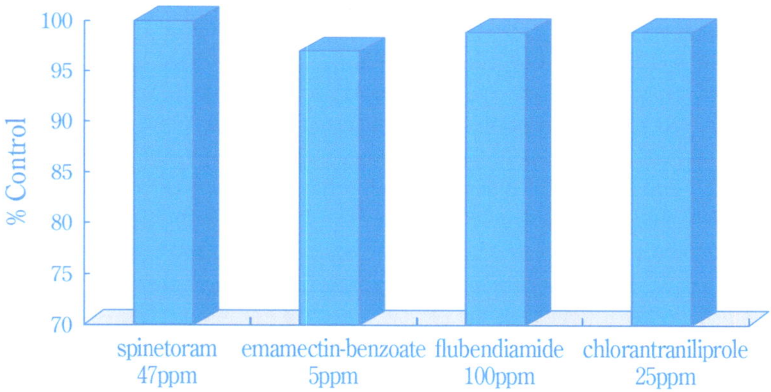


**Fig. 22.2** Efficacy against *H. undalis* on cabbage by foliar spray. (*Source*: Sumitomo Chemical Co. Ltd. 2012)

Spinetoram has also translaminar activity and is effective against soil-dwelling pests. Spinetoram has superior efficacy against *Hellula undalis* (Fabr.) over emamectin benzoate, flubendiamide and chlorantraniliprole (Fig. 22.2). It has also shown high activity against *Thrips tabaci* L. (Fig. 22.3).

## 22.8 Integrated Pest Management

Dependence on chemical pesticides has led to the problems such as insect pest resistance, resurgence and escalating cost of cultivation. Considering the ill-effects of chemical pesticides and the growing preference for chemical-free bio-products, efforts should be made to develop and popularize IPM technologies/strategies. Hence, the integration of all possible management strategies is thought and tested in several crops attacking different pests. Similarly, to control *C. punctiferalis*, many workers evaluated some of the available methods and reported the possibilities of

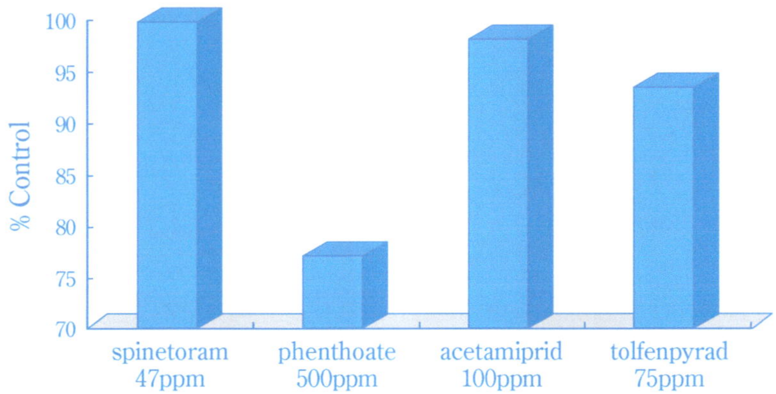


**Fig. 22.3** Efficacy against *T. tabaci* on cabbage by foliar spray. (*Source*: Sumitomo Chemical Co. Ltd. 2012)

using the same techniques in future as holistic approaches. Thyagaraj (2003) evaluated some of the strategies, viz. cultural, mechanical, botanicals and chemicals on *C. sahyadriensis* in cardamom plantations and found that individual methods failed to reduce the pest damage and recorded good control in combined strategies (Table 22.6). Chakravarthy and Khan (1987) reported the effectiveness of IPM tools in the management of *C. sahyadriensis*, and they found satisfactory reduction in pest population in combined and integrated manner. Rajagopal et al. (1987) evaluated some of the IPM tools for the management and found effective than individual methods. It has been reported that IPM tools are effective against cardamom thrips and shoot and fruit borer in Karnataka (Anonymous, 1990). Koya et al. (1991) suggested the similar pest management approaches for the management of ginger pests in Kerala. Hata et al. (1992) reported the IPM approaches as a novel method of pest management in red ginger in Hawaii. Devasahayam and Koya (1999) found IPM modules to be the most effective for cardamom pests in Kerala. Sarkar et al. (2016) evaluated bio-efficacy and non-target toxicity of an IPM compatible thiourea compound diafenthiuron against cardamom thrips and shoot and fruit borer and found effective and suggested that this can be included in the IPM schedule in other areas. Thrips and the borer are the major pests in all the cardamom cultivated areas in South India.

Decision-making in pest management requires a thorough analysis of agroecosystems. With increasing availability of information and understanding on how pests cause damage, new strategies are being devised to enhance protection that is possible. Plant breeding and biotechnology tools in combination are already providing new materials for better plant management. The pest management tools that have deployed have had a positive impact on the environment by reducing the amount of chemical pesticides applied to these crops. Farmers/growers' awareness in understanding the agroecosystems plays a very important role in practising IPM strategies. So, emphasis is now often laid on non-chemicals. An environmentally sustainable agricultural practice is recognized worldwide due to ecological imbalance caused by intensive agricultural practices. In order to address the adverse impact, integrated pest management has been evolved on ETL-based approach.

**Table 22.6** Effect of integrated management practices on cardamom thrips and shoot and fruit borer

| Treatment details | Thrips and shoot and fruit borer damage (%) | | | | | |
|---|---|---|---|---|---|---|
| | 2000–2001 | | | 2001–2002 | | |
| | (1)[a] | (2)[a] | (3)[b] | (1)[a] | (2)[a] | (3)[b] |
| Thrashing[c] | 18.50 | 5.50 | 6.50 | 19.69 | 6.00 | 7.00 |
| Timely application (February to March) of systemic insecticides monocrotophos 36%SL @ 15 ml (0.05%) in 10 l of water | 10.66 | 3.33 | 5.59 | 11.66 | 4.09 | 6.79 |
| Phosalone 35 % EC @ 20 ml/10 l of water 30 days after first spray + phosalone 35% EC (0.05%) @ 20 ml/l of water (30 days after 2nd spray) | 16.05 | 4.95 | 6.99 | 17.05 | 5.05 | 9.00 |
| Thrashing (February to March)+ timely application (March to April) of monocrotophos 36%SL @ 15 ml/10 l of water+phosalone 35%EC two sprays at 25–30 days interval after first spray (T1 + T2 + T3) | 4.39 | 2.00 | 2.50 | 5.99 | 2.05 | 3.05 |
| Control | 28.00 | 6.95 | 16.50 | 23.66 | 7.05 | 10.05 |
| CD @ 5% | 2.960 | 1.985 | 1.950 | 3.169 | 1.756 | 1.998 |

*Source*: Thyagaraj (2003)
[a]Mean of 1000 capsules of 5 harvests; 3 replications, [b]Calculated from 1000 randomly selected clumps, [c]Removal of dried and unproductive plant parts, *1* = Kajji (scabbed); *2* = Capsule damage; [3] = Shoot damage

Ecological engineering for pest management, a new paradigm, is gaining importance as a strategy for promoting biointensive integrated pest management. Since, there is rising public concern about the potential adverse effects of chemical pesticides on the human health, environment and biodiversity, IPM strategies are gaining importance. Crop protection practices, however, are advanced for the management of *Conogethes* spp. across different crops and locations.

**Simple strategies are:**

- Cultivate variety resistant/tolerant to *Conogethes* spp.
- Select healthy seeds/seedlings/planting materials for cultivation.
- Treat the seed/seedlings/planting materials with recommended soil pesticides.
- Follow appropriate spacing.
- Maintain good soil health, not too acidic/alkaline.
- Maintain required nutrients in the soil, balanced nutrition with biorational materials.
- Adopt proper irrigation systems like drip irrigation.
- Follow crop rotation, mixed cropping and intercropping.
- Use selective/recommended chemical molecules with need-based applications.
- Encourage natural enemies and inimical factors that suppress the borer populations.

*C. punctiferalis* management is really hard and tricky; the farmers/growers are facing difficulty. Unless they are made to know the weak stages or vulnerable conditions of this pest in different cropping situations, it is risky to grow crops. As it is already mentioned that *C. punctiferalis* is a potential and hazardous difficult-to-manage pest, large groups of workers throughout the world have tried their best by adopting multiple strategies to manage this pest. Such useful strategies have been discussed in this chapter in order to achieve sustainable crop yields. Information on the IPM of *Conogethes* is also discussed in other chapters of this book.

**Acknowledgement** Authors are thankful to the authorities of the University of Agricultural Sciences, GKVK, Bengaluru, Dr BR Ambedkar University, Agra and Dow Agri-Sci, Bombay, for the encouragement and the material used in this chapter.

## References

Akashe VB, Indi DV, Patil SR, Jadhav JD, Pawar PB (2015) Incidence of insect pest damage in castor in relation to meteorological parameters in the scarcity zone of Maharastra. J Agrometeorology 17(1):139–141

Ali MAA, Manoharan T, Kuttalam S (2014) Trends in Bioscience 7(22):3771–3773

Anonymous (1990) Improved cultivation practices for horticultural crops- hill region. UAS, Bangalore, pp 10–11

Armstrong K (2010) DNA bar-coding: a new module in New Zealand's plant biosecurity diagnostic toolbox. Bull OEPP/EPPO 40:91–100

Bhat VR (2000) Gameticidal effect of plant based pesticides in aercanut. Indian Farming 50(7):33–36

Boo KS (1998) Variation in sex pheromone composition of a few selected lepidopteran. J Asia Pac Entomol 1(1):17–23

CAB International (2007) Crop protection compendium. CAB International, Wallingford

Cai RX, Mu XL (1993) Trap trials for *Conogethes punctiferalis* with sex pheromones in citrus orchards. China Citrus 22(1):33

Chakravarthy AK (1988) Predation by golden backed wood pecker *Dinopium bengalensis* on cardamom shoot borer *Conogethes punctiferalis*. J Bombay Nat Hist Soc 85(2):427

Chakravarthy AK, Khan MM (1987) Innovative tools for protecting cardamom from herbivores-plea for crop management. Q News Lett Asia Pac Plant Prot Comm 30(34):12–17

Chakravarthy AK, Thyagaraj NE (1997) Response of cardamom shoot and fruit borer to different pheromone compounds. Insect Environ 2(4):127–128

Chakravarthy AK, Thyagaraj NE (1998) Evaluation of certain synthetic pheromones of the cardamom shoot and fruit borer *Conogethes punctiferalis* in Karnataka. Pest Manag Hortic Eco-syst 4(2):78–82

Chakravarthy AK, Thyagaraj NE (1999) Effect of different NPK fertilizer levels on the incidence of cardamom pests. Insect Environ 4(4):139–140

Chakravarthy AK, Shashank PR, Doddabasappa B, Thyagaraj NE (2012) Status of shoot and fruit borer, Conogethes spp (Lepidoptera: Crambidae) in the orient: biosystematics, biology and management. J Insect Sci 25(2):107–117

Chethan KS, Swamy BCH, Sowmya DK, Nagaraja A, Sathish R (2016) Evaluation of insecticide molecules against Shoot borer *Conogethes punctiferalis*. Environ Ecol 34(1A):225–228

Devasahayam S (2000) Evaluation of biopesticides for the management of *Conogethes punctiferalis* on ginger. In: Proceedings of the Centennial conference on spice and aromatic plants. Indian Society for Spices, Calicut, India, pp 276–277

Devasahayam S, Koya KMA (1999) Integrated management of insect pests in spice. Indian J Areca Nut, Spice Med Plants 1(1):19–23

Devasahayam S, Jacob TK, Abdul Koya KM, Sashikumar B (2010) Screening of ginger germplasm for resistance to *Conogethes punctiferalis*. J Med Aromat Plant Sci 32(2):137–138

Eapan SJ (1994) Effect of three granular insecticides on damage by thrips and borer in small cardamom. J Entomol Res 18(2):181–183

Gaur RK (2014) Diversity of insect pests of castor, *Ricinus communis* and their ecological interaction in South-West Haryana. Int J Farm Sci 4(4):147–152

Hata TY, Hara AH, Jang EB, Imaino AS, Hu BKS, Tenbrink VL (1992) Pest management before harvest and insecticidal dip after harvest as a system approach to quarantine security for red ginger. J Econ Entomol 85(6):2310–2316

He RQ (1997) Study on the control of yellow peach moths China. South –China-Fruits 26(4):33

Hou LX, Hou JC, Li ZX, Jonson JA, Wang SJ (2015) Validation of radio frequency treatments as alternative non-chemical methods for disinfecting chestnuts. J Stored Prod Res 63:75–79

Ingle YV, Vadaskar RM, Burgoni EB (2016) Host spectrum and relative virulence of entomological fungi *Nomuraea rileyi* and *Metarhizium anisopliae*. Curr Biotica 9(4):348–355

Jacob SA (1981) Biology of *Conogethes punctiferalis* on turmeric. J Plantation Crops 9(2):119–123

Jia XJ, Hao SD, Du YL, Zhang MZ, Quin ZC, Wang JZ, Wang HX, Ji WR (2015) cDNA cloning ,expression profiling and binding affinity assay of the pheromone binding protein Cpun –PBP1 in the yellow peach moth Conogethes punctiferalis. Acta Entomol Sin 58(11):1167–1176

Joseph Rajkumar A, Sainamole Kurian P, Backiyarani S, Murugan M (2002) Evaluation of botanicals against cardamom pests in cardamom plantations. J Spice Aromat Crops 11(2):132–134

Jyothi KN, Prasanna AL, Sighamony S, Prasad A, RandYadav JS (1996) Comparative electroantenogram study of *Parallelia algiria* and *Conogethes punctiferalis* to genera plant volatiles. J Entomol Res 20(3):189–195

Kimura T, Honda H (1999) Identification and possible functions of hair pencil scent of the yellow peach moth *Conogethes punctiferalis* in Japan. Appl Entomol Zool 34(1):147–153

Kondo A, Nagata K, Mochizuki (2008) Geographical difference in pheromone trap performance of the yellow peach moth *Conogethes punctiferalis*, occurring in Japanese peach orchards. Jpn Appl Entomol Zool Chugoku Branch 50:35–38

Konno Y, Honda H, Matsumoto Y (1980) Observation on the mating behavior and biology for the sex pheromones of yellow peach moth *Conogethes punctiferalis*. Appl Entomol Zool 15(3):321–327

Konno Y, Honda N, Matsumoto Y (1981) Mechanism of reproductive isolation between the fruit feeding and pin ace feeding type of yellow peach moth, Dichocrosis punctiferalis. Jpn J Appl Entomol Zool 25(4):253–258

Koya KMA, Devasahayam S, Kumar TP (1991) Insect pests of ginger and turmeric in India. J Plant Crop 19(1):1–3

Krishnamurthy KK, Avdani KK, Venkatesh J, Siddaramaih AL, Chakravathy AK, Gurumurthy BR (1989) Three decades of cardamom research, Regional Research Station, Mudigere (1958–1988). Station Technical Bulletin, pp 44–48

Kumaresan D (1982) Efficacy of modern synthetic insecticides for control of cardamom pests. Appl Entomol Zool 50(2):183–189

Kumaresan D (1983) Field evaluation of insecticides for control of cardamom pests. South-Indian Horticulture 2-3151-152

Li DY, Ali PP, Du YL, Sun SL, Zang MZ (2015) Effect of different host plants on the development and reproduction of yellow peach moth *Conogethes punctiferalis*. Aust Entomol 54(2):149–153

Liu MY, Tian Y, Li XY (1994) Identification of minor components of the sex pheromones of yellow peach moth *Conogethes punctiferalis*. Entomol Sin 1(2):150–155

Luo H, Honda H (2015) Olfactory and bio physical assessment of the oviposition stimulating potential of the host and non-host plants for the yellow peach moth *Conogethes punctiferalis*. Appl Entomol Zool 50(2):183–189

Mandal SC, Ghosh SN, Sinha RP (1978) Chemiocal control of castor capsule borer Conogethes punctiferalis. Indian J Entomol 40(4):460–462

Manikandan R, Muthukumar C, Ramalaxmi A, Balasubramani V, Udayasuriyan (2016) Screening of new isolates of Bacillus thuringiensis for cry1genes and testing of toxicity against *Conogethes punctiferalis*. Microbiol (Moscow) 85(2):191–197

Mori K, Watanable H, Fugiwahara M, Kuwahara S (1990) (E)-and(Z)-tetradecenyl formate, potent sex pheromone mimics against yellow peach moth. Pheromone Synth 122:1257–1259

Nari C, Kunikrishan, Zacharia PK, George KV, Nair MRGK, Kunhikrishan Nair C (1979) Field evaluation of newer insecticides for control of cardamom pests. Pesticides 13(8):49–50

Patel RK, Gangrade GA (1971) Note on the biology of castor capsule borer *Conogethes punctiferalis*. Indian J Agric Sci 41(5):443–444

Pervez R, Eapan SJ, Devasahayam S, Jacob TK (2014) Natural occurrence of Entomopathogenic nematodes associated with ginger eco-systems in India. Indian J Nematol 44(2):238–246

Rajabhaskar D, Regupathy R (2013) Plantation Journal of Biological Sciences, pp 1–8. ISSN:1028-8880. https://doi.org/10.3023/pjbs.2013

Rajagopal D, Chakravarthy AK, Gangappa E (1987) Role of biotic and abiotic factors in regulating the pests of cardamom. J Coffee Res 7(1):103–104

Renuka S, Regupathy A (2008) Acute and Selective toxicity of Profenofos against the shoot and capsule borer of the small cardamom. Resistant Pest Manag Newsl 17(2):33–35

Robinson GS, Tuck KR, Shaffer M (1994) A field guide to the smaller moths of South-East Asia. Malayan Nature Society and the Natural History Museum. Art Printing Works Sdn. Dhd., Kualalumpur

Sarkar PK, Roy D, Chakraraborthy G (2016) Bio-effectiveness and non-target toxicity of an IPM compatible thiourea compound diafenthiuron against cardamom pests under hill zone of the West Bengal 40(2):177–185

Sekiguchi K (1974) Morphology, biology and control of yellow peach moth Conogethes punctiferalis. Bulletin of the Ibaraki Horticultural Experiment Station, Special Issue, p 89

Sharma YR, Devasayam S, Anandaraj M (1992) Black pepper and cardamom: problems and prospects. In: Proceedings of national seminar on black pepper and cardamom, 17–18 May 1992, Calicut, Kerala

Sharma ML, Shukla A, Reddy RK (1995) Field testing of castor cultivars for field yield potential and resistance against shoot and fruit borer *Conogethes punctiferalis*. Crop Res Hissar 10(1):54–58

Shashank PR, Chakravarthy AK, Chandrashekaraiah R, Banu KRM (2014) Behavioural studies on shoot and fruit borer *Conogethes punctiferalis* host associated populations reveal occurrence of cryptic species. Entomol Generalis 35(2):103–115

Shashank PR, Kammar V, Mally R, Chakravarthy AK (2018) A new Indian species of shoot and capsule borer of the genus *Conogethes* (Lepidoptera: Crambidae), feeding on cardamom. Zootaxa 4374(2):215–234

Shetty GR, Hanchinamani V, Venkatesha J (2015) Correlation coefficients and occurrence of pests and diseases in turmeric cultivars under hill zone of Karnataka. J Environ Ecol 33(3):1201–1204

Singh SS, Rai AB, Mayank K, Kamal S (2009) Status, constraints and strategies of pest management in vegetable crops. Progress Hortic 41(1):46–53

Slon WJS (1945) Insect pests of grain sorghum. Queens Land Agric J 61(4):221–239

Solis MA (1999) Key to selected pyraloidea larvae intercepted at US ports of entry: revision of pyraloidea in keys to some frequently intercepted lepidopterous larvae by Weisman 1986. Proc Entomol Soc Wash 101(3):645–666

Sridharan S, Nagarajn N, Thamuraj S, Mohidin MK (1990) Effect of plating density on the capsule dame by cardamom major pests. South Indian Hortic 38(2):120–121

Sumitomo Chemical Co Ltd (2012) Development of the Novel Insecticide Spinetoram (DIANA®). A report

Thyagaraj NE (2003) Integrated management of some important cardamom pests in hill region of Karnataka, South India. Ph.D. thesis submitted to Dr. B.R. Ambedkar University, Agra, UP, p 226

Thyagaraj NE, Singh PK, Chakravarthy AK (2002) Effect of plat based insecticides on cardamom major pests infestation. Insect Environ 7(4):179–180

Tomomatsu S, Sekaguchi T, Ogino T, Misumi T, Kawakami F (1995) Methyl bromide fumigation for quarantine control of persimmon fruit moth and yellow peach moth on Japanese Persimon. Res Bull Plant Prot Serv 31:67–73

USDA (1957) Insects not known to in the United States. Yellow peach moth *Conogethes punctiferalis*, pp 37–38

Varadarasan S (2001) Insect pest management in cardamom –key to reduce cost of production. Spice India 14(7):19–22

Walsh BJ (1867) The apple worm and maggot. J Hortic 2:338–343

Wan NF, Zhang Y, Huang KH, Ji XY, Jiang JX (2016) Ecological engineering of trap cropping promotes biocontrol services in peach orchard ecosystems. Ecol Eng 90:427–430

Wang J, Zhang T-T, Wang Z-Y, He K-L, Liu Y, Jing L (2013) Molecular taxonomy of *Conogethes punctiferalis*, and *Conogethes pinicolalis* (Lepidoptera: Crambidae) based on mitochondrial DNA sequences. J Integr Agric 13(9). https://doi.org/10.1016/S2095-3119(3)60678-4

Wilson KI, Josseph D, Rehim MA, Nair MRGK (1978) Use of newer insecticides for control of cardamom pests. Agric Res J Kerala 15(2):192–194

MIX
Papier aus verantwortungsvollen Quellen
Paper from responsible sources
FSC® C105338

If you have any concerns about our products,
you can contact us on
**ProductSafety@springernature.com**

In case Publisher is established outside the EU,
the EU authorized representative is:
**Springer Nature Customer Service Center GmbH**
**Europaplatz 3, 69115 Heidelberg, Germany**

Printed by Libri Plureos GmbH
in Hamburg, Germany